Disease Mapping with WinBUGS and MLwiN

STATISTICS IN PRACTICE

Statistics in Practice is an important international series of texts which provide detailed coverage of statistical concepts, methods and worked case studies in specific fields of investigation and study.

With sound motivation and many worked practical examples, the books show in down-to-earth terms how to select and use an appropriate range of statistical techniques in a particular practical field within each title's special topic area.

The books provide statistical support for professionals and research workers across a range of employment fields and research environments. Subject areas covered include medicine and pharmaceutics; industry, finance and commerce; public services; the earth and environmental sciences, and so on.

The books also provide support to students studying statistical courses applied to the above areas. The demand for graduates to be equipped for the work environment has led to such courses becoming increasingly prevalent at universities and colleges.

It is our aim to present judiciously chosen and well-written workbooks to meet everyday practical needs. Feedback of views from readers will be most valuable to monitor the success of this aim.

A complete list of titles in this series appears at the end of the volume.

Disease Mapping with WinBUGS and MLwiN

Andrew B. Lawson
Department of Epidemiology and Biostatistics
University of South Carolina, USA

William J. Browne
School of Mathematical Sciences
University of Nottingham, UK

Carmen L. Vidal Rodeiro
Department of Epidemiology and Biostatistics
University of South Carolina, USA

WILEY

Telephone (+44) 1243 779777

Email (for orders and customer service enquiries): cs-bookswiley.co.uk
Visit our Home Page on www.wileyeurope.com or www.wiley.com

Other Wiley Editorial Offices

John Wiley & Sons Inc., 111 River Street, Hoboken, NJ 07030, USA

Jossey-Bass, 989 Market Street, San Francisco, CA 94103-1741, USA

Wiley-VCH Verlag GmbH, Boschstr. 12, D-69469 Weinheim, Germany

John Wiley & Sons Australia Ltd, 33 Park Road, Milton, Queensland 4064, Australia

John Wiley & Sons (Asia) Pte Ltd, 2 Clementi Loop #02-01, Jin Xing Distripark, Singapore 129809

John Wiley & Sons Canada Ltd, 22 Worcester Road, Etobicoke, Ontario, Canada M9W 1L1

Library of Congress Cataloging-in-Publication Data

Lawson, Andrew (Andrew B.)
Disease mapping with WinBUGS and MLwiN / Andrew B. Lawson, William J. Browne, Carmen L. Vidal Rodeiro.
p. cm. – (Statistics in practice)
Includes bibliographical references and index.
ISBN 0-470-85604-1 (hbk. : alk. paper)
1. Medical geography. 2. Medical geography – Maps – Data processing. 3. Epidemiology – Statistical methods. 4. Epidemiology – Data processing. 5. Public health surveillance. I. Browne, William J., Ph.D. II. Vidal Rodeiro, Carmen L. III. Title. IV. Statistics in practice (Chichester, England).

RA792.5.L388 2003
615.4′2′0727–dc21 2003053782

British Library Cataloguing in Publication Data

A catalogue record for this book is available from the British Library

ISBN 0–470–85604–1

Typeset in 10/12 pt by Kolam Information Services Pvt. Ltd, Pondicherry, India
Printed and bound in Great Britain by T J International Ltd, Padstow, Cornwall
This book is printed on acid-free paper responsibly manufactured from sustainable forestry in which at least two trees are planted for each one used for paper production.

Contents

Preface

The analysis of disease maps has seen considerable development over the last decade. This development has been reflected in a fast-increasing literature and has been matched by the development of software tools. The intersecting areas of spatial statistical methods development and geographical information systems (*GIS*) have both witnessed this growth. With increasing public health concerns about environmental risks and even bioterrorism, the need for good methods for analysing spatial health data is immediate. Two major software tools have now been developed, which allow the modelling of spatially-referenced small area health data. These tools, MLwiN and WinBUGS, both provide facilities for sophisticated modelling of realistically complex health data. WinBUGS was developed to allow the application of a wide range of hierarchical Bayesian models, exploiting modern computational advances, in particular Gibbs sampling. MLwiN was developed to allow the fitting of models to multilevel data where a natural parameter hierarchy exists. Originally, this was implemented using iterative likelihood and quasi-likelihood estimation methods. However, the most recent versions of the package have implemented Bayesian computational methodology and now have many parallel capabilities. Increasingly both packages are being used by researchers and also now there is a desire to be able to apply such methodology in practical public health applications. In response to this need, the authors have attempted to provide an introduction to the methods and types of applications where such modelling is feasible. We do not claim to provide a comprehensive text on disease mapping and have confined our attention to the main application of these methods to counted data, where numbers of cases are recorded within small areas.

This book is designed to be of interest to final-year undergraduate and graduate level statistics and biostatistics students but will also be of relevance to epidemiologists and public health workers both in higher education and beyond. The book provides in the introductory chapters (Chapters 1–5) general background to disease mapping, Bayesian hierarchical modelling and multilevel modelling approaches, and basic introductions to the use of WinBUGS and MLwiN. The latter part of the book is focused on application areas, and is divided between relative risk estimation (Chapter 6), focused clustering (Chapter 7), ecological analysis (Chapter 8), and finally spatial survival analysis (Chapter

9). Throughout the book we provide clear descriptions of the model programming execution and analysis of and interpretation of results. We have adopted the philosophy that we would attempt to demonstrate how MLwiN and WinBUGS approach the same data example, but also have included examples where either one or the other packages have limitations. We cannot necessarily hope to provide definitive answers to how modelling is to be approached in every case. However, we would hope that we provide useful pointers to the issues and potential benefits of the approaches described. As both MLwiN and WinBUGS are evolving packages, it is to be expected that features described here may vary in the future. However, we have done our best to describe the current or soon-to-be current form of the packages which is relevant to the potential audience for this published work. All the material described here is available in WinBUGS 1.4 (see Section 4.8.2 for download information and website *www.mrc-bsu.cam.ac.uk/bugs*), and in MLwiN (see section 5.6.1 and website *http://multilevel.ioe.ac.uk/index.html* for more details). Most datasets used in this book are available to download (with associated WinBUGS code) from the site *http://www.sph.sc.edu/alawson/*.

We would like to acknowledge the help and contribution of a number of people during the development of this work. First, we would like to acknowledge the help of the MLwiN project team, in particular Jon Rasbash, Harvey Goldstein, Amy Burch, Lisa Brennan, Fiona Steele and Min Yang. We would like to thank Allan Clark, Robin Puett, Lance Waller, Tom Richards, James Hebert, Alastair Leyland, Sudipto Banerjee, Robert McKeown, Ken Kleinman, Peter Rogerson, Dan Wartenburg and Martin Kulldorff for support and encouragement in the project. In addition, we would like to acknowledge the data availability afforded by the sophisticated online public access GIS layer system developed by, amongst others, Guang Zhao of the South Carolina Department of Health and Environmental Control. Finally, the continuing support and encouragement of Sian Jones and Rob Calver at Wiley Europe must be acknowledged and is much appreciated.

Andrew Lawson (Columbia, SC, USA)
William Browne (Nottingham, UK)
Carmen Vidal Rodeiro (Columbia, SC, USA)

March 2003

Notation

In complex random effects models there is often a myriad of different 'standard' notations to represent a statistical model. This is generally because the models were first discovered by many different authors at roughly the same time and each author had their own particular notation and style.

0.1 STANDARD NOTATION FOR MULTILEVEL MODELLING

In this book, in the multilevel modelling sections, as we will be using the MLwiN software package, we will use the notation used by this software package.

If we consider a three-level nested Normal model, then the standard multilevel model will be written as

$$y_{ijk} = X\beta + v_k + u_{jk} + e_{ijk}, \quad v_k \sim N(0, \sigma_v^2), \quad u_{jk} \sim N(0, \sigma_u^2), \quad e_{ijk} \sim N(0, \sigma_e^2)$$

Here the fixed effects are represented by β, X is a design matrix, and the random effects at levels 1, 2 and 3 are represented by e, u, and v respectively. Level 1 units are indexed i, level 2 units j and level 3 units k. There is a rather unfortunate notational clash as in disease mapping e is typically used to represent the expected counts. However, the level 1 residuals disappear from the equation in the Poisson response multilevel model which minimizes confusion. A three-level Poisson response model is typically written in MLwiN as

$$\begin{aligned} y_{ijk} &\sim Poisson(\pi_{ijk}), \\ \log(\pi_{ijk}) &= \log(e_{ijk}) + X\beta + v_k + u_{jk}, \\ v_k &\sim N(0, \sigma_v^2), \quad u_{jk} \sim N(0, \sigma_u^2). \end{aligned} \tag{1}$$

In standard disease mapping θ is often used rather than π, and the e_{ijk} is often put on the right-hand side of the equation.

0.2 SPATIAL MULTIPLE-MEMBERSHIP MODELS AND THE MMMC NOTATION

The disadvantage of the standard multilevel notation is that it relies on the nested structure of the model. Browne *et al.* (2001) consider more general random effect structures including crossed random effects and multiple-membership structures. Rather than give an index for each classification (level in a nested structure) they instead use mapping functions to define the unit in the classification that a particular observation belongs. For example let us consider the three-level Poisson model and assume that the levels are counties within regions within nations. Then in the notation of Browne *et al.* (2001) we can write Equation (1) as follows:

$$
\begin{aligned}
y_i &\sim Poisson(\pi_i),\\
\log(\pi_i) &= \log(e_i) + X\beta + u^{(3)}_{nation\,[i]} + u^{(2)}_{region[i]},\\
u^{(3)}_{nation[i]} &\sim N(0, \sigma^2_{u(3)}),\ u^{(2)}_{region[i]} \sim N(0, \sigma^2_{u(2)}).
\end{aligned}
$$

So here we define all terms with respect to the lowest (observation) level which is labelled i. The functions *nation*[i] and *region*[i] are mapping functions that return the nation and region respectively that observation i belongs to. As the random part of the model consists of a set of classifications which need not now be ordered in terms of nesting (and if the model contained crossed effects could not) we simply define each set of random effects with the letter u but include a superscript that gives the classification a number. We start numbering from 2 as 1 is reserved for the observation level.

The spatial multiple-membership models that we will consider later can be easily written in this notation as follows:

$$
\begin{aligned}
y_i &\sim Poisson(\pi_i),\\
\log(\pi_i) &= \log(e_i) + X\beta + \sum_{j\in neigh(i)} w^{(3)}_{i,j} u^{(3)}_j + u^{(2)}_{region[i]},\\
u^{(3)}_j &\sim N(0, \sigma^2_{u(3)}),\ u^{(2)}_{region[i]} \sim N(0, \sigma^2_{u(2)}).
\end{aligned}
$$

Here we have a set of region effects indexed by 2 and a set of neighbour effects that are indexed by 3.

0.3 STANDARD NOTATION FOR WinBUGS MODELS

In hierarchical models for disease maps, the notation commonly used is slightly different from that used in multilevel models. The basic Poisson likelihood model is defined as

$$y_i \sim Poisson(e_i\theta_i),$$

where e_i is the expected count and θ_i is the relative risk in the ith small area. Note that in the notation of multilevel models, $\pi_i = e_i\theta_i$. It is also common to use $\lambda_i = e_i\theta_i$, and this form is used in Chapter 7.

Modelling focuses on θ_i. Here we assume this notation for all the standard analysis within WinBUGS. In addition to region specific notation we also introduce space–time notation with a second subscript denoting the time period:

$$y_{ik} \sim Poisson(e_{ik}\theta_{ik}).$$

Here, k denotes the relevant time period and the expected count and relative risk are allowed to vary over time periods.

When random effects are introduced into models it is usual to denote region-specific uncorrelated heterogeneity as v_i, and correlated heterogeneity for the same unit as u_i. This differs slightly from the convention in multilevel models.

In each section the relevant notation for that section is introduced and it is hoped that any differences between sections will not create difficulties for the reader.

1

Disease Mapping Basics

The representation and analysis of maps of disease incidence data is now established as a basic tool in the analysis of regional public health. One of the earliest examples of disease mapping is the map of the addresses of cholera victims related to the locations of water supplies given by Snow (1854). In that case, the street addresses of victims were recorded and their proximity to putative pollution sources (water supply pumps) was assessed.

The subject area of disease mapping has developed considerably in recent years. This growth in interest has led to a greater use of geographical or spatial statistical tools in the analysis of data both routinely collected for public health purposes and in the analysis of data found within ecological studies of disease relating to explanatory variables. The study of the geographical distribution of disease can have a variety of uses. The main areas of application can be conveniently broken down into the following classes: (1) disease mapping, (2) disease clustering, and (3) ecological analysis. In the first class, usually the object of the analysis is to provide (estimate) the true *relative risk* of a disease of interest across a geographical study area (map): a focus similar to the processing of pixel images to remove noise. Applications for such methods lie in health services resource allocation, and in disease atlas construction (see, for example, Pickle *et al.*, 1999). The second class, that of disease clustering, has particular importance in public health surveillance, where it may be important to be able to assess whether a disease map is clustered and where the clusters are located. This may lead to examination of potential environmental hazards. A particular special case arises when a known location is thought to be a potential pollution hazard. The analysis of disease incidence around a putative source of hazard is a special case of cluster detection called focused clustering. The third class, that of ecological analysis, is of great relevance within epidemiological research, as its focus is the analysis of the geographical distribution of disease in relation to explanatory covariates, usually at an aggregated spatial level. Many issues relating to disease mapping are also found in this area, in addition to issues relating specifically to the incorporation of covariates.

Disease Mapping with WinBUGS and MLwiN A. Lawson, W. Browne and C. Vidal Rodeiro
 ISBN: 0-470-85604-1 (HB)

In this volume, we focus on the issues of modelling. While the focus here is on *statistical* methods and issues in disease mapping, it should be noted that the results of such statistical procedures are often represented visually in mapped form. Hence, some consideration must be given to the purely cartographic issues that affect the representation of geographical information. The method chosen to represent disease intensity on the map, be it colour scheme or symbolic representation, can dramatically affect the resulting interpretation of disease distribution. It is not the purpose of this review to detail such cognitive aspects of disease mapping, but the reader is directed to some recent discussions of these issues (MacEachren, 1995; Monmonier, 1996; Pickle and Hermann, 1995; Walter, 1993).

1.1 DISEASE MAPPING AND MAP RECONSTRUCTION

To begin, we consider two situations which commonly arise in studies of the geographic distribution of disease. These situations are defined by the form of the mapped data which arises in such studies. First a study area or window is defined and within this area for a fixed period of time the locations of cases of a specified disease are recorded. These locations are usually residential addresses (street address or, at a higher spatial scale, zip code (USA) or post code unit (UK)). When such addresses are known it is possible to proceed by direct analysis of the case locations. This is termed *case-event* analysis. Often this analysis requires the use of point process models and associated methodology. This form of analysis is reviewed in Lawson (2001, Chapters 4 and 5) and elsewhere (see, for example, Elliott *et al.* (2000, Chapter 6)). Due to the requirements of medical confidentiality, it is often not possible to obtain data at this level of resolution and so resort must be made to the analysis of *counts* of cases within small areas within the study window. These small areas are arbitrary regions usually defined for administrative purposes, such as census tracts, counties, municipalities, electoral wards or health district regions. Data of this type consist of counts of cases within tracts and the analysis of this data is termed *tract count* analysis. In this volume we focus exclusively on tract count analysis. An example of the analysis of case-event data with a Bernouilli model using WinBUGS is given in Congdon (2003, Chapter 7).

Essentially the count is an aggregation of all the cases within the tract. By aggregation, the individual case spatial references (locations) are lost and therefore any georeference of the count is related to the tract 'location'. Often this is represented by the tract centroid. In a chosen study window there is found to be m tracts. Denote the counts of disease within the m tracts as $\{y_i\}$, $i = 1, \ldots, m$. Figure 1.1 displays a tract count example.

This example is of the 46 counties of South Carolina in which were collected the congenital abnormality death counts for the year 1990.

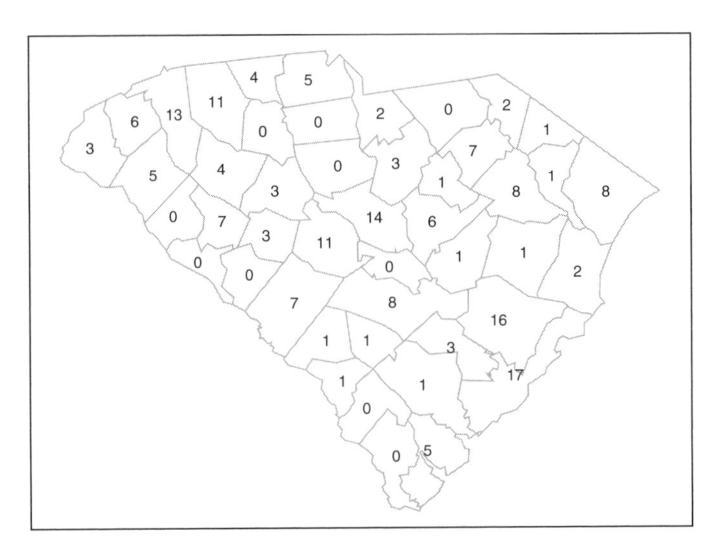

Figure 1.1 South Carolina congenital abnormality deaths 1990 by counties.

1.2 DISEASE MAP RESTORATION

1.2.1 Simple statistical representations

The representation of disease-incidence data can vary from pictorial representation of counts within tracts, to the mapping of estimates from complex models purporting to describe the structure of the disease events. In this section, we describe the range of mapping methods from simple representations to model-based forms. The geographical incidence of disease has as its fundamental unit of observation, the address location of cases of disease. The residential address (or possibly the employment address) of cases of disease contains important information relating to the type of exposure to environmental risks. Often, however, the exact address locations of cases are not directly available, and one must use instead counts of disease in arbitrary administrative regions, such as census tracts or postal districts.

1.2.1.1 Crude representation of disease distribution

The simplest possible mapping form is the depiction of disease rates at specific sets of locations. For counts within tracts, this is a pictorial representation of the number of events in the tracts plotted at a suitable set of locations (e.g., tract centroids). The locations of case-events within a spatially heterogeneous population can display a small amount of information concerning the overall pattern of disease events within a window. However, any interpretation of the structure of these events is severely limited by the lack of information concerning the spatial distribution of the background population which might be 'at risk' from the disease of concern and which gave rise to the cases of disease. This population also has a spatial distribution and failure to take account of this spatial variation severely limits the ability to interpret the resulting case-event map. In essence, areas of high density of 'at risk' population would tend to yield high incidence of case-events and so, without taking account of this distribution, areas of high disease intensity could be spuriously attributed to excess disease risk.

In the case of counts of cases of disease within tracts, similar considerations apply when crude count maps are constructed. Here, variation in population density also affects the spatial incidence of disease. It is also important to consider how a count of cases could be depicted in a mapped representation. Counts within tracts are totals of events from the whole tract region. If tracts are irregular, then a decision must be made to either 'locate' the count at some tract location (e.g. tract centroid, however defined) with suitable symbolization, or to represent the count as a fill colour or shade over the whole tract (choropleth thematic map). In the former case, the choice of location will affect interpretation. In the latter case, symbolization choice (shade and/or colour) could distort interpretation also, although an attempt to represent the whole tract may be attractive.

In general, methods that attempt to incorporate the effect of background 'at risk' population (termed: *at risk background*) are to be preferred. These are discussed in the next section.

1.2.1.2 Standardized mortality/morbidity ratios and standardization

To assess the status of an area with respect to disease incidence, it is convenient to attempt to first assess what disease incidence should be locally 'expected' in the tract area and then to compare the observed incidence with the 'expected' incidence. This approach has been traditionally used for the analysis of counts within tracts. Traditionally, the ratio of observed to expected counts within tracts is called a Standardized Mortality/Morbidity Ratio (SMR) and this ratio is an estimate of *relative risk* within each tract (i.e., the ratio describes the odds of being in the disease group rather than the background group). The justification for the use of SMRs can be supported by the analysis of likelihood models with multiplicative expected risk (see, for example, Breslow and Day, 1987). In Section 1.2.3.1, we explore further the connection between likelihood models and tract-based estimators of risk. Figure 1.2 displays the SMR thematic map for congenital abnormality deaths within South Carolina, USA, for the year 1990 based on expected rates calculated from the South Carolina 1990–1998 statewide rate per 1000 births.

Define y_i as the observed count of the case disease in the ith tract, and e_i as the expected count within the same tract. Then the SMR is defined as:

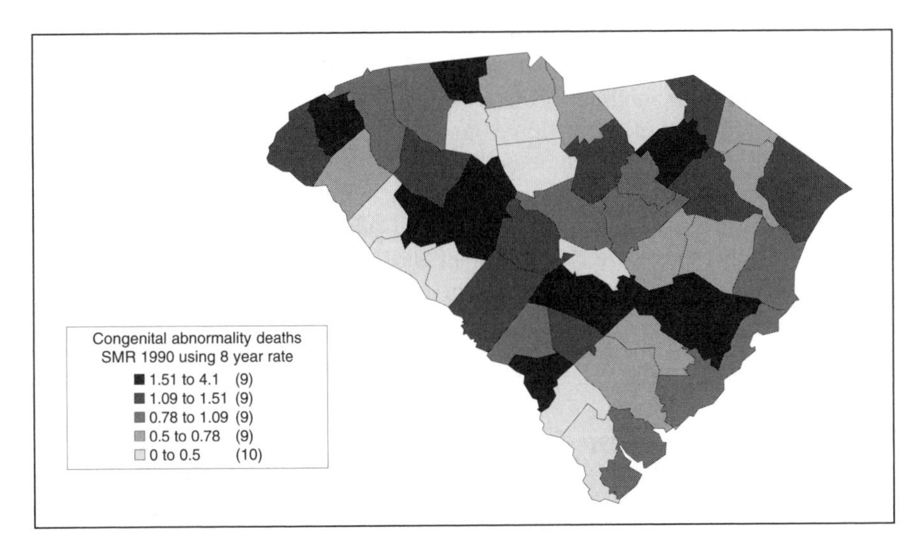

Figure 1.2 South Carolina congenital abnormality deaths 1990: SMRs.

$$\widehat{\theta}_i = \frac{y_i}{e_i}. \tag{1.1}$$

In this case it must be decided whether to express the $\widehat{\theta}_i$ as fill patterns in each region, or to locate the result at some specified tract location, such as the centroid. If it is decided that these measures should be regarded as continuous across regions then some further interpolation of $\widehat{\theta}_i$ must be made (see, for example, Breslow and Day, 1987, pp. 198–9).

SMRs are commonly used in disease map presentation, but have many drawbacks. First, they are based on ratio estimators and hence can yield large changes in estimate with relatively small changes in expected value. In the extreme, when a (close to) zero expectation is found the SMR will be very large for any positive count. Also the zero SMRs do not distinguish variation in expected counts, and the SMR variance is proportional to $1/e_i$. The SMR is essentially a saturated estimate of relative risk and hence is not parsimonious.

1.2.2 Informal methods

To circumvent the problems associated with SMRs a variety of methods have been proposed. Some of these are relatively informal or nonparametric and others highly parametric. In the rest of this volume we will concentrate on the model-based relative risk estimation methods. However, it is useful here to present briefly some notes on alternative methods.

One approach to the improvement of relative risk estimation is to employ smoothing tools on SMRs to reduce the noise. These tools could be based on interpolation methods, or more commonly on nonparametric smoothers such as kernel regression (Nadaraya-Watson, local linear) (Bowman and Azzalini, 1997), and partition methods (Ferreira *et al.*, 2002). A variety of exploratory data analysis (EDA) methods have also been advocated (see, for example, Cressie, 1993). These methods usually require the estimation of a smoothing constant which describes the overall behaviour of the relative risk surface. Some local methods are also available. Generalized additive models have also been proposed and these have the advantage of allowing the incorporation of covariates (see, for example, Kelsall and Diggle, 1998).

1.2.3 Basic models

When more substantive hypotheses and/or greater amounts of prior information are available concerning the problem, then it may be advantageous to consider a model-based approach to disease map construction. Model-based approaches can also be used in an exploratory setting, and if sufficiently general models are employed then this can lead to better focusing of subsequent hypothesis generation. In what follows, we consider first likelihood models for case event data and then discuss the inclusion of extra information in the form of random effects.

1.2.3.1 Likelihood models

Usually the basic model for case-event data is derived from the following assumptions:

(1) Individuals within the study population behave independently with respect to disease propensity, after allowance is made for observed or unobserved confounding variables.
(2) The underlying at risk background intensity has a continuous spatial distribution, within a specified boundary.
(3) The case-events are unique, in that they occur as single spatially separate events.

Assumption (1) above allows the events to be modelled via a likelihood approach, which is valid conditional on the outcomes of confounder variables. Further, assumption (2), if valid, allows the likelihood to be constructed with a background continuous modulating intensity function representing the 'at risk' background. The uniqueness of case-event locations is a requirement of point process theory (the property called orderliness: see, for example, Daley and Vere-Jones, 1988), which allows the application of Poisson-process models in this analysis. Assumption (1) is generally valid for non-infectious diseases. It may also be valid for infectious diseases if the information about current infectives were known at given time points. Assumption (2) will be valid at appropriate scales of analysis. It may not hold when large areas of a study window include zones of zero population (e.g. harbours/industrial zones). Often models can be restricted to exclude these areas however. Assumption (3) will usually hold for relatively rare diseases but may be violated when households have multiple cases and these occur at coincident locations. This may not be important at more aggregate scales, but could be important at a fine spatial scale. Remedies for such non-orderliness are the use of de-clustering algorithms (which perturb the locations by small amounts), or analysis at a higher aggregation level. Note that it is also possible to use a conventional case-control approach to this problem (Diggle *et al.*, 2000).

In the case of observed counts of disease within tracts, the Poisson-process assumptions given above mean that the counts are Poisson distributed with, for each tract, a different expectation. Often at this point a simplifying assumption is made where the *i*th tract count expectation is regarded as being a function of a parameter within a model hierarchy, without considering the spatial continuity of the intensity. This assumption leads to considerable simplifications and the distribution of the tract counts is often assumed to be

$$y_i \sim Poisson(e_i\theta_i),$$

where θ_i is assumed to be a constant relative risk parameter. In this definition the expected value of the count is a multiplicative function of the expected count/rate (e_i) and a relative risk. This is the classic model assumed in many disease mapping studies. The log-likelihood associated with this model is, bar a constant, given by:

$$l = \sum_{i=1}^{m} y_i \ln(e_i\theta_i) - \sum_{i=1}^{m} e_i\theta_i.$$

Note that by differentiation the saturated maximum likelihood estimator of θ_i is just y_i/e_i, the SMR.

This model makes a number of assumptions. First it is assumed that any excess risk in a tract will be expressed beyond that described by e_i. For example, the expected rate (e_i) can be estimated in a variety of ways. Often external standardization is used, where known supra-regional rates for different age × sex groups are applied to the local population in each tract. The use of external standardization alone to estimate the expected counts/rates within tracts may provide a different map from that provided by a combination of external standardization and measures of tract-specific deprivation (e.g. deprivation indices (Carstairs, 1981)). If any confounding variables are available and can be included within the estimate of the at risk background, then these should be considered for inclusion. Examples of confounding variables could be found from national census data, particularly relating to socioeconomic measures. These measures are often defined as 'deprivation' indicators, or could relate to lifestyle choices. For example, the local rate of car ownership or percentage unemployed within a census tract or other small area, could provide a surrogate measure for increased risk, due to correlations between these variables and poor housing, smoking lifestyles, and ill-health. Hence, if it is possible to include such variables, then any resulting map will display a close representation of the 'true' underlying risk surface. When it is not possible to include such variables, it is sometimes possible to adapt a mapping method to include covariates of this type within regression setting.

1.2.3.2 *Fixed effects*

Usually the focus of attention when more sophisticated models are applied in disease mapping is the relative risk. Hence, all the models we will examine in this volume will be models for the $\{\theta_i\}$. One simple model for the relative risks would be to suppose that there could be a spatial trend or long-range variation over the study area. To do this we can construct a model which is a function of the spatial coordinates of the tract centroids: $\{x_{1i}, x_{2i}\}$ representing eastings and northings, say. Simple forms of spatial trend can be modelled by using the centroid coordinates or functions of the coordinates as covariates and assuming a regression-type model. As the relative risks must be positive it is usual to model the logarithm of the relative risk as a linear function. Hence, in this case we could have:

$$\theta_i = \exp\{\beta_0 + \beta_1 x_{1i} + \beta_2 x_{2i}\}. \tag{1.2}$$

This model includes a constant rate ($\exp\{\beta_0\}$) which captures the overall rate across the whole study region, and two linear parameters: β_1, β_2. This model describes a planar trend across the study region, and can be easily extended to include higher-order trend surfaces by adding power functions of the coordinates. Here we have used centroid locations as covariates, and indeed this model can be generalized simply when you observe other covariates measured within the tracts. For example it may be possible to include deprivation scores for each tract or census variables such as percentage unemployed or percentage car ownership. In general, assume that the intercept (constant rate) term is defined for a variable x_{0i} which is 1 for each tract. Hence we can specify the model compactly as

$$\theta_i = \exp\{\mathbf{x}_i\boldsymbol{\beta}\},$$

where $\mathbf{x}$ is a $m \times p$ matrix consisting of $p - 1$ covariates, $\boldsymbol{\beta}$ is a $p \times 1$ parameter vector and $\mathbf{x}_i$ denotes the ith observation row of $\mathbf{x}$.

This type of fixed effect model can be fitted in conventional statistical packages which allow Poisson regression or log-linear modelling. The *glm* function in R or S-Plus with a log link and log offset of the $\{e_i\}$ can be used, for example.

1.2.3.3 Random effects

In the sections above some simple approaches to mapping counts within tracts have been described. These methods assume that once all known and observable confounding variables are included then the resulting map will be clean of all artefacts and hence depicts the true excess risk surface. However, it is often the case that unobserved effects could be thought to exist within the observed data and that these effects should also be included within the analysis. These effects are often termed *random* effects, and their analysis has provided a large literature both in statistical methodology and in epidemiological applications (see, for example, Manton *et al.*, 1981; Tsutakawa, 1988; Breslow and Clayton, 1993; Clayton, 1991; Best and Wakefield, 1999; Lawson, 2001; Richardson, 2003). Within the literature on disease mapping, there has been a considerable growth in recent years in modelling random effects of various kinds. In the mapping context, a random effect could take a variety of forms. In its simplest form, a random effect is an extra quantity of variation (or variance component) which is estimable within the map and which can be ascribed a defined probabilistic structure. This component can affect individuals or can be associated with tracts or covariates. For example, individuals vary in susceptibility to disease and hence individuals who become cases could have a random component relating to different susceptibility. This is sometimes known as *frailty*. Another example is the interpolation of a spatial covariable to the locations of case events or tract centroids. In that case, some error will be included in the interpolation process, and could be included within the resulting analysis of case or count events. Also, the locations of case-events might not be precisely known or subject to some random shift, which may be related to uncertain residential exposure. Finally, within any predefined spatial unit, such as tracts or regions, it may be expected that there could be components of variation attributable to these different spatial units. These components could have different forms depending on the degree of prior knowledge concerning the nature of this extra variation. For example, when observed counts, thought to be governed by a Poisson distribution, display greater variation than expected (i.e. variance $>$ mean), it is sometimes described as overdispersion. This overdispersion can occur for various reasons. Often it arises when clustering occurs in the counts at a particular scale. It can also occur when considerable numbers of cells have zero counts (sparseness), which can arise when rare diseases are mapped. In spatial applications, it is important furthermore to distinguish two basic forms of extra variation. First, as in the aspatial case, a form of independent and spatially uncorrelated extra variation can be assumed. This is often called *uncorrelated heterogeneity* (see, for example, Besag *et al.*, 1991). Another form of random effect is that which arises from a model where it is thought that the spatial unit (such as case-events, tracts or regions) is correlated with neighbouring spatial units. This is often termed *correlated heterogeneity*. Essentially, this form of extra variation implies that there exists spatial autocorrelation between spatial units (see, for example, Cliff and Ord (1981) for an accessible introduction to spatial autocorrelation). This autocorrelation could arise for a variety of reasons. First, the disease of concern could be naturally clustered in its spatial distribution at the scale of observation. Many infectious diseases display such spatial clustering, and a number of apparently non-infectious diseases also cluster (see, for example, Cuzick and Hills, 1991; Glick, 1979). Second, autocorrelation can be induced in spatial disease patterns by the existence of unobserved environmental or frailty effects. Hence the extra variation observed in any application could arise from confounding variables that have not been included in the analysis. In disease mapping examples, this could easily arise when simple mapping methods are used on SMRs with just basic age–sex standardization.

In the discussion above on heterogeneity, it is assumed that a global measure of heterogeneity applies to a mapped pattern. That is, any extra variation in the pattern can be captured by including a general heterogeneity term in the mapping model. However, often spatially-specific heterogeneity may arise where it is important to consider local effects as well as, or instead of, general heterogeneity. To differentiate these two approaches, we use the term *specific* and *nonspecific* heterogeneity. Specific heterogeneity implies that spatial locations are to be modelled locally; for example, clusters of disease are to be detected on the map. In contrast, 'nonspecific' describes a global approach to such modelling, which does not address the question of the location of effects. In this definition, it is tacitly assumed that the locations of clusters of disease can be regarded as random effects themselves. Hence, there are strong parallels between image processing tasks and the tasks of disease mapping.

Random effects can take a variety of forms and suitable methods must be employed to provide correctly estimated maps under models including these effects. In this section, we discuss simple approaches to this problem from a frequentist, multilevel and Bayesian viewpoint.

A frequentist approach. In what follows, we use the term 'frequentist' to describe methods that seek to estimate parameters within a hierarchical model structure. The methods *do* assume that the random effects have mixing (or prior) distributions. For example a common assumption made when examining tract counts is that $y_i \sim Poisson(e_i\theta_i)$ independently, and that $\theta_i \sim Gamma(\alpha, \beta)$. This latter distribution is often assumed for the Poisson relative risk parameter and provides for a measure of overdispersion relative to the Poisson distribution itself, depending on the α, β values used. The joint distribution is now given by the product of a Poisson likelihood and a gamma distribution. At this stage a choice must be made concerning how the random intensities are to be estimated or otherwise handled. One approach to this problem is to average over the values of θ_i to yield what is often called the *marginal* likelihood. Having averaged over this density, it is then possible to apply standard methods such as maximum likelihood. This is usually known as marginal maximum likelihood (see, for example, Bock and Aitkin, 1981; Aitkin, 1996b). In this approach, the parameters of the gamma distribution are estimated from the integrated likelihood. A further development of this approach is to replace the gamma density with a finite mixture. This approach is essentially nonparametric and does not require the complete specification of the parameter distribution (see, for example, Aitkin, 1996a).

Although the example specified here concerns tract counts, the method described above can equally be applied to case-event data, by inclusion of a random component in the intensity specification.

A Bayesian approach. It is natural to consider modelling random effects within a Bayesian framework. First, random effects naturally have prior distributions and the joint density discussed above is proportional to the posterior distribution for the parameters of interest. Hence, the development of full Bayes and empirical Bayes (posterior approximation) methods has progressed naturally in the field of disease mapping. The prior distribution(s) for the ($\boldsymbol{\theta}$, say) parameters in the intensity specification $e_i\theta_i$, have hyperparameters (in the Poisson–gamma example above, these were α, β). These hyperparameters can also have hyperprior distributions. The distributions chosen for these parameters depend on the application. In the full Bayesian approach, inference is based on the posterior distribution of $\boldsymbol{\theta}$ given the data. However, as in the frequentist approach above, it is possible to adopt an intermediate approach where the posterior distribution is approximated in some way, and subsequent inference may be made via 'frequentist-style' estimation of parameters or by computing the approximated posterior distribution. In the tract-count example, approximation via intermediate prior-parameter estimation would involve the estimation of α and β, followed by inference on the estimated posterior distribution (see, for example, Carlin and Louis, 1996, pp. 67–8).

For count data, a number of examples exist where independent Poisson distributed counts (with constant within-tract rate) are associated with prior distributions of a variety of complexity. The earliest examples of such a Bayesian mapping approach can be found in Manton *et al.* (1981) and Tsutakawa (1988). Also, Clayton and Kaldor (1987) developed a Bayesian analysis of a Poisson likelihood model where y_i has expectation $e_i\theta_i$, and found that with a prior distribution given by $\theta_i \sim Gamma(\alpha, \beta)$, the Bayes estimate of θ_i is the posterior expectation:

$$\frac{y_i + \alpha}{e_i + \beta}. \tag{1.3}$$

Hence one could map directly these Bayes estimates. Now, the distribution of θ_i conditional on y_i is $Gamma(y_i + \alpha,\ e_i + \beta)$ and a Bayesian approach would

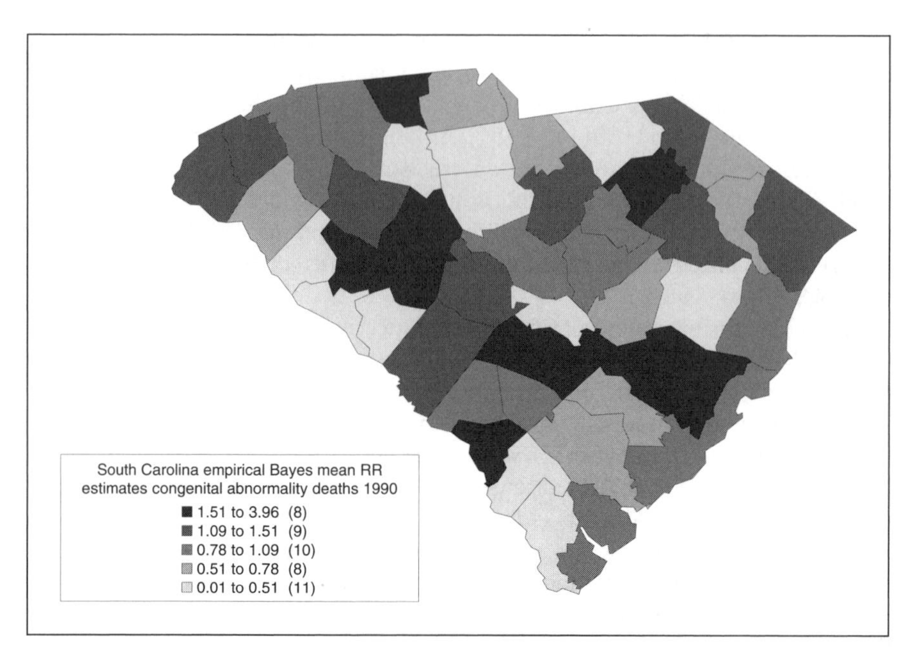

Figure 1.3 Empirical Bayes mean relative risk (RR) estimates.

require summarization of θ_i from this posterior distribution. In practice, this is often obtained by generation of realizations from this posterior and then the summarizations are empirical (e.g. Markov Chain Monte Carlo (MCMC) methods). Figure 1.3 displays the empirical Bayes estimates under the Poisson–gamma model with α and β estimated as in Clayton and Kaldor. Note that in contrast to the SMR map (Figure 1.2), Figure 1.3 presents a smoother relative risk surface.

Other approaches and variants in the analysis of simple mapping models have been proposed by Tsutakawa (1988), Marshall (1991) and Devine and Louis (1994). In the next section, more sophisticated models for the prior structure of the parameters of the map are discussed.

1.2.4 Advanced Bayesian models

Many of the models discussed above can be extended to include the specification of prior distributions for parameters and hence can be examined via Bayesian methods. In general, we distinguish here between empirical Bayes methods and full Bayes methods, on the basis that any method which seeks to approximate the posterior distribution is regarded as empirical Bayes (Bernardo and Smith, 1994). All other methods are regarded as full Bayes. This latter category includes maximum a posteriori estimation, estimation of posterior functionals, as well as posterior sampling.

1.2.4.1 *Empirical Bayes methods*

The methods encompassed under the definition above are wide-ranging, and here we will only discuss a subset of relevant methods. The first method considered by the earliest workers was the evaluation of simplified (constrained) posterior distributions. Manton *et al.* (1981) used a direct maximization of a constrained posterior distribution, Tsutakawa (1988) used integral approximations for posterior expectations, while Marshall (1991) used a method of moments estimator to derive shrinkage estimates. Devine and Louis (1994) further extended this method by constraining the mean and variance of the collection of estimates to equal the posterior first and second moments.

The second type of method which has been considered in the context of disease mapping is the use of likelihood approximations. Clayton and Kaldor (1987) first suggested employing a quadratic normal approximation to a Poisson likelihood, with gamma prior distribution for the intensity parameter of the Poisson distribution and a spatial correlation prior. Extensions to this approach lead to simple generalized least squares (GLS) estimators for a range of likelihoods (Lawson, 1994; 1997).

A third type is the Laplace asymptotic integral approximation, which has been applied by Breslow and Clayton (1993) to a generalized linear modelling framework in a disease mapping example. This integral approximation method allows the estimation of posterior moments and normalizing integrals (see, for example, Bernardo and Smith, 1994, pp. 340–4). A further, but different, integral approximation method is where the posterior distribution is integrated across the parameter space: that is, the nuisance parameters are 'integrated out' of the model. In that case the method of nonparametric maximum likelihood (NPML) can be employed (Bock and Aitkin, 1981; Aitkin, 1996b; Clayton and Kaldor, 1987). Another possibility is to employ Linear Bayes methods (Marshall, 1991).

1.2.4.2 *Full Bayes methods*

Full posterior inference for Bayesian models has now become available, largely because of the increased use of MCMC methods of posterior sampling. The first full sampler reported for a disease mapping example was a Gibbs sampler applied to a general model for intrinsic autoregression and uncorrelated heterogeneity by Besag *et al.* (1991). Subsequently, Clayton and Bernardinelli (1992), Breslow and Clayton (1993) and Bernardinelli *et al.* (1995) have adapted this approach to mapping, ecological analysis and space–time problems.

This has been facilitated by the availability of general Gibbs sampling packages such as BEAM and BUGS (GeoBUGS and WinBUGS). Such Gibbs sampling methods can be applied to putative source problems as well as mapping/ecological studies. Alternative, and more general, posterior sampling methods, such as the Metropolis–Hastings algorithm, are currently not separately available in a packaged form, although these methods can accommodate considerable variation in model specification. WinBUGS does provide such estimators when non-convex posterior distributions are encountered. Metropolis–Hastings algorithms have been applied in comparison to approximate maximum a posteriori (MAP) estimation by Lawson *et al.* (1996) and Diggle *et al.* (1998); hybrid Gibbs–Metropolis samplers have been applied to space–time problems by Waller *et al.* (1997). In addition, diagnostic methods for Bayesian MCMC sample output have been discussed for disease mapping examples by Zia *et al.* (1997). Developments in this area have been reviewed recently (Lawson *et al.*, 1999; Elliott *et al.*, 2000; Lawson, 2001).

1.2.5 Multilevel modelling approaches

An alternative to the above specification can be considered where a log-linear form is specified:

$$\theta_i = \exp\{\beta_0 + v_i\},$$

where the random term has a zero mean Gaussian distribution, i.e. $v_i \sim N(0, \sigma_v^2)$ and σ_v^2 is the variance of the random effects v.

This model may be rewritten (in terms of counts rather than rates) as

$$y_i \sim Poisson(\mu_i),\ log(\mu_i) = log(e_i) + \beta_0 + v_i.$$

Here $\mu_i = e_i\theta_i$ and the $log(e_i)$ are treated as known 'offset' terms.

Generally, multilevel models (see, for example, Goldstein, 1995) are fitted to data that possess levels of clustering in their structure. In disease mapping and geographical applications in general such levels would be different levels of geographical aggregation, for example, census tracts nested within counties nested within countries. For each level of geography we could then fit normally distributed random effects so for example if we had data on census tracts nested within counties we could fit

$$y_{ij} \sim Poisson(\mu_{ij}),\ log(\mu_{ij}) = log(e_{ij}) + \beta_0 + v_j + u_{ij},$$

where both the county and tract random effects have Gaussian distributions, i.e. $v_j \sim N(0, \sigma_v^2)$ and $u_{ij} \sim N(0, \sigma_u^2)$.

Poisson response multilevel models can be fitted using either frequentist or Bayesian approaches. Frequentist approaches generally involve some approximations, for example the software package MLwiN (Rasbash *et al.*, 2000) uses quasi-likelihood methods that involve Taylor series approximations (Goldstein, 1991; Goldstein and Rasbash, 1996) to transform the problem so that it can be fit using the iterative general least squares algorithm (IGLS; Goldstein (1986)). Other common frequentist approaches include Laplace approximations (Raudenbush *et al.*, 2000) and Gaussian quadrature (e.g. Rabe-Hesketh *et al.*, 2001). Bayesian estimation involves extending the models to include prior distributions and fitting using MCMC estimation.

Fitting higher-level random effects may account for some spatial auto-correlation in the data but multilevel models can also more directly account for spatial correlations via their extension to multiple-membership models (Hill and Goldstein, 1998). Although originally used to account for missing unit identifiers, Langford *et al.* (1999) showed how to use such models to fit spatial data. The model can be described as

$$y_i \sim Poisson(\mu_i),\ log(\mu_i) = log(e_i) + \beta_0 + v_i + \sum_{j \in neigh(i)} w_{ij} u_j,$$

where $u_i \sim N(0,\ \sigma_u^2)$ and $v_i \sim N(0,\ \sigma_v^2)$. Here we have, for each observation, an unstructured random effect v_i and a group of (weighted) 'neighbour' random effects u_j. Browne *et al.* (2001) consider Bayesian extensions of this model as a member of the family of models that they call multiple-membership multiple-classification (MMMC) models. Langford *et al.* (1999) also introduce a correlation between the two sets of random effects giving u and v a multivariate Gaussian distribution. Multilevel approaches will be discussed in greater detail in Chapter 3.

2

Bayesian Hierarchical Modelling

In this chapter we introduce the basic ideas behind Bayesian modelling in relation to disease mapping applications. In the previous chapter we have highlighted some simple approaches to relative risk estimation and have stressed the shortcoming of the SMR as a relative risk estimator. To circumvent these shortcomings a variety of estimators have been proposed. One approach to improving the estimation process is to control the behaviour of the $\{\theta_i\}$. To this end it is appropriate to extend our model by assuming that the parameters have distributions. These distributions control the form of the parameters and are specified by the investigator based, usually, on their prior belief concerning their behaviour. These distributions are termed *prior* distributions. We will denote a prior distribution by $g(\theta)$. In the disease mapping context a commonly assumed prior distribution for θ is the gamma distribution and the resulting model is the Poisson–gamma model.

2.1 LIKELIHOOD AND POSTERIOR DISTRIBUTIONS

Prior distributions and likelihood provide two sources of information about any problem. The likelihood informs about the parameter via the data, while the prior distributions inform via prior beliefs or assumptions. When there are large amounts of data, i.e. the sample size is large, the likelihood will contribute more to the relative risk estimation. When the example is data poor then the prior distributions will dominate the analysis.

Denote the likelihood of data $\{y_i\}$ given the parameters $\{\theta_i\}$ as $L(\mathbf{y}|\boldsymbol{\theta})$ and the log-likelihood as $l(\mathbf{y}|\boldsymbol{\theta})$. Note that $\boldsymbol{\theta}$ does not have to be the same dimension as $\mathbf{y}$. The product of the likelihood and the prior distributions is called the posterior distribution. This distribution describes the behaviour of the parameters after

Disease Mapping with WinBUGS and MLwiN A. Lawson, W. Browne and C. Vidal Rodeiro
 ISBN: 0-470-85604-1 (HB)

the data are observed and prior assumptions are made. The posterior distribution is defined as:

$$p(\boldsymbol{\theta}|\mathbf{y}) \propto L(\mathbf{y}|\boldsymbol{\theta})\mathbf{g}(\boldsymbol{\theta}),$$

where $\mathbf{g}(\boldsymbol{\theta})$ is the joint prior distribution of the $\boldsymbol{\theta}$ vector.

A simple example of this type of model in disease mapping occurs when the data likelihood is Poisson and there is a common relative risk parameter with a single gamma prior distribution:

$$p(\boldsymbol{\theta}|\mathbf{y}) \propto L(\mathbf{y}|\theta)g(\theta),$$

where $g(\theta)$ is a gamma distribution with parameters α, β, i.e. $Gamma(\alpha, \beta)$, and $L(\mathbf{y}|\theta) = \prod_{i=1}^{m}\{(e_i\theta)^{y_i}\exp(e_i\theta)\}$ bar a constant only dependent on the data. A compact notation for this model is:

$$\begin{aligned} y_i|\theta &\sim Poisson(e_i\theta), \\ \theta &\sim Gamma(\alpha, \beta). \end{aligned}$$

2.2 HIERARCHICAL MODELS

In the previous section a simple example of a likelihood and prior distribution was given. In that example the prior distribution for the parameter also had parameters controlling its form. These parameters (α, β) can have assumed values, but more usually an investigator will not have a strong belief in the prior parameter values. The investigator may therefore want to estimate these parameters from the data. Alternatively and more formally, as parameters within models are regarded as stochastic (and thereby have probability distributions governing their behaviour), then these parameters must also have distributions. These distributions are known as hyperprior distributions, and the parameters are known as hyperparameters.

The idea that the values of parameters could arise from distributions is a fundamental feature of Bayesian methodology and leads naturally to the use of models where parameters arise within hierarchies. In the Poisson–gamma example there is a two level hierarchy: θ has a $Gamma(\alpha, \beta)$ distribution at the first level of the hierarchy and α will have a hyperprior distribution (h_α) as will β (h_β), at the second level of the hierarchy. This can be written as:

$$\begin{aligned} y_i|\theta &\sim Poisson(e_i\theta), \\ \theta|\alpha, \beta &\sim Gamma(\alpha, \beta), \\ \alpha|\nu &\sim h_\alpha(\nu), \\ \beta|\rho &\sim h_\beta(\rho). \end{aligned}$$

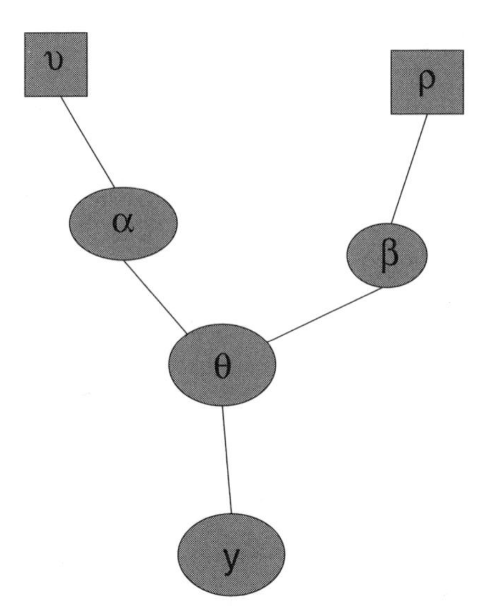

Figure 2.1 Directed acyclic graph for the two-level Poisson–gamma model.

For these types of models it is also possible to use a graphical tool to display the linkages in the hierarchy. This is known as a directed acyclic graph or DAG for short. On such a graph lines connect the levels of the hierarchy and parameters are nodes at the ends of the lines. Clearly it is important to terminate a hierarchy at an appropriate place, otherwise one could assume an infinite hierarchy of parameters. Usually the cut-off point is chosen to lie where further variation in parameters will not affect the lowest-level model. At this point the parameters are assumed to be fixed. For example, in the Poisson–gamma model if you assume α and β were fixed then the gamma prior would be fixed and the choice of α and β would be uninformed. The data would not inform about the distribution at all. However, by allowing a higher level of variation, i.e. hyperpriors for α, β, then we can fix the values of ν and ρ without heavily influencing the lower-level variation. Figure 2.1 displays the DAG for the simple two-level Poisson–gamma model just described.

2.3 POSTERIOR INFERENCE

When a simple likelihood model is employed, often maximum likelihood is used to provide a point estimate and associated variability for parameters. This is true for simple disease mapping models. For example, in the model $y_i|\theta \sim Poisson(e_i\theta)$ the maximum likelihood estimate of θ is the overall rate for the study region, i.e. $\sum y_i / \sum e_i$. On the other hand, the SMR is the maximum likelihood estimate for the model $y_i|\theta_i \sim Poisson(e_i\theta_i)$.

When a Bayesian hierarchical model is employed it is no longer possible to provide a simple point estimate for any of the θ_is. This is because the parameter is no longer assumed to be fixed but is assumed to arise from a distribution of possible values. Given the observed data, the parameter or parameters of interest will be described by the posterior distribution, and hence this distribution must be found and examined. It is possible to examine the expected value (mean) or the mode of the posterior distribution to give a point estimate for a parameter. Just as the maximum likelihood estimate is the mode of the likelihood, then the *maximum a posteriori* estimate is that value of the parameter or parameters at the mode of the posterior distribution. More commonly the expected value of the parameter or parameters is used. This is known as the posterior mean (or Bayes estimate). For simple unimodal symmetrical distributions, the modal and mean estimates coincide.

For some simple posterior distributions it is possible to find the exact form of the posterior distribution and to find explicit forms for the posterior mean or mode. However, it is commonly the case that for reasonably realistic models within disease mapping, it is not possible to obtain a closed form for the posterior distribution. Hence it is often not possible to derive simple estimators for parameters such as the relative risk. In this situation resort must be made to posterior sampling, i.e. using simulation methods to obtain samples from the posterior distribution which can then be summarized to yield estimates of relevant quantities. In the next section we discuss the use of sampling algorithms for this purpose.

An exception to this situation where a closed form posterior distribution can be obtained is the Poisson–gamma model with relative risk parameters $\{\theta_i\}$, where α, β are fixed. In this case, the relative risks have posterior distribution given by:

$$\theta_i \sim Gamma(y_i + \alpha, e_i + \beta),$$

and the posterior expectation of θ_i is $(y_i + \alpha)/(e_i + \beta)$. Of course if α and β are not fixed and have hyperprior distributions then the posterior distribution is more complex. Clayton and Kaldor (1987) use an approximation procedure to obtain estimates of α and β from a marginal likelihood based on the assumption that α and β had uniform hyperprior distributions. These estimates are those displayed in Figure 1.3. Note that these are not the full posterior expected estimates of the parameters from within a two-level model hierarchy.

2.4 MARKOV CHAIN MONTE CARLO METHODS

Often in disease mapping, realistic models for maps have two or more levels and the resulting complexity of the posterior distribution of the parameters requires the use of sampling algorithms. In addition, the flexible modelling of disease could require switching between a variety of relatively complex models. In this case, it is convenient to have an efficient and flexible posterior sampling method which could be applied across a variety of models. Efficient algorithms for this purpose were developed within the fields of physics and image processing to handle large-scale problems in estimation. In the late 1980s and early 1990s these methods were developed further, particularly for dealing with Bayesian posterior sampling for more general classes of problems (Gilks *et al.* 1993; 1996). Now posterior sampling is commonplace and a variety of packages (including WinBUGS and MLwiN) have incorporated these methods. For general reviews of this area the reader is referred to Casella and George (1992), and Robert and Casella (1999). Markov chain Monte Carlo (MCMC) methods are a set of methods which use iterative simulation of parameter values within a Markov chain. The convergence of this chain to a stationary distribution, which is assumed to be the posterior distribution, must be assessed.

Prior distributions for the p components of $\boldsymbol{\theta}$ are defined as $g_i(\theta_i)$ for $i = 1, \ldots, p$. The posterior distribution of $\boldsymbol{\theta}$ and $\mathbf{y}$ is then defined as:

$$P(\boldsymbol{\theta}|\mathbf{y}) \propto L(\mathbf{y}|\boldsymbol{\theta}) \prod_i g_i(\theta_i). \tag{2.1}$$

The aim is to generate a sample from the posterior distribution $P(\boldsymbol{\theta}|\mathbf{y})$. Suppose we can construct a Markov chain with state space $\boldsymbol{\theta}_c$, where $\boldsymbol{\theta} \in \boldsymbol{\theta}_c \subset \mathcal{R}^k$. The chain is constructed so that the equilibrium distribution is $P(\boldsymbol{\theta}|\mathbf{y})$, and the chain should be easy to simulate from. If the chain is run over a long period, then it should be possible to reconstruct features of $P(\boldsymbol{\theta}|\mathbf{y})$ from the realized chain values. This forms the basis of the MCMC method, and algorithms are required for the construction of such chains. A selection of recent literature on this area is found in Ripley (1987), Gelman *et al.* (1995), Smith and Roberts (1993), Besag and Green (1993), Cressie (1993), Smith and Gelfand (1992), Tanner (1996) and Robert and Casella (1999).

The basic algorithms used for this construction are:

(1) the Metropolis algorithm and its extension the Metropolis–Hastings algorithm;
(2) the Gibbs Sampler algorithm.

2.5 METROPOLIS AND METROPOLIS–HASTINGS ALGORITHMS

In all MCMC algorithms, it is important to be able to construct the correct *transition probabilities* for a chain which has $P(\boldsymbol{\theta}|\mathbf{y})$ as its equilibrium distribution. A Markov chain consisting of $\boldsymbol{\theta}^1, \boldsymbol{\theta}^2, \ldots, \boldsymbol{\theta}^t$ with state space Θ and equilibrium distribution $P(\boldsymbol{\theta}|\mathbf{y})$ has transitions defined as follows.

Define $q(\boldsymbol{\theta}, \boldsymbol{\theta}')$ as a transition probability function, such that, if $\boldsymbol{\theta}^t = \boldsymbol{\theta}$, the vector $\boldsymbol{\theta}^t$ drawn from $q(\boldsymbol{\theta}, \boldsymbol{\theta}')$ is regarded as a proposed possible value for $\boldsymbol{\theta}^{t+1}$.

2.5.1 Metropolis updates

In this case choose a symmetric proposal $q(\boldsymbol{\theta}, \boldsymbol{\theta}')$ and define the transition probability as

$$p(\boldsymbol{\theta}, \boldsymbol{\theta}') = \begin{cases} \alpha(\boldsymbol{\theta}, \boldsymbol{\theta}')q(\boldsymbol{\theta}, \boldsymbol{\theta}') & \text{if } \boldsymbol{\theta}' \neq \boldsymbol{\theta}, \\ 1 - \sum_{\theta''} q(\boldsymbol{\theta}, \boldsymbol{\theta}'')\alpha(\boldsymbol{\theta}, \boldsymbol{\theta}'') & \text{if } \boldsymbol{\theta}' = \boldsymbol{\theta}, \end{cases}$$

where

$$\alpha(\boldsymbol{\theta}, \boldsymbol{\theta}') = \min\left\{1, \frac{P(\boldsymbol{\theta}'|\mathbf{y})}{P(\boldsymbol{\theta}|\mathbf{y})}\right\}.$$

In this algorithm a proposal is generated from $q(\boldsymbol{\theta}, \boldsymbol{\theta}')$ and is accepted with probability $\alpha(\boldsymbol{\theta}, \boldsymbol{\theta}')$. The acceptance probability is a simple function of the ratio of posterior distributions as a function of $\boldsymbol{\theta}$ values. The proposal function $q(\boldsymbol{\theta}, \boldsymbol{\theta}')$ can be defined to have a variety of forms but must be an irreducible and aperiodic transition function. Specific choices of $q(\boldsymbol{\theta}, \boldsymbol{\theta}')$ lead to specific algorithms.

2.5.2 Metropolis–Hastings updates

In this extension to the Metropolis algorithm the proposal function is not confined to symmetry and

$$\alpha(\boldsymbol{\theta}, \boldsymbol{\theta}') = \min\left\{1, \frac{P(\boldsymbol{\theta}'|\mathbf{y})q(\boldsymbol{\theta}', \boldsymbol{\theta})}{P(\boldsymbol{\theta}|\mathbf{y})q(\boldsymbol{\theta}, \boldsymbol{\theta}')}\right\}.$$

Some special cases of chains are found when $q(\boldsymbol{\theta}, \boldsymbol{\theta}')$ has special forms. For example, if $q(\boldsymbol{\theta}, \boldsymbol{\theta}') = q(\boldsymbol{\theta}', \boldsymbol{\theta})$ then the original Metropolis method arises and further, with $q(\boldsymbol{\theta}, \boldsymbol{\theta}') = q(\boldsymbol{\theta}')$, (i.e. when no dependence on the previous value is assumed) then

$$\alpha(\boldsymbol{\theta}, \boldsymbol{\theta}') = \min\left\{1, \frac{w(\boldsymbol{\theta}')}{w(\boldsymbol{\theta})}\right\},$$

where $w(\boldsymbol{\theta}) = P(\boldsymbol{\theta}|\mathbf{y})/q(\boldsymbol{\theta})$ and $w(.)$ are importance weights. One simple example of the method is $q(\boldsymbol{\theta}') \sim$ *Uniform* $(\boldsymbol{\theta}_a, \boldsymbol{\theta}_b)$ and $g_i(\theta_i) \sim$ *Uniform* $(\theta_{ia}, \theta_{ib})$ $\forall i$, this leads to an acceptance criterion based on a likelihood ratio. Hence the original Metropolis algorithm with uniform proposals and prior distributions leads to a stochastic exploration of a likelihood surface. This, in effect, leads to the use of prior distributions as proposals. However, in general, when the $g_i(\theta_i)$ are not uniform this leads to inefficient sampling. The definition of $q(\boldsymbol{\theta}, \boldsymbol{\theta}')$ can be quite general in this algorithm and, in addition, the posterior distribution only appears within a ratio as a function of $\boldsymbol{\theta}$ and $\boldsymbol{\theta}'$. Hence, the distribution is only required to be known up to proportionality.

2.5.3 Gibbs updates

The Gibbs Sampler has gained considerable popularity, particularly in applications in medicine, where hierarchical Bayesian models are commonly applied (see, for example, Gilks *et al.*, 1993). This popularity is mirrored in the availability of software which allows its application in a variety of problems (e.g. WinBUGS, MLwiN, BACC). This sampler is a special case of the Metropolis–Hastings algorithm where the proposal is generated from the conditional distribution of θ_i given all other $\boldsymbol{\theta}$s, and the resulting proposal value is accepted with probability 1.

More formally, define

$$q(\theta_j, \theta_j') = \begin{cases} p(\theta_j^*|\theta_{-j}^{t-1}) & \text{if } \theta_{-j}^* = \theta_{-j}^{t-1}, \\ 0 & \text{otherwise}, \end{cases}$$

where $p(\theta_j^*|\theta_{-j}^{t-1})$ is the conditional distribution of θ_j given all other $\boldsymbol{\theta}$ values (θ_{-j}) at time $t-1$. Using this definition it is straightforward to show that

$$\frac{q(\boldsymbol{\theta}, \boldsymbol{\theta}')}{q(\boldsymbol{\theta}', \boldsymbol{\theta})} = \frac{P(\boldsymbol{\theta}'|\mathbf{y})}{P(\boldsymbol{\theta}|\mathbf{y})}$$

and hence $\alpha(\boldsymbol{\theta}, \boldsymbol{\theta}') = 1$.

2.5.4 Metropolis–Hastings (M-H) versus Gibbs algorithms

There are advantages and disadvantages to M-H and Gibbs methods. The Gibbs Sampler provides a *single* new value for each $\boldsymbol{\theta}$ at each iteration, but requires the evaluation of a conditional distribution. On the other hand the M-H step does not require evaluation of a conditional distribution but does not guarantee the acceptance of a new value. In addition, block updates of parameters are available in M-H, but not usually in Gibbs steps (unless joint conditional distributions are available). If conditional distributions are difficult to obtain or computationally expensive, then M-H can be used and is usually available.

In summary, the Gibbs Sampler may provide faster convergence of the chain if the computation of the conditional distributions at each iteration is not time-consuming. The M-H step will usually be faster at each iteration, but will not necessarily guarantee exploration. In straightforward hierarchical models where conditional distributions are easily obtained and simulated from, then the Gibbs Sampler is likely to be favoured. In more complex problems, such as many arising in spatial statistics, resort may be required to the M-H algorithm.

A simple M-H example Assume that for m regions, the count $y_i, i = 1, \ldots, m$ is observed. In addition, the expected count in the ith region, e_i is also observed. Assume also that the counts are independently distributed and have a Poisson distribution with $E(y_i) = e_i\theta$, where θ is a constant parameter describing the relative risk over the whole study window. The likelihood in this case, bar a constant, is given by

$$L(\theta) = \exp\left(-\theta \sum_{i=1}^{m} e_i\right) \prod_{i=1}^{m} (\theta e_i)^{y_i}. \tag{2.2}$$

Assuming a flat prior distribution for θ, then the M-H sampler for this problem reduces to a stochastic exploration of the likelihood surface. Hence the following sampler criterion is found for the θ parameter in this case:

$$\frac{L(\theta')}{L(\theta)} = \exp\{s_e(\theta - \theta')\}\left(\frac{\theta'}{\theta}\right)^{s_y},$$

where

$$s_e = \sum_{i=1}^{m} e_i \text{ and } s_y = \sum_{i=1}^{m} y_i.$$

2.5.5 Special methods

Alternative methods exist for posterior sampling when the basic Gibbs or M-H updates are not feasible or appropriate. For example, if the range of the parameters are restricted then slice sampling can be used (Robert and Casella, 1999, Chapter 7; Neal, 2003). When exact conditional distributions are not available but the posterior is log-concave then adaptive rejection sampling algorithms can be used. The most general of these algorithms (adaptive rejection sampling (ARS) algorithm; Robert and Casella (1999, pp. 57–9)) has wide applicability for continuous distributions, but may not be efficient for specific cases. Block updating can also be used to effect in some situations. When generalized linear model components are included then block updating of the covariate parameters can be effected via multivariate updating.

2.5.6 Convergence

MCMC methods require the use of diagnostics to assess whether the iterative simulations have reached the equilibrium distribution of the Markov chain. Sampled chains require to be run for an initial burn-in period until they can be assumed to converge to the posterior distribution of interest. This burn-in period can vary considerably between different problems. In addition, it is important to ensure that the chain manages to explore the parameter space properly so that the sampler does not 'stick' in local maxima of the surface of the distribution. Hence, it is crucial to ensure that a burn-in period is adequate for the problem considered. Judging convergence has been the subject of much debate and can still be regarded as art rather than science: a qualitative judgement has to be made at some stage as to whether the burn-in period is long enough.

There are a wide variety of methods now available to assess convergence of chains within MCMC. Robert and Casella (1999) and Chen *et al.* (2000) provide recent reviews. The available methods are largely based on checking the distributional properties of samples from the chains.

2.5.6.1 Single-chain methods

First, global methods for assessing convergence have been proposed which involve monitoring functions of the posterior probability at each iteration. These methods look for stabilization of the probability value. This value forms a time series, and special cusum methods have been proposed (Yu and Mykland, 1998). Other related methods seek to examine summary statistics based on distance evaluations (Brooks–Draper diagnostic) and on a binary control model where the batch size is successively evaluated for a two-state Markov chain (Raftery–Lewis diagnostic). Section 8.4 in Robert and Casella (1999) discusses in detail the advantages and disadvantages of these measures.

Second, graphical methods have been proposed which allow the comparison of the whole distribution of successive samples. Quantile–quantile plots of successive lengths of single variable output from the sampler can be used for this purpose.

2.5.6.2 Multi-chain methods

Single-chain methods can, of course, be applied to each of a multiple of chains. In addition, there are methods that can only be used for multiple chains. The Gelman–Rubin statistic was proposed as a method for assessing the convergence of multiple chains via the comparison of summary measures across chains (Gelman and Rubin, 1992; Brooks and Gelman, 1998; Robert and Casella,

1999, Chapter 8). There is some debate about whether it is useful to run one long chain as opposed to multiple chains with different start points. The advantage of multiple chains is that they provide evidence for the robustness of convergence across different subspaces. However, as long as single chains sample the parameter space adequately, then these have benefits. The reader is referred to Robert and Casella (1999, Chapter 8) for a thorough discussion of diagnostics and their use.

2.6 RESIDUALS AND GOODNESS OF FIT

2.6.1 Model goodness-of-fit measures

As a model choice criterion the *Bayesian Information Criterion (BIC)* is widely used in Bayesian and hierarchical models. It asymptotically approximates a Bayes factor. In a model with log-likelihood $l(\theta)$ the BIC value is estimated from the output of an MCMC algorithm by

$$2\hat{l}(\theta^*) - p \ln n,$$

where p is the number of linearly independent parameters, n is the number of data points and

$$\hat{l}(\theta^*) = \frac{1}{G}\sum_{i=1}^{G} l(\theta_i^*),$$

the averaged log-likelihood over G posterior samples of θ. Recently, the *Deviance Information Criterion (DIC)* (Spiegelhalter *et al.*, 2002a) has been proposed. This is defined as

$$DIC = 2E_{\theta|x}\{D\} - D\{E_{\theta|x}(\theta)\},$$

where $D(.)$ is the deviance of the model and x is the observed data. This uses the average of the posterior samples of θ to produce an expected value of θ. This value can also be computed from a sample output from a chain.

2.6.2 General residuals

The analysis of residuals and summary functions of residuals forms a fundamental part of the assessment of model goodness of fit in any area of statistical application. In the case of disease mapping there is no exception, although full residual analysis is seldom presented in published work in the area. Often goodness-of-fit measures are aggregate functions of piecewise residuals, while measures relating to individual residuals are also available. A variety of methods are available when full residual analysis is to be undertaken. We define a piecewise residual as the standardized difference between the observed value and the fitted model value. Usually the standardization will be based on a measure of the variability of the difference between the two values.

It is common practice to specify a residual as:

$$r_{1i} = \{y_i - \widehat{y}_i\}$$

or

$$r_{2i} = \{y_i - \widehat{y}_i\}/\sqrt{var(y_i - \widehat{y}_i)}, \tag{2.3}$$

where $\widehat{y}_i$ is a fitted value under a given model. When complex spatial models are considered, it is often easier to examine residuals, such as $\{r_{1i}\}$ using Monte Carlo methods. In fact it is straightforward to implement a Parametric Bootstrap (PB) approach to residual diagnostics for likelihood models. The simplest case, is that of tract count data, where for each tract an observed count can be compared to a fitted count. In general, when Poisson likelihood models are assumed with $y_i \sim Poisson\{e_i\theta_i\}$ then it is straightforward to employ a PB by generating a set of simulated counts $\{y^*_{ij}\}$ $j = 1, \ldots, J$, from a Poisson distribution with mean $e_i\widehat{\theta}_i$. In this way, a tract-wise ranking, and hence p-value, can be computed by assessing the rank of the residual within the pooled set

$$(y_i - e_i\widehat{\theta}_i; \{y^*_{ij} - e_i\widehat{\theta}_i\}, j = 1, \ldots, J). \tag{2.4}$$

Denote the observed standardized residual as r_{2i} and the simulated residuals as $\{r^*_{2ij}\}$. Note that it is now possible to compare functions of the residuals as well as making direct comparisons. For example, in a spatial context, it may be appropriate to examine the spatial autocorrelation of the observed residuals. Hence, a Monte Carlo assessment of degree of residual autocorrelation could be made by comparing Moran's I statistic for the observed residuals, say, $M(\{r_{2i}\})$, to that found for the simulated count residuals $M(\{r^*_{2ij}\})$.

2.6.3 Bayesian residuals

In a Bayesian setting it is natural to consider the appropriate version of Equation (2.3). Carlin and Louis (1996) describe a Bayesian residual as:

$$r_i = y_i - \frac{1}{G}\sum_{g=1}^{G} E(y_i|\theta_i^{(g)}), \tag{2.5}$$

where $E(y_i|\theta_i)$ is the expected value from the posterior predictive distribution, and (in the context of MCMC sampling) $\{\theta_i^{(g)}\}$ is a set of parameter values sampled from the posterior distribution.

In the tract count modelling case, this residual can therefore be approximated, when a constant tract rate is assumed, by:

$$r_i = y_i - \frac{1}{G}\sum_{g=1}^{G} e_i \theta_i^{(g)}. \tag{2.6}$$

This residual averages over the posterior sample. An alternative possibility is to average the parameters within $\{\theta_i^{(g)}\}$ and estimate the relative risk, $\widehat{\theta}_i$ say, using 'plug-in' average parameters, to yield a posterior expected value of y_i, say $\widehat{y}_i = e_i\widehat{\theta}_i$, and to form $r_i = y_i - \widehat{y}_i$. A further possibility is to simply form r_{2i} at each iteration of a posterior sampler and to average these over the converged sample (Spiegelhalter *et al.*, 1996). These residuals can provide pointwise goodness-of-fit (gof) measures as well as global gof measures, and can be assessed using Monte Carlo methods.

Deletion residuals and residuals based on conditional predictive ordinates (CPOs) can also be defined for tract counts (Stern and Cressie, 2000). To further assess the distribution of residuals, it would be advantageous to be able to apply the equivalent of a PB in the Bayesian setting. With convergence of a MCMC sampler, it is possible to make subsamples of the converged output. If these samples are separated by a distance (h) which will guarantee approximate independence, then a set of J such samples could be used to generate data and the residual computed from the data can be compared to the set of J residuals computed from the generated data. In turn, these residuals can be used to assess functions of the residuals and gof measures. The choice of J will usually be 99 or 999 depending on the level of accuracy required.

3

Multilevel Modelling

When data are collected in many application areas, for example education, medicine and public health, there is an inherent structure to the units of the population from which the data come. Often this structure is nested (or hierarchical) with the individual observations being taken from clustered 'higher level' units. For example, in education, data is often collected on individual students who are nested within classrooms within schools. Similarly, in public health, observations may be on individuals who are nested within various levels of geography (wards, counties, regions etc.).

Although it is plausible that within a particular clustered unit, observations are independent, we may expect that observations within a particular cluster are more alike than observations from different clusters, due to the shared social and geographical environment experienced within the cluster. When we wish to model such data it is therefore important to take account of the underlying structure and in particular the correlation between observations from the same cluster.

Multilevel modelling is a term used generically to describe the fitting of statistical models to data that exhibit levels of clustering. It is one of many terms used for such models and alternatives include hierarchical models and random effects models. In this chapter we intend to show how we can use multilevel modelling in the analysis of disease mapping data. We will begin with a history of multilevel modelling and an introduction to modelling of continuous responses. Note that although disease mapping data are not generally continuous responses the methods used to fit multilevel models to count data are extensions of the methods used for continuous data so we consider such models first. We will then consider the standard count data that is met in practice. We will finally show extensions that allow the incorporation of spatial effects and higher-level effects.

Disease Mapping with WinBUGS and MLwiN A. Lawson, W. Browne and C. Vidal Rodeiro
 ISBN: 0-470-85604-1 (HB)

3.1 CONTINUOUS RESPONSE MODELS

Multilevel modelling has been used in the field of education since the 1980s. The early works of Aitkin *et al.* (1981) and Aitkin and Longford (1986) fuelled interest in the community of statisticians working in education and within the next decade several statistical software packages were born that were designed specifically to fit multilevel models. Harvey Goldstein and his team of researchers at the Institute of Education produced a progression of software packages of increasing complexity (ML2, ML3 and MLN) which were the forerunners of the current MLwiN package (Rasbash *et al.*, 2000). In the USA, Bryk, Raudenbush and colleagues developed the package HLM (Bryk *et al.*, 1988) at the University of Chicago and Nick Longford produced the package VARCL (Longford, 1988).

In education the data collected often have a natural hierarchical structure, for example pupils nested within classrooms or schools, or marks measured over time nested within individual children. Generally the responses considered were attainment scores which although being marks out of a given total were treated as continuous measures. Let us for example consider a dataset with children nested within schools as shown in Figure 3.1, and as a response let us consider each child's exam mark at the end of their schooling. To predict such a response we will have student characteristics, for example their gender, their parents' social class and occupation and their marks on earlier tests. We will also have

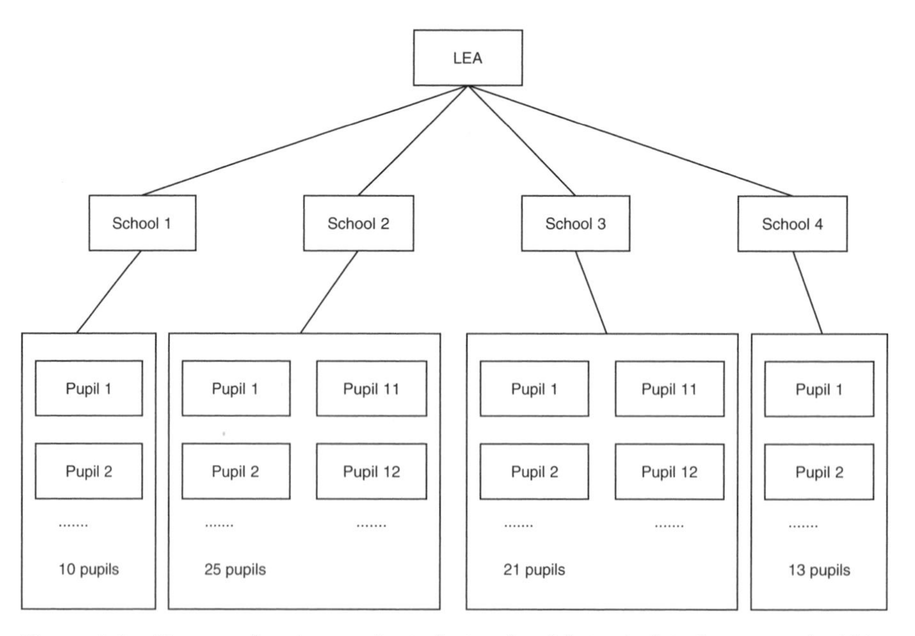

Figure 3.1 Diagram showing a subset of a two-level dataset of students nested within schools in a single local education authority (LEA).

school-level predictors (including the school identifier) such as the size and type of the school.

Perhaps the simplest multilevel model that we can fit is a *variance components* model which is also known as a random intercepts model. This model is an extension of a linear model where the effects of the school each student attends are treated as random variables. We can write such a model as follows:

$$y_{ij} = \beta_0 + X_{ij}^{(1)}\beta^{(1)} + X_j^{(2)}\beta^{(2)} + u_j + e_{ij}. \qquad (3.1)$$

Here we have a global intercept, β_0, a set of student level effects, $\beta^{(1)}$, for the student level predictors, e.g. gender, a set of school level effects, $\beta^{(2)}$, for the school level predictors, e.g. school type, a set of school level residuals, u_j, and a set of pupil level residuals, e_{ij}. The first three terms are referred to as the *fixed* part of the model and are often combined to a single term. The final two residual terms are referred to as the *random* part of the model and additional distributional assumptions are required for these terms. The model is therefore more commonly written:

$$y_{ij} = X_{ij}\beta + u_j + e_{ij}, \quad u_j \sim N(0, \sigma_u^2), \quad e_{ij} \sim N(0, \sigma_e^2). \qquad (3.2)$$

Normal distributions are assumed for both the school and student residuals and these residuals have variances σ_u^2 and σ_e^2 respectively. Interest often lies in the intra-class (intra-school) correlation or ICC which is the percentage of the variance in the data that is explained by the higher-level (school) residuals:

$$ICC = \frac{\sigma_u^2}{\sigma_u^2 + \sigma_e^2}. \qquad (3.3)$$

Typical values of the ICC will range from 10% to 20% for educational problems and the bulk of the variation is still unexplained and resides at the pupil level.

3.1.1 Applying continuous response models to disease mapping data

Although continuous response variables occur often in education, in disease mapping our responses are generally counts and such data are treated differently. Leyland and McLeod (2000), however, considered treating the Standardized Mortality ratio (SMR) as a continuous response variable. The SMR is the observed count/expected count for a particular area, in this case the areas are the administrative districts of England and Wales. The dataset that they analysed consists of all deaths in England and Wales (counted yearly) over the period 1979 to 1992 for each of the 403 districts within England and Wales.

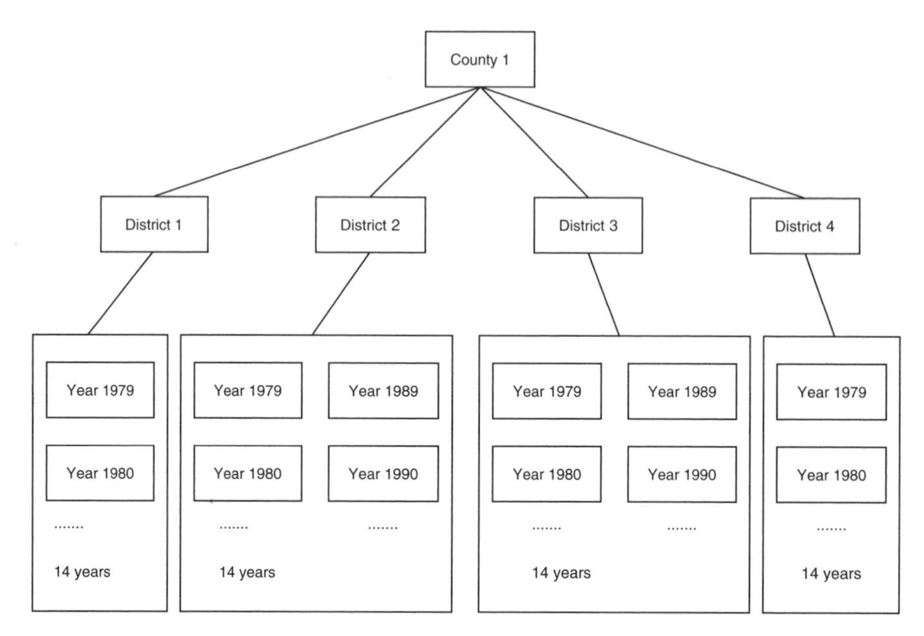

Figure 3.2 Diagram showing the structure of a subset of the data for the England and Wales mortality data.

Although modelling children's progress in school and modelling deaths in the population of England and Wales would appear very different, the underlying structure of the data is very similar. We will firstly consider the SMRs for the 14 years to be nested within each district. We can represent this structure in the data by a diagram (see Figure 3.2) which is almost identical to Figure 3.1 for the education example.

Data structures where we have observations taken at various times (years) nested within individuals (districts) are often referred to as *repeated measures* data. The multilevel modelling approach is particularly suited to fitting such data and has the added advantage that such models can easily cater with missing observations (assuming they are missing at random). For example, if there were districts in our dataset which did not start recording deaths until 1985, we could still include them in the model.

A simple multilevel model that we could fit to this data is the following:

$$\begin{aligned} SMR_{ij} &= \beta_0 + \beta_1(year_{ij} - 1979) + u_j + e_{ij}, \\ u_j &\sim N(0, \sigma_u^2), \;\; e_{ij} \sim N(0, \sigma_e^2). \end{aligned} \tag{3.4}$$

This model is a variance components model where we assume a linear trend for the SMRs over time with random intercepts (u_j) for each district. Here we use i to index year (within district) and j to index district. We have subtracted 1979 from the predictor $year_{ij}$ and it is common practice to either centre predictor variables or to subtract a constant so that the intercepts in our model have a more meaningful interpretation. For our example β_0 is now the average SMR value in year 1979.

Interestingly in repeated measures models we often find that we get large ICC values as more variability typically exists between individuals (districts) than the time points (after adjusting for the trend). For the mortality dataset 82% of the variability is found to be between districts.

A linear trend may not be appropriate for this dataset and so we could include in the model higher-order polynomial functions of time. In fact Leyland and McLeod (2000) find a quadratic term gives a better model fit. This model will result in a series of quadratic curves, one for each district that are all parallel but which have differing intercepts, hence the name *random intercepts* model. It is also possible that other predictors may have different effects for each school, for example it may not be a plausible assumption that districts have parallel curves and we may want to investigate fitting the slope (linear) term as random across districts.

This will result in the model

$$\begin{aligned}
SMR_{ij} &= \beta_{0ij} + \beta_{1j}(year_{ij} - 1979) + \beta_2(year_{ij} - 1979)^2, \\
\beta_{0ij} &= \beta_0 + u_{0j} + e_{0ij}, \\
\beta_{1j} &= \beta_1 + u_{1j}, \\
u_j &\sim MVN(0, \Omega_u), \; e_{0ij} \sim N(0, \sigma_e^2).
\end{aligned} \tag{3.5}$$

Here we fit a quadratic regression to all districts but both the intercept and slope terms are allowed to differ for each district. The two district-level residuals $u_j = (u_{0j}, u_{1j})^{\mathrm{T}}$ representing the intercept and slope are assumed to have a multivariate normal distribution with variance matrix

$$\Omega_u = \begin{bmatrix} \sigma_{u0}^2 & \sigma_{u01} \\ \sigma_{u01} & \sigma_{u1}^2 \end{bmatrix}.$$

When we allow the slopes to vary across districts such a model is often referred to as a *random slopes* regression model or more generally for predictor variables a *random coefficient* model. Note it would also be feasible to allow the quadratic term to also vary across districts.

Of course not all disease mapping data is longitudinal and we will often have cross-sectional data where we have total counts over a fixed period for a set of geographical areas. We may in this case ask how we can use a multilevel model for such data? In fact the subset of the mortality dataset shown in the earlier figure is a set of districts that all come from one county in the UK and there are in total 54 counties in the dataset. The 54 counties are also nested within 10 regions so in reality our dataset has four levels with yearly SMRs nested within districts nested within counties nested within regions. To include these extra levels of geography in our model we would include random effects for these higher levels. For example if we were to expand our last model to additionally include county effects we would have a three-level model that can be written:

$$\begin{aligned}SMR_{ijk} &= \beta_{0ijk} + \beta_{1jk}(year_{ijk} - 1979) + \beta_2(year_{ijk} - 1979)^2,\\ \beta_{0ij} &= \beta_0 + v_{0k} + u_{0jk} + e_{0ijk},\\ \beta_{1j} &= \beta_1 + u_{1jk},\\ v_k &\sim N(0, \sigma^2_v),\ u_{jk} \sim MVN(0, \Omega_u),\ e_{0ijk} \sim N(0, \sigma^2_e).\end{aligned} \quad (3.6)$$

Here the additional subscript k defines the county number and v_k is the effect for county k. We have allowed both the intercept and the slope of our SMR curves to vary across districts and additionally allowed the intercept to vary across counties. We may be interested in how important the county effects are and to do this we can compare the corresponding size of the variances σ^2_v and σ^2_{u0}. It should be noted that comparisons between the levels of geography are easier to interpret in a variance components model (with the random district level slopes omitted) when we could work out the ICC values for each level.

If we were to consider just a cross-section of the dataset for example the counts for 1980 we would still have three nested levels of geography with district SMRs nested within counties nested within regions. In the models considered so far we have included level 1 (year) predictors (the *year* itself and $year^2$) but we can also consider fitting higher-level predictor variables. In the mortality dataset we have a categorical variable that categorizes the districts into six types. The districts were classified into whether they were in London, and if not whether they were rural, urban, mining, mature (areas with large retired populations) or prospering. Leyland and McLeod (2000) considered fitting this categorical predictor and found that in fact a classification into three types, mining (with highest SMR), urban and London, and rural, mature and prospering (with lowest SMR) produced a better model.

Leyland and McLeod (2000) also did not use the highest-level classification, region in their modelling. This raises an interesting issue with multilevel modelling as to when a classification should be fitted as a random effect. There are two issues as to whether a particular categorical predictor (classification) should be treated as a set of random or fixed effects.

The first issue is due to the definition of random effects, in that a classification should only be treated as random if its units can be thought of as a (random) sample from some population of units, for example the districts chosen here could be thought of as a sample from the population of districts in the UK whereas a predictor such as gender should never be treated as a set of random effects as there are only two genders and they are not two genders from a population of genders! It should be noted here that in the case of districts, even if we choose all the districts in the UK, it could still be argued that districts should be treated as random effects. This is because we are not primarily interested in the individual districts and the boundaries that define each district are fairly arbitrary and so selecting district boundaries could be seen as sampling from the population of districts. The other issue is a practical issue in that if we have too few units in a classification it may be firstly difficult to find variability between the units and secondly the estimate of this variability will have larger uncertainty the fewer units we have. In the mortality example there are only 10 regions and so it would probably be better to fit this classification as fixed effects.

3.2 ESTIMATION PROCEDURES FOR MULTILEVEL MODELS

In the above section we have described several potential (continuous response) multilevel models that can be fitted to a disease mapping dataset. In practice multilevel models cannot be estimated using simple matrix formulae like linear models and so more complicated estimation procedures are required. In this section we will describe some of the more commonly used methods and also discuss model comparison. There are two main approaches to fitting multilevel models, a likelihood-based (frequentist) approach and a Bayesian approach. The frequentist approach involves finding maximum likelihood estimates for the unknown parameters in the model. The frequentist approaches pre-date the Bayesian approaches as they are less computationally intensive and so were feasible earlier. There are several algorithms to find maximum likelihood estimates. Here we will describe the method that MLwiN uses, as we will be using it in later chapters, but we will also give references to other methods.

3.2.1 Iterative generalized least squares estimation

Goldstein (1986) introduced the use of the iterative generalized least squares (IGLS) algorithm for estimation of normal response multilevel models. It is a sequential refinement procedure based on generalized least squares estimation that produces maximum likelihood (ML) estimates. The algorithm can fit all normal response (nested) multilevel models and can be found in more details in Goldstein (1995). Basically, if we consider our first example,

$$y_{ij} = X_{ij}\beta + u_j + e_{ij},\ u_j \sim N(0, \sigma_u^2),\ e_{ij} \sim N(0, \sigma_e^2), \tag{3.7}$$

then we can write this as $y_{ij} = X_{ij}\beta + e^*_{ij},\ i = 1, \ldots, n_j,\ j = 1, \ldots, J$ where $e^*_{ij} = u_j + e_{ij}$ and we can similarly rewrite any other nested model in this way with different substitutions. If we consider the vector $e^* = (e_{1,1}, e_{2,1}, \ldots, e_{n_J,J})$ then this has a multivariate normal distribution with mean 0 and covariance matrix V. We now require estimates for the fixed effects β and the variance matrix V. If we knew the matrix V then the problem becomes a use of GLS estimation and we could hence find an estimate of β, $\hat{\beta}$.

Similarly if we knew the value of β then we could construct the residuals $\tilde{y}_{ij} = y_{ij} - X_{ij}\beta$. We can construct the matrix $Y^* = \tilde{Y}\tilde{Y}^T$ where $\tilde{Y}$ is the vector formed from stacking $\tilde{y}_{ij}$. If we also stack the columns of Y^* we can form the column vector $Y^*_{(s)}$. Then we can also transform the V matrix into a stacked column vector and match elements so that we have $Y^*_{(s)} = Z^*\theta + \epsilon$ where θ in our example contains the elements σ^2_u and σ^2_e i.e. $\theta = (\sigma^2_u, \sigma^2_e)^T$ and Z^* is a (known) design matrix. We now have a set of equations of size N^2 where N is the number of level 1 units from which we can (using GLS) calculate an estimate of θ, $\hat{\theta}$.

The IGLS algorithm begins by evaluating starting values for β (assuming no higher-level random effects and using ordinary least squares). Then the two GLS steps are iterated between until the estimates produced by consecutive iterations are within (predefined) tolerance limits.

3.2.2 Restricted iterative generalized least squares (RIGLS)

As with other ML procedures, the IGLS algorithm produces biased estimates when sample sizes are small. When fitting multilevel models this bias is noticeable in the higher-level variance estimates where the sample size is determined by the number of higher-level units. Goldstein (1989) describes how, if we define the residuals as $\tilde{y}_{ij} = y_{ij} - X_{ij}\hat{\beta}$ and $Y^* = \tilde{Y}\tilde{Y}^T$, then

$$E(\hat{Y}^*) = V - X(X^TV^{-1}X)^{-1}X^T. \tag{3.8}$$

Consequently we can adjust the ML estimates (for bias) by adding the second term on the right-hand side of the above equation to $\hat{Y}^*$ at each iteration. This technique is known as restricted IGLS (or RIGLS) and is equivalent to restricted maximum likelihood (REML) in normal response models. Estimated standard errors in both the REML and ML approaches are based on the final (converged) values of the covariance matrices of both $\hat{\beta}$ and $\hat{\theta}$ and expressions for these can be found in Goldstein (1995).

3.2.3 Other maximum likelihood approaches

There are other methods that can be used to produce maximum likelihood estimates for multilevel models. Longford (1987) introduced a procedure that was based on a 'Fisher scoring' algorithm and is used in the software package VARCL (Longford, 1988). This method has since been shown to be formally equivalent to the IGLS algorithm. Bryk and Raudenbush (1992) describe the use of the EM algorithm to provide empirical Bayes estimates for multilevel

models and this algorithm is used in the software package HLM (
1988). In the case of normally distributed responses these empiri
estimates are equivalent to the maximum likelihood estimates obtaine

3.2.4 Model comparison

To compare two (nested) normal response models is straightforwardly do
the likelihood ratio test. For each model the deviance (−2* log-likelihood)
calculated. The difference in deviance has a chi-squared distribution
degrees of freedom equal to the number of additional parameters in the
complex model. As we are effectively fitting a very large multivariate no
model, counting the number of parameters is easy. For example, if we wis
compare the two earlier models (Equations (3.4) and (3.5)) then in the seco
model we have added a quadratic fixed effect, and a set of random effe
associated with time. We therefore have one additional parameter due to t
fixed effect and two due to the additional variance and covariance associate
with the set of random effects and so the second model is better if the differenc
in deviance is significant when compared with a χ^2 distribution with three
degrees of freedom. The likelihood ratio test assumes ML estimates and so
cannot be used with the RIGLS method or more importantly with the quasi-likelihood methods that MLwiN uses for Poisson response models.

3.2.5 Bayesian estimation

For each multilevel model we are considering we can create equivalent Bayesian models by incorporating prior distributions for each of the unknown parameters in the model and performing inference on the resulting posterior distributions. We can pick some 'default' choice of 'diffuse' priors so that we can have a one-to-one correspondence with the likelihood approach or alternatively try several models to test sensitivity to the choice of prior distribution. When we have completely specified our Bayesian model we can then fit it using MCMC estimation. For the normal response case, if we use conjugate priors, we can use a form of MCMC estimation known as Gibbs sampling. Gibbs sampling is described in detail in the chapter on Bayesian hierarchical modelling (Chapter 2). Here we will consider the Bayesian extension of the standard variance components model (Equation (3.7))

$$\begin{gathered} y_{ij} = X_{ij}\beta + u_j + e_{ij},\ u_j \sim N(0, \sigma_u^2),\ e_{ij} \sim N(0, \sigma_e^2), \\ i = 1, \ldots, n_j,\ j = 1, \ldots, J,\ \sum n_j = N, \\ \beta \sim N(\mu_\beta, S_\beta),\ \sigma_u^2 \sim \chi^{-2}(\nu_u, s_u^2),\ \sigma_e^2 \sim \chi^{-2}(\nu_e, s_e^2) \end{gathered} \tag{3.9}$$

ate out the u_j like the IGLS algorithm but ariables, and hence we have a four-step s consists of sampling in turn from the full $u, \sigma_u^2, \sigma_e^2)$, $p(u|y, \beta, \sigma_u^2, \sigma_e^2)$, $p(\sigma_u^2|y, \beta, u, \sigma_e^2)$, and e following forms:

$$
\begin{aligned}
&N\left[\hat{D}_\beta\left(\sum_{ij}\frac{(X_{ij})^T(y_{ij}-u_j)}{\sigma_e^2}+S_\beta^{-1}\mu_\beta\right),\hat{D}_\beta\right]\\
&N\left[\frac{\hat{D}_j}{\sigma_e^2}\sum_{i=1}^{n_j}(y_{ij}-X_{ij}\beta),\hat{D}_j\right]\\
&\sim \Gamma^{-1}\left[\frac{J+\nu_u}{2},\frac{1}{2}\left(\nu_u s_u^2+\sum_{j=1}^{J}u_j^2\right)\right]\\
&\sigma_u^2) \sim \Gamma^{-1}\left[\frac{N+\nu_e}{2},\frac{1}{2}\left(\nu_e s_e^2+\sum_{ij}e_{ij}^2\right)\right]
\end{aligned}
\qquad (3.10)
$$

$$
\hat{D}_\beta=\left(\sum_{ij}\frac{(X_{ij})^T X_{ij}}{\sigma_e^2}+S_\beta^{-1}\right) \text{ and } \hat{D}_j=\left(\frac{n_j}{\sigma_e^2}+\frac{1}{\sigma_u^2}\right)^{-1}.
$$

ing from these distributions in turn will produce, after a suitable burn- eriod in which the chains are converging to the joint posterior distri- on, chains of parameter estimates from the joint posterior distribu- n. These chains can then be used to produce both point estimates and full istributional summaries. Model comparison in Bayesian models is also achieved by different means including the DIC diagnostic (Spiegelhalter *et al.*, 2002a) which is also described in the chapter on Bayesian hierarchical modelling.

3.3 POISSON RESPONSE MODELS

Although we have considered fitting normal response models thus far, generally the responses that are considered in disease mapping are counts of deaths in particular areas. It is therefore more appropriate to fit a distribution that is specifically designed for count data for example the Poisson distribution. If we continue with the mortality dataset from Leyland and McLeod (2000) then we can consider fitting Poisson models to the number of deaths rather than normal models to the SMR. We do of course want to account for the differences in expected counts in the various areas and so we treat the logarithm of the expected counts as an offset variable. If we consider model (3.4) then an equivalent Poisson model is the following:

$$
\begin{aligned}
(y_{ij} \mid \pi_{ij}) &\sim Poisson(\pi_{ij}),\\
\log(\pi_{ij}) &= \log(e_{ij}) + \beta_0 + \beta_1(year_{ij} - 1979) + u_j,\\
u_j &\sim \mathrm{N}(0, \sigma_u^2).
\end{aligned} \tag{3.11}
$$

Note here that we are using e_{ij} to represent the expected count and previously we used e_{ij} to represent the level 1 residuals in normal response models. This is an unfortunate notation clash but, as for the remainder of this book we will not refer to normal response models, these clashes should be minimal. The above model states that the observed number of deaths in a particular region over a particular time period follow a Poisson distribution whose parameter is a function of the expected number of deaths, the year and the county. Just as with the Normal model case we can add complexity to the model for example in an analogous expansion to model (3.5), we can form the Poisson model

$$
\begin{aligned}
(y_{ij} \mid \pi_{ij}) &\sim Poisson(\pi_{ij}),\\
\log(\pi_{ij}) &= \log(e_{ij}) + \beta_{0ij} + \beta_{1j}(year_{ij} - 1979) + \beta_2(year_{ij} - 1979)^2,\\
\beta_{0ij} &= \beta_0 + u_{0j},\\
\beta_{1j} &= \beta_1 + u_{1j},\\
u_j &\sim MVN(0,\ \Omega_u).
\end{aligned} \tag{3.12}
$$

Here as with the equivalent expansion for the normal models we have included an interaction between the year trend effect and the individual counties and added a quadratic effect to the global trend of deaths. Estimating parameters of Poisson models is not as easy as the Normal case due to the nonlinear relationship between the response and predictors and consequently we need to use some alternative estimation procedures which will be outlined below.

3.3.1 Marginal and penalized quasi-likelihood

The IGLS and RIGLS algorithms are designed to fit normally distributed multilevel models and so in order to fit multilevel models to non-Normal responses for example the Poisson responses models we have just discussed we need to use different estimations procedures. It is possible to calculate exact maximum likelihood estimates by using Gauss–Hermite quadrature estimation (see, for example, Hedeker and Gibbons, 1994, and Rabe-Hesketh *et al.*, 2001) but typically these methods are time-consuming and are sensitive to amongst other things the number of quadrature points (Lesaffre and Spiessens, 2001). The alternative is to use some form of approximation to give quasi-likelihood estimates rather than maximum likelihood estimates. There are several types of approximation and we will here describe those built on Taylor series expansion as these are used in the MLwiN package. It is, however, worth noting that a method based on Laplace approximations (Raudenbush *et al.*, 2000) is also available in the HLM software package. Goldstein and Rasbash (1996) describe quasi-likelihood algorithms for binary response models but the algorithms are similar for Poisson response models. If we consider a general two-level Poisson model then we can write this as

$$(y_{ij}|\pi_{ij}) \sim Poisson(\pi_{ij}) \text{ with } \pi_{ij}/e_{ij} = f(X_{ij}\beta + Z_{ij}u_j), \tag{3.13}$$

where $f(l) = \exp(l)$ a nonlinear function. To approximate this model using the standard IGLS algorithm we first need to linearize this function so that it assumes the form of a standard two-level normal model and then apply quasi-likelihood estimation using the Poisson distribution assumption to define the level 1 variation. We linearize the model via Taylor series expansion with H_t chosen as a suitable value to perform the expansion of the function $f(.)$ around. We have the equation

$$\begin{aligned} f(H_{t+1}) = f(H_t) + X_{ij}(\hat{\beta}_{t+1} - \hat{\beta}_t)f'(H_t) + \\ Z_{ij}u_j f'(H_t) + \frac{1}{2}(Z_{ij}u_j)^2 f''(H_t) \end{aligned} \tag{3.14}$$

The simplest choice for H_t is the fixed-part predicted value of f, $H(t) = X_{ij}\hat{\beta}_t$. This leads to the marginal quasi-likelihood (MQL) algorithm. A better approximation is often found by expanding around the entire predicted value for the ijth unit, $H_t = X_{ij}\hat{\beta}_t + Z_{ij}\hat{u}_j$ where $\hat{u}_j$ are the current estimated random effects. If we combine this choice for H_t with an improved approximation of Equation (3.14)

$$\begin{aligned} f(H_{t+1}) = f(H_t) + X_{ij}(\hat{\beta}_{t+1} - \hat{\beta}_t)f'(H_t) + \\ Z_{ij}(u_j - \hat{u}_j)f'(H_t) + \frac{1}{2}(Z_{ij}(u_j - \hat{u}_j))^2 f''(H_t), \end{aligned} \tag{3.15}$$

then the result is the method called the penalized or predictive quasi-likelihood (PQL) algorithm. Both the MQL and PQL estimation algorithms are usually referred to with an order which refers to the number of terms involved in the Taylor series expansion underlying the approximation. Here we have described the second-order algorithms which are commonly called MQL2 and PQL2. Although the PQL method generally gives better estimates than the MQL method it is more prone to convergence problems and so a typical strategy is to try MQL estimation first and if this succeeds then move onto PQL estimation using the MQL estimates as starting values for the PQL procedure.

3.3.2 Bayesian estimation

Gibbs sampling in a random effects Poisson response model is not as easy as in normal response models. For example, in perhaps the simplest model

$$\begin{aligned} (y_{ij} \mid \pi_{ij}) &\sim Poisson(\pi_{ij}), \\ \log(\pi_{ij}) &= \log(e_{ij}) + \beta + u_j, \quad u_j \sim N(0, \sigma_u^2). \end{aligned} \tag{3.16}$$

If we assume a uniform prior for β then the full conditional for β is

$$p(\beta \mid y, u, \sigma_u^2) \propto \prod_{ij} (\exp(\beta + u_j))^{y_{ij}/e_{ij}} \exp(-\exp(\beta + u_j)). \tag{3.17}$$

This distribution does not lend itself to direct sampling and consequently there are many ways to proceed. The WinBUGS package employs the adaptive rejection (AR) sampler (Gilks and Wild, 1992) which will work on any log-concave posterior distribution and involves constructing an envelope function around the posterior, sampling from this envelope and applying a form of rejection sampling to the resulting sampled values. The MLwiN package uses a hybrid Metropolis–Gibbs approach. Here the variance parameters are updated via Gibbs sampling whilst the fixed effects and residuals are updated using univariate random walk Metropolis sampling (see Browne (1998) for more details).

3.3.3 Including overdispersion at the lowest level

The Poisson distribution has the underlying property that the variance of a Poisson response variable is equal to its mean. Often, however, we will find that this assumption is not true and that there is in fact more variation in a dataset than a Poisson distribution suggests. One of the motivations of fitting multilevel models is that the higher-level random effects, in our example the county effects, will explain some of this additional variation and make the Poisson assumption (conditional on the county effects) more plausible. The quasi-likelihood methods in fact can estimate the degree of overdispersion (extra variation) by including a scale factor parameter in the linearization procedure. This scale factor will then give a multiplicative estimate of the overdispersion as we will see in examples in later chapters.

If the higher level effects do not explain all the overdispersion or we do not have higher-level effects (and hence a multilevel structure) then we can fit additional (normally distributed) random effects for each observation. This can be thought of as additive overdispersion and a simple example is the following model:

$$(y_i \mid \pi_i) \sim Poisson(\pi_i),$$
$$\log(\pi_i) = \log(e_i) + \beta + u_i, \; u_i \sim N(0, \sigma_u^2). \qquad (3.18)$$

Here the variance σ_u^2 will give the amount of variance not explained by the Poisson assumption and the effects u_i will be described as unstructured random effects in later chapters. This model can, in fact be thought of as a multilevel model. Let us assume our observations in the above model are deaths in counties of a particular area. Then if we are prepared to have level 1 represent the individual observations and level 2 the counties as in Equation (3.16) but with each county having only one observation so in fact each level 2 unit (county) has only one level 1 unit. This trick allows us to fit this model easily using both quasi-likelihood and MCMC methods in the MLwiN software package.

3.4 INCORPORATING SPATIAL INFORMATION

The standard multilevel model structure does not explicitly incorporate spatial structure, although through the use of higher levels of geography as additional levels in the model we can indirectly incorporate spatial clustering effects. One extension to the standard multilevel model is the multiple-membership model.

3.4.1 Multiple-membership (MM) models

This family of models was first used in Hill and Goldstein (1998) to solve the problem of missing school identification in an educational dataset. In their problem they had a two-level data structure with pupils nested within schools, but for some pupils the exact school was not known, although the set of possible schools they could belong to was a subset of the whole population of schools. Rather than throw away the data for these pupils they were included in the model but with a set of (weighted) random effects for the possible schools they could have attended. Perhaps a better example of the use of this model is where pupils move schools during their education and so we cannot identify just one school for such pupils when adding school effects to a model.

We can write a multiple-membership model as follows (using the notation of Browne *et al.* (2001) rather than that originally used by Hill and Goldstein (1998)):

$$y_i = X_i\beta + \sum_{j \in School[i]} w_{i,j}^{(2)} u_j^{(2)} + e_i,$$
$$i = 1, \ldots, N, \; School[i] \subset (1, \ldots, J),$$
$$u_j^{(2)} \sim N(0, \sigma_{u(2)}^2), \; e_i \sim N(0, \sigma_e^2).$$

Here as we do not have a straight nested structure we use the index i to represent the lowest level (student) and then use the function *School[i]* to refer to the set of schools attended by pupil i. The weights, $w_{i,j}^{(2)}$ give the relevant importance of the various schools the pupil attended, and typically are constrained to sum to 1 ($\Sigma_{j \in School[i]} w_{i,j}^{(2)} = 1 \; \forall i$). Possible weighting formulations are equal weights for each school attended, or weights proportional to the time spent in each school. This model can be fitted using either an adaptation of the algorithm for cross-classified models based on IGLS given in Rasbash and Goldstein (1994) or by using MCMC estimation as described in Browne *et al.* (2001). As with standard multilevel models if we have non-normal data then we can fit a model with Poisson responses and the same random effects structure by either using quasi-likelihood methods or MCMC estimation. Langford *et al.* (1999) describe how to use multiple-membership models in a spatial context by using an extension to the basic multiple-membership model described by Browne *et al.* (2001) as a multiple-membership multiple-classification (MMMC) model. To illustrate such a model consider a dataset with observed counts of a disease over time for a set of counties with a known neighbourhood structure. Then we can consider a model

$$
\begin{aligned}
(y_i \mid \pi_i) &\sim Poisson(\pi_i), \\
\log(\pi_i) &= \log(e_i) + X_i\beta + u_{area[i]}^{(2)} + \sum_{j \in Neighbour[i]} w_{i,j}^{(3)} u_j^{(3)}, \qquad (3.19) \\
u_j^{(2)} &\sim N(0, \sigma_{u(2)}^2), \; u_j^{(3)} \sim N(0, \sigma_{u(3)}^2).
\end{aligned}
$$

Here i indicates an observed count, *area[i]* the area from which that observed count was taken, and *Neighbour[i]* the set of neighbouring areas to the area from which the count was taken. So in this model specification the observed count is affected by various predictor variables, the area where the count was taken and the neighbouring areas. Typically the weights in this model are such that $\Sigma_{j \in School[i]} w_{i,j}^{(2)} = 1 \; \forall i$ and generally all neighbours are given equal weights so that in fact $w_{i,j}^{(2)} = 1/n_i$ where n_i is the number of neighbours to *area[i]*. As there is a one-to-one correspondence between the set of area and neighbour residuals Langford *et al.* (1999) also consider fitting a joint multivariate Normal distribution to these two sets of residuals. Browne *et al.* (2001) consider the Bayesian equivalent of MMMC models (excluding the model with correlated area and neighbour residuals) and also compare these models with the more commonly used CAR spatial models.

3.5 DISCUSSION

In this chapter we have described some of the background to multilevel modelling. We have given examples of both normal and Poisson multilevel models and shown how such models can be applied to disease mapping problems. We have also given some details of the estimation algorithms used to fit multilevel models both from a likelihood and Bayesian point of view. In Chapter 5 we will apply the estimation methods discussed here on real data problems and we will also use some of the methods used here in the three final chapters. Multilevel models are widely used in many applied fields and there is a vast literature which the interested reader can explore. For a longer introduction to the use of multilevel modelling in health applications, Leyland and Goldstein (2001) is to be recommended.

4

WinBUGS Basics

There has been considerable development of Geographical Information Systems software which can manipulate and display spatial data and can provide a useful presentation tool for disease maps. The common purpose of these packages is to produce thematic maps of relative risk estimates (for example SMRs). Often, the statistical capabilities of these packages are severely limited.

The statistical procedures needed to produce disease maps can be found in a number of statistical packages (for example R, S-Plus or SAS). Specialist Bayesian modelling of disease maps is facilitated by the use of WinBUGS and GeoBUGS. WinBUGS allows the fitting of full Bayesian models to a wide range of data. GeoBUGS is intended to provide a GIS-style presentation of output from spatial analysis in the WinBUGS program.

4.1 ABOUT WinBUGS

Bayesian inference **U**sing **G**ibbs **S**ampling (BUGS) is a program that carries out Bayesian inference on statistical problems using Markov chain Monte Carlo (MCMC) methods, such as Gibbs sampling. It is intended for complex models for which there is no exact analytic solution or for which even standard approximation techniques have difficulties. BUGS assumes a Bayesian probability model, in which all unknown parameters are treated as random variables. The model consists of a defined joint distribution over all unobserved (parameters and missing data) and observed quantities (data); then it is necessary to condition on the data in order to obtain a posterior distribution for the parameters and unobserved data. Empirical summary statistics can be obtained from samples of the posterior and are used to draw inferences for the quantities of interest.

Typical problems that can be handled using BUGS include generalized linear mixed models with spatial or temporal random effects, censored data, missing data problems and analyses in which prior information needs to be incorporated. However, there are some restrictions on the class of models that can be

Disease Mapping with WinBUGS and MLwiN A. Lawson, W. Browne and C. Vidal Rodeiro
 ISBN: 0-470-85604-1 (HB)

analysed using BUGS; also the fact that BUGS is a general-purpose program means that it is inevitable that many types of models, such as spatial smoothing, could be more efficiently implemented in special-purpose software.

The software also provides limited facilities for detecting convergence, summarizing the accumulated samples and performing model diagnostics; these procedures, however, are best handled externally to the program.

Currently, the most advanced version of BUGS, released in January 2003, is WinBUGS 1.4 running under Windows. The program is currently developed by the Medical Research Council (MRC) Biostatistics Unit (Cambridge, UK) and the Department of Epidemiology and Public Health of the Imperial College School of Medicine at St Mary's Hospital (London). It can be downloaded from the website *http://www.mrc-bsu.cam.ac.uk/bugs/*. The free version of WinBUGS is a restricted version; you need to email the BUGS project for a key that will let you use the full version. WinBUGS 1.4 also includes a new version of GeoBUGS and a scripting facility which permits batch running and also, therefore, the ability to run WinBUGS from other programs.

The WinBUGS Manual (Spiegelhalter *et al.*, 2002b) is available online. It gives brief instructions on WinBUGS but it also helps to refer back to earlier manuals for more elaborate descriptions and for issues in modelling using MCMC methods. Two volumes of WinBUGS examples are also available online. A good way to attack a problem is to go to the volumes of examples and find one that is like your problem; you can then adapt the code to run on your problem. The BUGS website provides additional links to sites of interest, some of which provide extensive examples and tutorial materials.

4.1.1 Markov chain Monte Carlo techniques

As noted previously, BUGS is intended for use on problems for which there is no exact analytic solution, and for which standard approximation techniques have difficulties. Statistical analysis is conducted using Markov chain Monte Carlo which is increasingly being used as an approach for dealing with such problems (Gilks *et al.*, 1996).

Classic BUGS used Gibbs sampling to obtain the posterior distributions of the parameters of interest. In this technique, at each iteration a new value for every unobserved stochastic node is sampled from the corresponding parameter's full conditional distribution.

WinBUGS has new features that were not part of the classic BUGS package: a more general Metropolis sampler and the ability to determine the sampling method for the estimation of the target distributions. In cases where a node's full conditional distribution is available in closed form WinBUGS can identify that closed form and implement the appropriate specialized sampling method. Where a node's full conditional is not available in closed form, the software examines the circumstances and chooses an appropriate sampling method. (If it

Table 4.1 Sampling method hierarchy used by WinBUGS.

Target distribution	Sampling method
Discrete	Inversion of cumulative distribution function
Conjugate	Direct sampling
Log-concave	Derivative-free adaptive rejection sampling
Restricted range	Slice sampling
Unrestricted range	Metropolis–Hastings

is unable to do it, an error message will be returned). WinBUGS is able to recognize conjugate specifications (e.g. Poisson–gamma) so that the corresponding nodes can be updated via direct sampling using standard algorithms. Free adaptive rejection sampling is used for nodes having a log-concave full conditional distribution. For nodes with other distributions, WinBUGS uses Metropolis–Hastings techniques. Table 4.1 shows the five types of sampling methods currently used by WinBUGS and the types of distribution for which they are employed. Each method is only used if no previous method in the hierarchy is appropriate.

WinBUGS 1.4 has tools for advanced control of the MCMC algorithms (**update options** from the **options menu**) but the user needs to be familiar with MCMC techniques to modify the default options.

4.2 START USING WinBUGS

To start WinBUGS, click on the WinBUGS icon or program file (WinBUGS14.exe). You will get a message about the licence conditions that you can read and close (Figure 4.1).

4.2.1 Main menus on WinBUGS

WinBUGS 1.4 has 13 menus of which the following are commonly used.

The **File** menu allows you to open existing files or to start a new file to program your own examples. WinBUGS documents can be printed from this menu.

The **Model** menu permits you to check and compile models expressed as either doodles or in the BUGS language.

The **Inference** menu controls the monitoring, display and summary of the simulated variables. From this menu it is possible to use tools for model comparison such as the Deviance Information Criterion (Spiegelhalter *et al.*, 2002a).

The **Info** menu provides a log of the run and information about the nodes in the model.

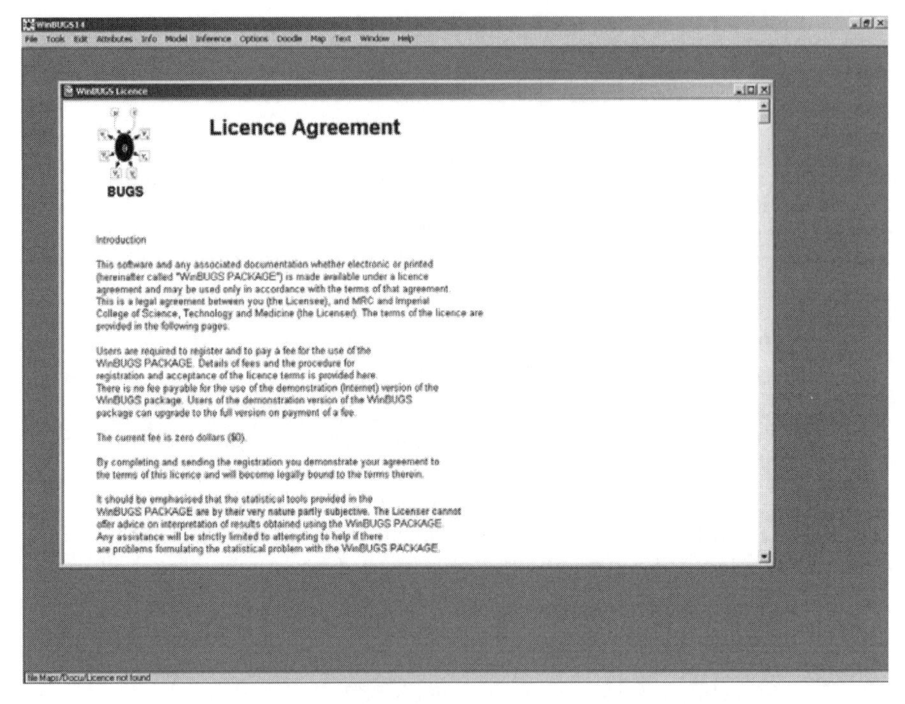

Figure 4.1 WinBUGS 1.4 program start screen.

The **Doodle** menu provides the tools to start a file for a new model specified in graphical format.

The **Options** menu allows the user some control over the output and the available MCMC algorithms.

The **Map** menu provides tools that produce maps of the output of spatial models and allows the creation and manipulation adjacency matrices that are required for spatial smoothing.

The **Help** menu allows the user to open the program manuals and the examples. These are written in something called a compound document format that allows programs, graphs and explanations to be together.

4.2.2 Compound documents and doodles

The WinBUGS software has been designed so that it can get its input from a compound document and produce output directly to a compound document.

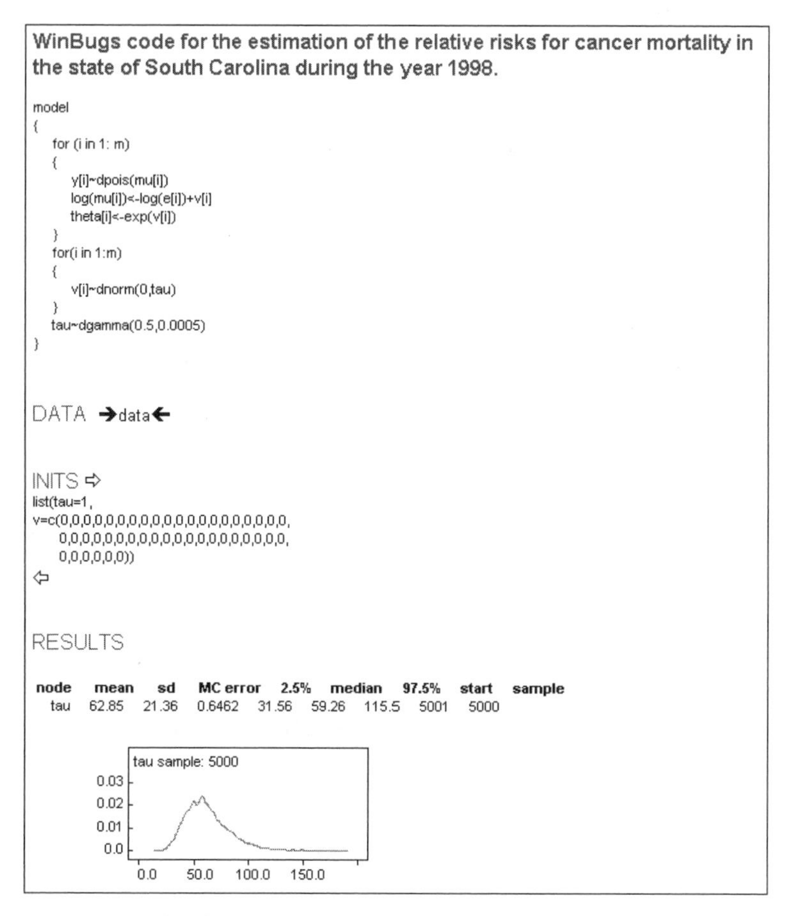

WinBugs code for the estimation of the relative risks for cancer mortality in the state of South Carolina during the year 1998.

```
model
{
   for (i in 1: m)
   {
      y[i]~dpois(mu[i])
      log(mu[i])<-log(e[i])+v[i]
      theta[i]<-exp(v[i])
   }
   for(i in 1:m)
   {
      v[i]~dnorm(0,tau)
   }
   tau~dgamma(0.5,0.0005)
}
```

DATA ➔data🡐

INITS ⇨

```
list(tau=1,
v=c(0,0,0,0,0,0,0,0,0,0,0,0,0,0,0,0,0,0,0,0,0,0,0,
   0,0,0,0,0,0,0,0,0,0,0,0,0,0,0,0,0,0,0,0,0,0,0,
   0,0,0,0,0,0))
```

⇦

RESULTS

node	mean	sd	MC error	2.5%	median	97.5%	start	sample
tau	62.85	21.36	0.6462	31.56	59.26	115.5	5001	5000

Figure 4.2 Example of a compound document.

Compound documents resemble word-processor documents that contain various types of information (text, tables, formulae, plots, graphs, etc.) displayed in a single window and stored in a single file with the extension **.odc**. However, it is more usual to have the model code, the data, etc. in separate files. Figure 4.2 is an example of a compound document.

The WinBUGS software also works with doodles, which allow statistical models to be described in terms of graphs, but there are many features in the BUGS language that cannot be fully expressed with doodles. **DoodleBUGS** is a specialized graphics editor and is fully described in Spiegelhalter *et al.* (2002b). Figure 4.3 is an example of a doodle.

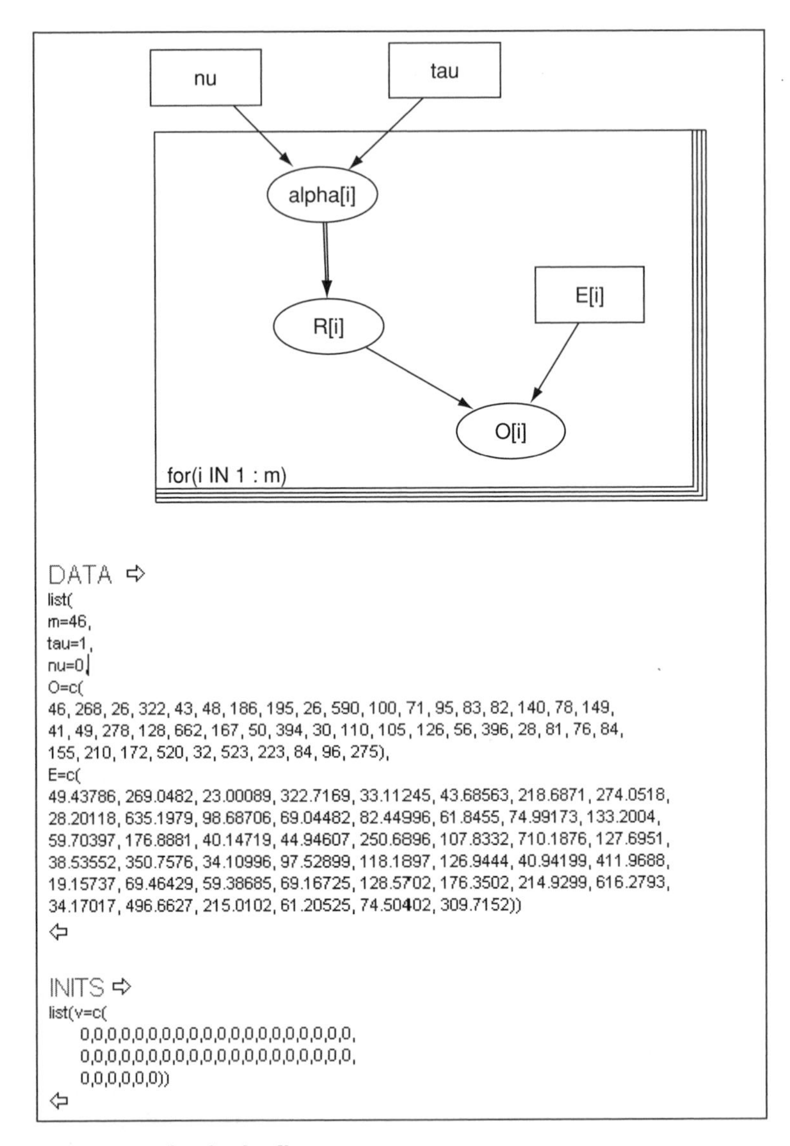

Figure 4.3 Example of a doodle.

4.3 SPECIFICATION OF THE MODEL

The first step in any analysis using WinBUGS should be the construction of the model. In WinBUGS terminology, data, variables, etc. are called nodes. Nodes can be of three types:

- **Stochastic nodes** are variables that are given a distribution. These nodes can be observed, in which case they are data, or may be unknown quantities, censored observations or missing data and hence are treated as parameters.
- **Constant nodes** are fixed quantities and must be specified in a data file.
- **Logical nodes** are logical functions of stochastic nodes and/or constant nodes.

4.3.1 Graphical models: DoodleBUGS editor

In graphical modelling, each quantity in the model is represented by a node and nodes are connected by lines or arrows to show direct dependence. Knowledge of each node's parents, i.e. the nodes upon which it directly depends, is enough to construct the full model.

Doodles consists of three elements:

- *Nodes*: Stochastic and logical nodes are denoted as ellipses in the graph and may be parents or descendants. Constants are denoted as rectangles and they do not have parents.
- *Edges*: These are links between nodes and can be of two types. A solid arrow indicates stochastic dependence while a hollow arrow indicates a logical function.
- *Plates*: These are used to represent repeated parts of the graphs.

4.3.1.1 *Creating a directed graph*

In this section we are going to create the directed graph for a model that estimates the relative risk of a disease in m regions.

Let θ_i be the unknown relative risk for the ith area, $i = 1, \ldots, m$. Let $(y_1, \ldots, y_m)$ and $(e_1, \ldots, e_m)$ denote the number of deaths and the expected number of deaths, respectively, from the disease during the study period. When the disease is non-contagious and rare, the numbers of deaths in each area are assumed to be mutually independent and to follow Poisson distributions

$$y_i \sim Poisson(e_i \theta_i) \quad \forall i$$

and the log of the relative risk can be modelled as

$$\log \theta_i = v_i, \\ v_i \sim N(0, \tau),$$

where τ is a fixed value.

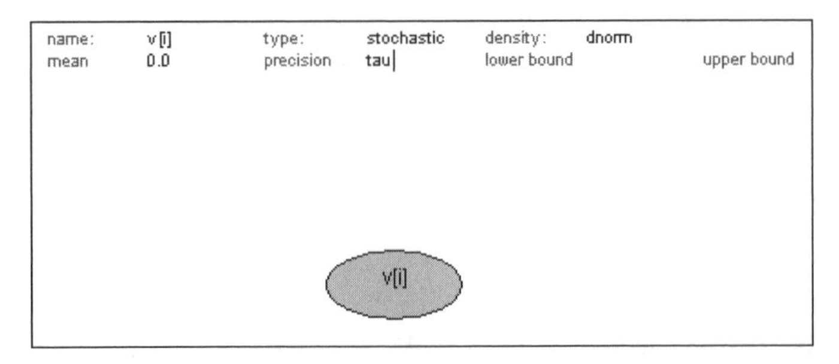

Figure 4.4 Definition of a node using DoodleBUGS.

To create the directed graph for this model, select the **Doodle menu** from the menu bar and click on **New**. This provides you with a window in which to construct the doodle, which can be built using the mouse and the keyboard.

To draw a node, click in an empty region of the doodle window. A new node appears, along with its properties (Figure 4.4).

Nodes are by default defined as stochastic and are associated with a Normal distribution. To change this distribution click on **density** and select the appropriate one from the folder list (distributions available in WinBUGS are described in Section 4.3.2.1). For each density, the name of its parameters will then be displayed. For some densities, default values of parameters will be displayed next to the parameter name.

The type of the node (stochastic, logical or constant) can be changed by clicking on the word **type** at the top of the doodle. Logical nodes can be given a link function and a value to be evaluated. Constants can be given a name (in this case they must be specified in a data file) or a numerical value. Vectors and matrices are denoted using squared brackets, with indices separated by commas.

To add an edge to the graph, select the node into which the edge should point and then click on its parent while holding down the CTRL key. In Figure 4.5 the solid arrows between the nodes indicate stochastic dependence. The hollow arrow indicates that the node θ_i is a logical function of the stochastic node v_i.

To create a plate for the range (i *in* $1 : m$) point the mouse cursor to an empty region of the window and click while holding the CTRL key. Figure 4.6 displays the final doodle for the model described above.

After constructing a doodle, it can be moved into a compound document where data and initial values can be added.

4.3.2 Text-based BUGS language

As a more flexible alternative to the doodle representation, the model can be specified using the text-based BUGS language. Figure 4.7 shows the code in the BUGS language for the doodle in Figure 4.6.

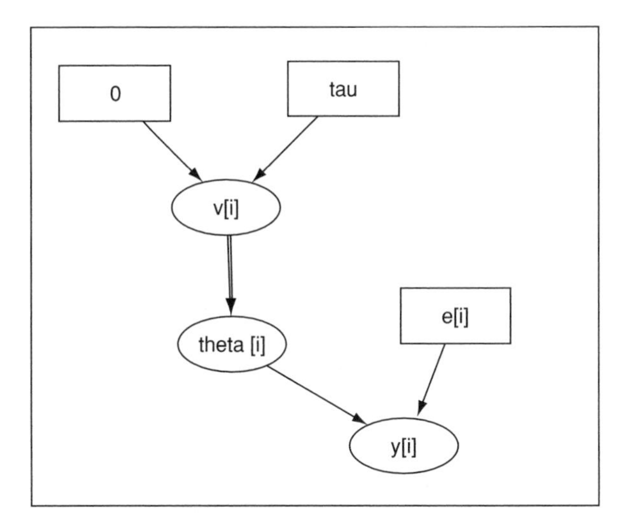

Figure 4.5 Nodes and edges for model described in Section 4.3.1.1.

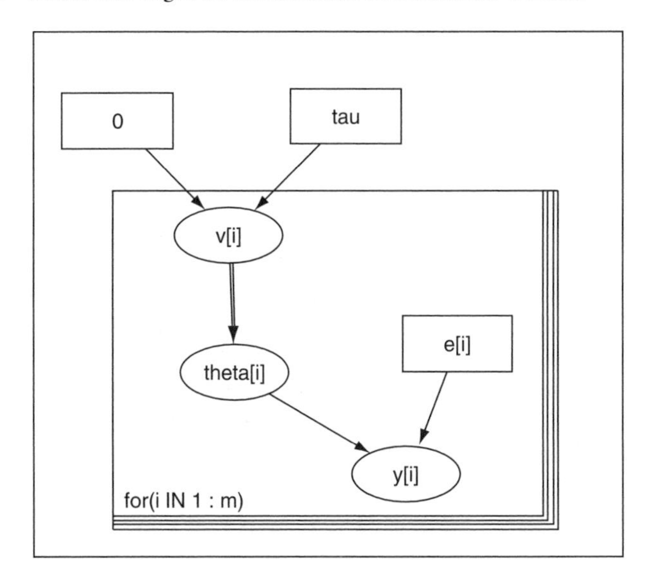

Figure 4.6 Doodle for model described in Section 4.3.1.1.

We now explain the notation used to write the model in Figure 4.7:

- Arrays are indexed by terms within squared brackets. The four basic operators $+, -, *, /$ and brackets are allowed to calculate an integer function of an index.
- The $\sim$ notation denotes *distributed as*.
- The notation $<-$ identifies logical relationships.

```
model
{
for (i in 1:m)
{
    y[i]~dpois(mu[i])
    log(mu[i])<-log(e[i])+v[i]
    theta[i]<-exp(v[i])
    v[i]~dnorm(0,tau)
}
}
```

Figure 4.7 Text-based BUGS code for model specified in Figure 4.6.

- $1 : m$ represents $1, 2, \ldots, m$.
- Repeated structures are specified using a '*for*' loop. The syntax for this is

for (*i in* 1 : *m*)
{
list of statements to be repeated for values of loop variable *i*
}

4.3.2.1 *Stochastic nodes*

Stochastic nodes are represented by the node name followed by a tilde symbol and its distribution. The parameters of a distribution must be explicit nodes and may not be function expressions. Tables 4.2 and 4.3 describe the distributions that can be used in WinBUGS. It is possible to use sampling distributions that are not included in the list of standard distributions; for details, see section 'Tricks: Advanced Use of the BUGS language' in the WinBUGS documentation.

In the new version of WinBUGS, WinBUGS 1.4, spatial models have been moved to GeoBUGS and new spatial distributions have been implemented (Thomas *et al.*, 2002). See Section 4.7.2 in this volume for details about these.

4.3.2.2 *Logical nodes*

Logical nodes are represented by the name followed by a left-pointing arrow followed by a logical expression which can be built using the following operators: plus, multiplication, minus, division and unitary minus. A link function (*log*, *logit*, *cloglog*) can also be specified acting on the left-hand side of a logical node. Table 4.4 shows a summary of functions that can also be used in logical expressions.

Logical nodes cannot be given data or initial values.

Table 4.2 Univariate distributions available in WinBUGS (Spiegelhalter *et al.*, 2002b).

Name	Usage	Distribution
Bernoulli	$r \sim$ dbern(p)	$p^r(1-p)^{1-r};\ r = 0, 1$
Beta	$p \sim$ dbeta(a, b)	$p^{a-1}(1-p)^{b-1}\frac{\Gamma(a+b)}{\Gamma(a)\Gamma(b)};\ 0 < p < 1$
Binomial	$r \sim$ dbin(p, n)	$\frac{n!}{r!(n-r)!}p^r(1-p)^{n-r};\ r = 0, \ldots, n$
Negative binomial	$x \sim$ dnegbin(p, r)	$\frac{(x+r-1)!}{x!(r-1)!}p^r(1-p)^x;\ x = 0, 1, 2, \ldots$
Categorical	$r \sim$ dcat(p[])	$p[r];\ r = 1, 2, \ldots, \dim(p); \sum_i p[i] = 1$
Chi-squared	$x \sim$ dchisqr(k)	$\frac{2^{-\frac{k}{2}}x^{\frac{k}{2}-1}e^{-\frac{x}{2}}}{\Gamma(\frac{k}{2})};\ x > 0$
Exponential	$x \sim$ dexp(*lambda*)	$\lambda \cdot e^{-\lambda x};\ x > 0$
Double exponential	$x \sim$ ddexp(*mu*, *tau*)	$\frac{\tau}{2}\exp(-\tau\lvert x-\mu\rvert);\ -\infty < x < \infty$
Gamma	$x \sim$ dgamma(r, *mu*)	$\frac{\mu^r x^{r-1}e^{-\mu x}}{\Gamma(r)};\ x > 0$
Generalized gamma	$x \sim$ gen.gamma(r, *mu*, *beta*)	$\frac{\beta}{\Gamma(r)}\mu^{\beta \cdot r}\exp\left(-(\mu x)^{\beta}\right);\ x > 0$
Logistic	$x \sim$ dlogis(*mu*, *tau*)	$\tau \cdot e^{r(x-\mu)}/(1+e^{r(x-\mu)})^2;\ -\infty < x < \infty$
Log-normal	$x \sim$ dlnorm(*mu*, *tau*)	$\sqrt{\frac{\tau}{2\pi}}\frac{1}{x}e^{-\frac{\tau}{2}(\log x-\mu)^2};\ x > 0$
Normal	$x \sim$ dnorm(*mu*, *tau*)	$\sqrt{\frac{\tau}{2\pi}}e^{-\frac{\tau}{2}(x-\mu)^2};\ -\infty < x < \infty$
Pareto	$x \sim$ dpar(*alpha*, c)	$\alpha c^{\alpha}x^{-(\alpha+1)};\ x > c$
Poisson	$x \sim$ dpois(*lambda*)	$\frac{\lambda^r}{r!}e^{-\lambda};\ r = 0, 1, \ldots$
Student-t	$x \sim$ dt(*mu*, *tau*, *nu*)	$\frac{\Gamma(\frac{\nu+1}{2})}{\Gamma(\frac{\nu}{2})}\sqrt{\frac{\tau}{\nu\pi}}\left[1+\frac{\tau}{\nu}(x-\mu)^2\right]^{-\frac{\nu+1}{2}};\ -\infty < x < \infty$
Uniform	$x \sim$ dunif(a, b)	$\frac{1}{(b-a)};\ a < x < b$
Weibull	$x \sim$ dweib(r, *lambda*)	$r\lambda x^{r-1}x^{-\lambda x^r};\ x > 0$

Table 4.3 Multivariate distributions available in WinBUGS (Spiegelhater *et al.*, 2002b).

Name	Usage	Distribution
Dirichlet	$p[] \sim$ ddirch ($alpha[]$)	$\frac{\Gamma(\sum_i \alpha_i)}{\prod_i \Gamma(\alpha_i)} \prod_i p_i^{\alpha_i - 1}$; $0 < p_i < 1$; $\sum_i p_i = 1$
Multivariate normal	$x[] \sim$ dmnorm($mu[]$, $tau[,]$)	$(2\pi)^{-\frac{d}{2}} \lvert\tau\rvert^{\frac{1}{2}} e^{-\frac{1}{2}(x-\mu)^T \tau(x-\mu)}$; $-\infty < x_j < \infty, j = 1, \ldots, dim(x) = d$
Multivariate Student-t	$x[] \sim$ dmt($mu[]$, $T[,]$, k)	$\frac{\Gamma((k+d)/2)}{\Gamma\frac{(k)}{2}k^{\frac{d}{2}}\pi^{\frac{d}{2}}} \lvert T\rvert^{\frac{1}{2}}[1 + \frac{1}{k}(x-\mu)'T(x-\mu)]^{-\frac{k+d}{2}}$; $-\infty < x < \infty$; $k \geq 2$
Multinomial	$r[] \sim$ dmulti($p[]$, N)	$N! \prod_i \frac{p_i^{r_i}}{r_i!}$; $0 < p_i < 1$; $\sum_i p_i = 1$; $\sum_i r_i = N$
Wishart	$x[,] \sim$ dwish($R[,]$, k)	$\lvert R\rvert^{\frac{k}{2}} \lvert x\rvert^{(k-p-1)/2} \exp(-\frac{1}{2} Tr(Rx))$; x-symmetric and positive-definite

Table 4.4 Summary of functions used in logical expressions.

Function	Value
abs(x)	$\|x\|$
cos(x)	*cosine* (x)
cloglog(x)	$\ln(-\ln(1-x))$
exp(x)	$exp(x)$
log(x)	$ln(x)$
logit(x)	$ln(\frac{x}{1-x})$
max(x, y)	x if $x > y$, else y
mean(v)	sum of components of v divided by the number of components
min(x, y)	x if $x < y$, else y
phi(x)	standard normal cdf
pow(x, y)	x^y
sqrt(x)	$x^{\frac{1}{2}}$
sin(x)	$sine(x)$
sd(v)	standard deviation of components of v
sum(v)	sum of the components of v

4.3.3 Introducing the data

All the variables in a data file must be defined in the model description. There are two ways to get the data into WinBUGS: data can be in S-Plus format or in rectangular format. Missing values are represented by the two-character string NA.

S-Plus format. Data are named and given values in a single structure headed by the keyword *list*. For example, if we want to specify the data for the example in Figure 4.6 (a scalar m and vectors y and e) we could use the format shown in Figure 4.8.

For a two-dimensional array e with 10 rows and 5 columns, we could use the following format:

$$\begin{aligned}\text{list}(e = \text{structure}(.\text{Data} = \text{c}(&123, 12, 34, 23, 34,\\ &17, 123, 45, 67,\\ &\ldots\ldots\ldots\ldots\ldots\\ &67, 45, 56),\ .Dim = c(10, 5)))\end{aligned}$$

When specifying data in S-Plus format note that WinBUGS reads data into an array by filling the rightmost index first.

Rectangular format. The columns for the data in rectangular format have to be headed by the array name. The arrays need to be of equal size, their names must be followed by squared brackets and the whole array must be specified in the same file. In Figure 4.9 we display the data for vectors y and e (as above) in rectangular format.

The version 1.4 of WinBUGS requires the use of the END command for rectangular format so, the data file must end with an END statement.

```
list(
m=46,
y=c(0,7,1,5,1,1,5,16,0,17,4,0,0,1,1,7,1,3,0,0,8,2,
 13,7,0,8,0,3,2,4,1,11,0,1,2,3,3,8,6,14,3,11,6,0,1,5),
e=c(1.129778827,6.667008775,0.650279674,6.988864371,0.95571406,
 1.123210345,5.908349156,8.539026017,0.601016062,18.92051111,
 2.272694617,1.73736337,2.019808077,1.688099759,1.747216093,
 3.221840201,1.835890594,5.221942834,0.978703751,1.254579976,
 6.407553754,2.676656232,16.57884744,3.077333607,1.087083697,
 7.606301637,1.018114641,2.15774619,2.844152512,2.955816698,
 0.985272233,9.22871658,0.38097193,1.855596038,1.579719813,
 1.579719813,2.647098065,4.791707292,4.144711859,15.70852363,
 0.765228101,11.32077795,6.256478678,1.500898035,2.085492893,
 7.297583004)
)
```

Figure 4.8 Data in S-Plus format.

```
y[]  e[]
0    1.129778827
7    6.667008775
1    0.650279674
5    6.988864371
1    0.955714066
1    1.123210345
5    5.908349156
16   8.539026017
0    0.601016062
17   18.92051111
..............................
6    4.144711859
14   15.70852363
3    0.765228101
11   11.32077795
6    6.256478678
0    1.500898035
1    2.085492893
5    7.297583004
END
```

Figure 4.9 Data in rectangular format.

Multidimensional arrays can be specified by explicit indexing. The first position of the array must always be empty. For example:

$y[,1]$	$y[,2]$	$y[,3]$
12	11	23
24	12	1
...	...	...

It is possible to load a mixture of rectangular and S-Plus format data files for the same model.

Other formats. The command to input data in WinBUGS is not very flexible and in many situations data is not in any of the formats described above. However, there are macros developed by WinBUGS users that allow conversion of formats into BUGS format. For example, there are Excel add-ins that convert an Excel data sheet to a text file (S-Plus list format) readable by WinBUGS. Also, SAS macros that output a SAS data set into a rectangular data format are available.

For more information about these macros and to download them, visit the web *http://www.mrc-bsu.cam.ac.uk/bugs/overview/list.shtml.*

4.4 MODEL FITTING

4.4.1 Specifying the model

The first stage in model fitting is to specify the model. This can either be done graphically using DoodleBUGS or in the text-based BUGS language. Once the model and the data are specified it is possible to run the model in WinBUGS. From the **model menu** you can select a **specification tool** (Figure 4.10). The following steps can then be followed:

Check model. It is necessary to check that the model description defines a valid statistical model. Highlight the keyword *model* at the beginning of the model description (or click anywhere in the area of a doodle to make it active) and click on the **check model** button. Any error messages will appear in the status bar at the bottom of the screen and the cursor will be placed where the error was found. If there are no errors in the code, you will get the message '*model is syntactically correct*'. You will also see that the compile and load data buttons will become active.

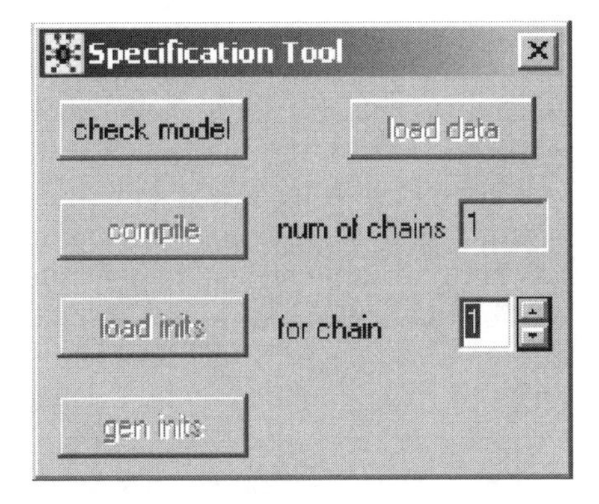

Figure 4.10 Specification tool.

Load data. You can use the **load data** button to read the data in WinBUGS. Data can be entered either in list format or in rectangular format (see examples in Section 4.3.3) and can be entered in two ways:

(a) If the data are part of a compound document, either the key word *list* (if the data are in S-Plus format), or the first array name (if the data are in rectangular format) must be highlighted and the data will then be read from this point on.

(b) If the data are in a separate file, the window containing that file needs to be in the focus view when the load data command is used.

Errors or data inconsistencies are displayed in the status line. Corrections can be made to the data without returning to the check model stage. When the data have been successfully loaded '*Data loaded*' should appear in the status line. The load data button ceases to be active once the model has been successfully compiled.

The number of chains to be simulated can be entered into the text field **number of chains**. This field can be input after the model has been checked and before the model has been compiled. By default, one chain is simulated. In practice, you may wish to run the model using a single chain to check that the model compiles and runs. After that you can re-run it using multiple chains.

Compile. Now you are ready to compile your model using the **compile** button. This button builds the internal data structures needed to carry out the sampling and chooses the specific MCMC updating algorithm to be used by WinBUGS for that particular model. The model is checked for completeness and consistency with the data. When the model has been successfully compiled, '*model compiled*' should appear in the status line.

Load initial values. All nodes that are not given as data, or derived from other nodes, need to be given initial values. This is done either by setting them from values in a file (**load inits**) or by generating a random value from the prior distribution (**gen inits**). These buttons become active once the model has been successfully compiled. Like the data, initial values should be in the form of an S-Plus object or rectangular array and should be consistent with any previously loaded data. If after loading the initial values, the model is fully initialized this will be reported in the status bar by the message '*initial values loaded: model initialized*'. Otherwise the status line will show the message '*initial values loaded: model contains uninitialized nodes*'. In this case, further initial values can be loaded, or the **gen inits** button can be pressed to generate initial values for all the uninitialized nodes. The **load inits** button can still be executed once Gibbs sampling has been started. It will have the effect of starting the sampler on a new trajectory.

The initial values for each stochastic node can be arbitrary but convergence can be poor if inappropriate values are chosen. If you are running the model using multiple chains, a different set of initial values is needed for each chain.

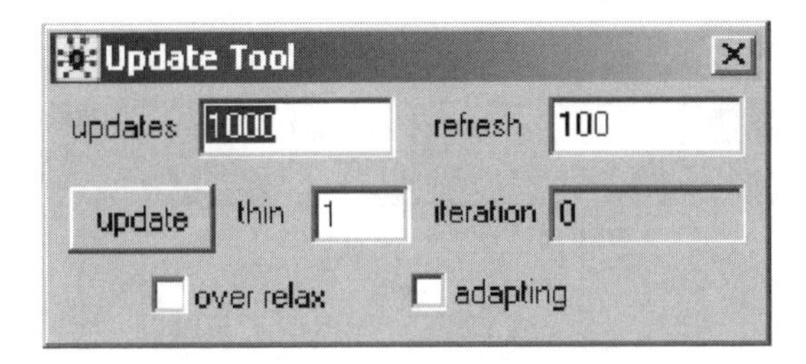

Figure 4.11 Update tool.

4.4.2 Generating samples

Once the model has been compiled and initialized, the update tool becomes active and you are now ready to generate samples and to examine the simulated output. To start the sampler, go to the **model menu** and then select **update** and you will get an **updating tool** (Figure 4.11).

You can select how many updates you get for each press of the **update** button (the default value is 1000) and how often the screen is refreshed to show how sampling is proceeding.

The samples from every kth iteration can be stored, where k is the value of the field **thin**. Setting $k > 1$ can help to reduce the autocorrelation in the sample, but there is no real advantage in thinning except to reduce storage requirements when very long runs are being carried out and to reduce Monte Carlo errors.

Clicking the **over-relax** box allows you to select an over-relaxed form of MCMC (Neal, 1998). The time per iteration will be increased, but the within-chain correlations should be reduced and hence fewer iterations may be necessary.

When the updates are finished, the message '*updates took #s*' will appear in the status bar.

4.4.3 Storing values and summarizing results

In MCMC methods, one usually wants to run the sampler for some time to be sure it has converged before the storage of values. The update tool performs the sampling, but does not store any values.

There are two options for monitoring parameters in WinBUGS, both in the inference menu: sample monitor and summary monitor.

4.4.3.1 Sample monitor

Choosing sample monitor tells WinBUGS to store every value of the chosen parameters. This option is needed to view trace plots of samples or to obtain

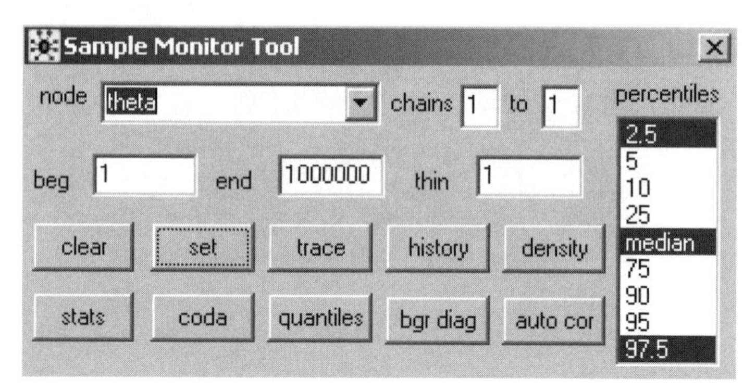

Figure 4.12 Sample monitoring tool.

exact posterior quantiles. The **sample monitoring tool** (Figure 4.12) is located in the **inference menu**.

To use the sample tool we need to enter the parameter names to monitor in the **node** box and for each one press **set**. If the variable is an array, slices of the array can be selected using the notation *variable*[*lower*0: *upper*0, *lower*1: *upper*1, ...]. The names are retained in a pull down list. A star (∗) can be entered in the node text file as shorthand for all the stored samples. WinBUGS automatically sets a logical node to measure the *deviance* of the model; this node may be accessed in the same way as any other variable of interest.

The **beg** and **end** numerical fields are used to select a subset of the stored sample for the analysis.

The field **thin** can be used to select every kth iteration of each chain to contribute to the statistics being calculated, where k is the value of the field.

To select the chains which contribute to the statistics being calculated specify them in the fields **chains ... to**.

Clear removes the stored values of the variable from the computer memory.

Stats gives summary statistics including the parameter mean, standard deviation, median and percentiles (Figure 4.13). These can be used for credible intervals. It is also possible to obtain a Monte Carlo error for the mean that will indicate how well the mean of the posterior has been estimated from the samples.

Trace plots the variable value against the number of iterations and **History** plots out a complete trace for the variable.

Density gives a kernel density estimate of the marginal posterior distribution.

Quantiles plots out the running mean with running 95% confidence intervals against the number of iterations.

The **auto cor** button plots the autocorrelation function of the variable up to lag 50. (See Figure 4.14.)

node	mean	sd	MC error	2.5%	median	97.5%	start	sample
theta[1]	0.9705	0.09428	0.001503	0.7958	0.9679	1.166	5001	5000
theta[2]	0.9961	0.05529	7.17E-4	0.891	0.9949	1.108	5001	5000
theta[3]	1.041	0.1166	0.00151	0.8314	1.034	1.292	5001	5000
theta[4]	0.9981	0.05145	6.996E-4	0.9022	0.9974	1.102	5001	5000
theta[5]	1.114	0.1166	0.001781	0.9065	1.107	1.371	5001	5000
theta[6]	1.045	0.1029	0.001522	0.8545	1.042	1.257	5001	5000
theta[7]	0.8861	0.05628	8.669E-4	0.7801	0.8855	0.9983	5001	5000

Figure 4.13 Summary statistics.

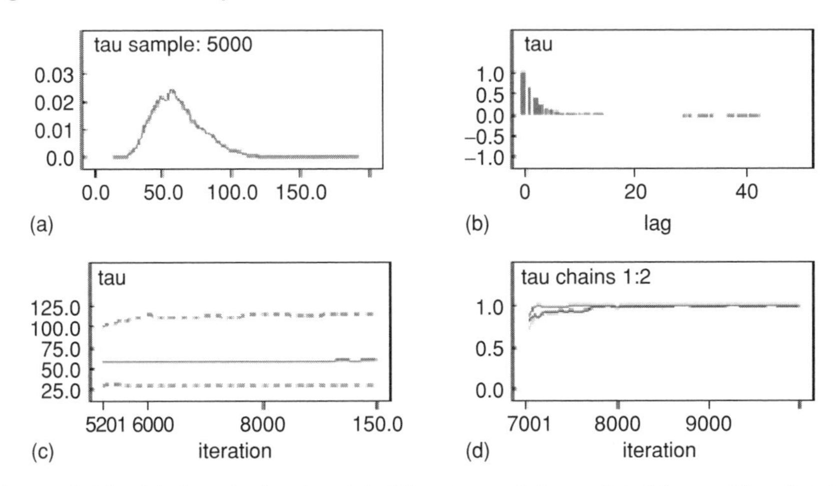

Figure 4.14 (a) Sample density plot, (b) autocorrelations plot, (c) quantiles plot, (d) Gelman and Rubin convergence diagnostic plot.

All WinBUGS graphics have a set of properties (margins, axes, titles, labels, fonts, colours, ...) that can be modified via the **plot properties options**. For details about graphics in WinBUGS see Spiegelhalter *et al.* (2002b).

If you are going to use some external software for monitoring convergence, an output file for each chain showing the iteration number and value can be produced by clicking on the **coda** button. A file containing a description of which lines of the output file correspond to which variable is also produced.

bgr diag calculate the Gelman and Rubin convergence statistic, as modified by Brooks and Gelman (1998), which is based on running parallel chains from overdispersed starting values.

4.4.3.2 Summary monitor

Using this tool WinBUGS store the mean and standard deviation for the parameters and the approximate quantiles. The **summary tool** (Figure 4.15) is located in the **inference menu**.

To use the summary tool we enter the parameter names to monitor in the **node** box and for each one press the **set** button.

Figure 4.15 Summary tool.

The **stats** button displays the mean and standard deviation for each node while the **means** button displays the means for each node in a comma delimited form.

4.5 SCRIPTS

As an alternative to the menu/dialog interface of WinBUGS a scripting language has been provided in the latest version of the software. This language can be useful for automating routine analysis.

To run a script from WinBUGS, open the file containing the commands and select **script** from the **model** menu.

4.5.1 A simple example

In Figure 4.16 we display the commands to perform a simple analysis of the model in Figure 4.2.

```
display('log')
check('sc-lognormal-code.txt')
data('sc-data.txt')
data('data-sc-1990.txt')
compile(2)
inits(1,'sc-inits1.txt')
inits(2,'sc-inits2.txt')
update(2000)
set(theta)
set(tau)
set(deviance)
dic.set()
update(4000)
stats(*)
trace(*)
gr(*)
dic.stats()
```

Figure 4.16 Script commands in WinBUGS.

A minimum of four files are required to make use of the scripting language: the script itself, a file containing the BUGS code for the model (*sc-lognormal-code.txt*), a file or several files containing the data (*sc-data.txt, data-sc-1990.txt*) and for each simulated chain a file containing the initial values (*sc-inits1.txt, sc-inits2.txt*).

This script performs the following actions:

- checks the model syntax,
- loads the data,
- specifies the number of chains (2) and compiles the model,
- loads the initial values for each chain,
- starts the simulation (1000 iterations),
- sets sample to store values of some parameters,
- carries out further updates,
- summarizes samples of the parameters: statistics, sample trace plots, Gelman and Rubin convergence diagnostics, deviance and DIC.

The output of this script is displayed in a log file. It can be saved in a file if we add the line **save(file)** at the end of the script.

A list of all currently implemented commands in the scripting language along with their menu equivalents appears in the WinBUGS 1.4 documentation.

The shortcut **BackBUGS** has been set up to run the commands contained in a file called **script.odc** located in the root directory of WinBUGS. Thus, a WinBUGS session may be embedded within any software that can execute the **BackBUGS** shortcut.

4.6 CHECKING CONVERGENCE

WinBUGS uses an iterative estimation algorithm that starts from arbitrary initial values (that can be generated from priors, based on frequentist estimates, chosen to represent extreme regions of the parameter space, etc.) and eventually converges to a target value. Software such as CODA – *Convergence Diagnostic and Output Analysis* – (Best *et al.*, 1995) may be used to perform convergence diagnosis and statistical and graphical analysis of the samples. This is a menu-driven suite of S-Plus functions for investigating MCMC output in general and WinBUGS output in particular. The program produces:

- summary statistics, such as means, quantiles, standard errors, etc.,
- sample traces and density plots,
- sample autocorrelations,
- statistics and plots for convergence diagnosis.

The program and manual are available over the website *http://www.mrc-bsu.cam.ac.uk/bugs/classic/coda04/readme.shtml.*

An S-Plus/R revision of CODA called BOA – *Bayesian Output Analysis* – (Smith, 1999) is also available for carrying out convergence diagnostic and statistical and graphical analysis of Monte Carlo sampling output. BOA can be used as an output processor for WinBUGS or any other program which produces sampling output. It is freely available over the web at *http://www.public-health.iowa.edu/boa.*

It is recommended to use a combination of convergence diagnostics plus visual inspection of the trace plots and summary statistics to look for evidence of when the simulation appears to have stabilized. For models with many parameters it is not practical to check the convergence for every parameter so a choice must be made of relevant parameters to monitor. In this way, convergence of the WinBUGS output can be assessed with a reasonable degree of confidence.

It is also possible to check the convergence of the algorithm using the sample traces plots and the Gelman–Rubin convergence diagnostic implemented in WinBUGS. As mentioned in Section 4.4.3, this test is based on running parallel chains from different starting values. The output is a plot where the width of the central 80% interval of the pooled runs is green, the average width of the 80% intervals within the individual runs is blue, and their ratio $R =$ (pooled/within) is red. R would generally be expected to be greater than 1 if the starting values are suitable overdispersed. Brooks and Gelman (1998) emphasize that one should be concerned both with convergence of R to 1, and with convergence of both the pooled and the within interval widths to stability.

The plots of Figures 4.17–4.19 are examples of chains for which convergence looks reasonable and chains which have not reached convergence.

After convergence, further iterations are needed to obtain samples for posterior inference. Running for further iterations will produce more accurate posterior estimates, but how many iterations are needed after convergence? Accuracy of the posterior estimates can be assessed by using the Monte Carlo standard error for each parameter. WinBUGS reports the Monte Carlo standard error of the mean, i.e. the standard deviation of the difference between the mean of the sampled values and the true posterior mean. To reach efficiency, the Monte Carlo error should be small in relation to the posterior standard deviation.

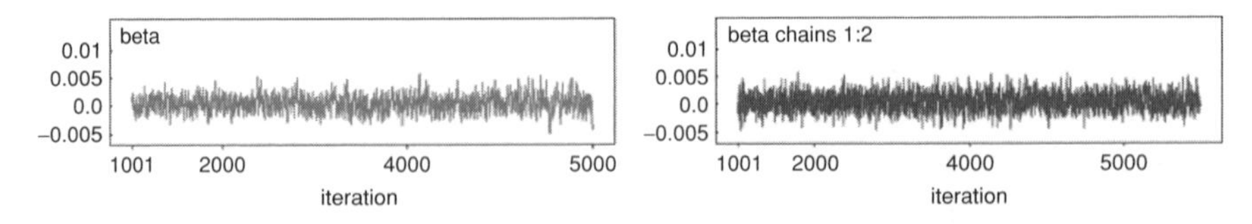

Figure 4.17 Sample traces of chains for which convergence looks reasonable.

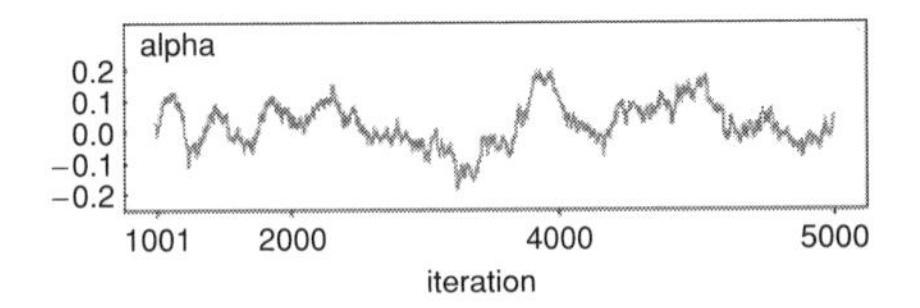

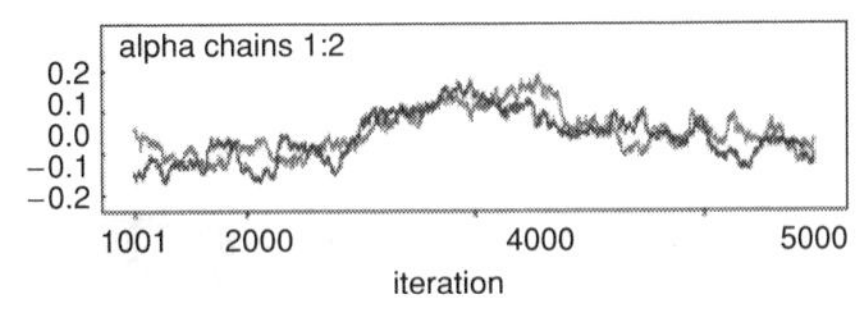

Figure 4.18 Sample traces of chains which have not reached convergence.

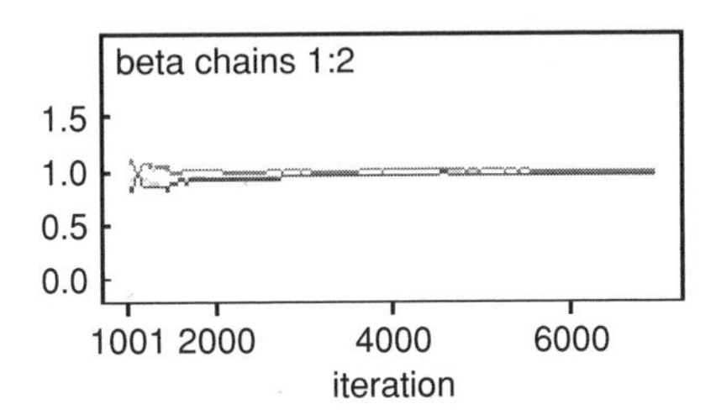

Figure 4.19 Gelman and Rubin diagnostic test for two chains for which convergence looks reasonable.

4.7 SPATIAL MODELLING: GeoBUGS

GeoBUGS is an add-on module to WinBUGS that fits spatial models and produces maps as output. It also provides an interface for creating and manipulating adjacency matrices that are required for carrying out spatial smoothing. It has been developed by a team at the Department of Epidemiology and Public Health of the Imperial College School of Medicine at St Mary's Hospital (London).

Currently, the version 1.1 of GeoBUGS can produce maps for

- districts in Scotland,
- wards in a London Health Authority,
- counties in Great Britain,
- *départements* in France,
- *nomoi* in Greece,
- districts in Belgium,
- communes in Sardinia,
- subquarters in Munich.

GeoBUGS 1.1 also has facilities for importing user-defined maps reading polygon formats from ArcInfo, EpiMap and S-Plus. This can be done using the import facilities in the **map menu** of WinBUGS (Figure 4.20). The different GeoBUGS import formats are designed to follow as closely as possible the format in which S-Plus, ArcInfo and EpiMap export polygons. However, some editing of

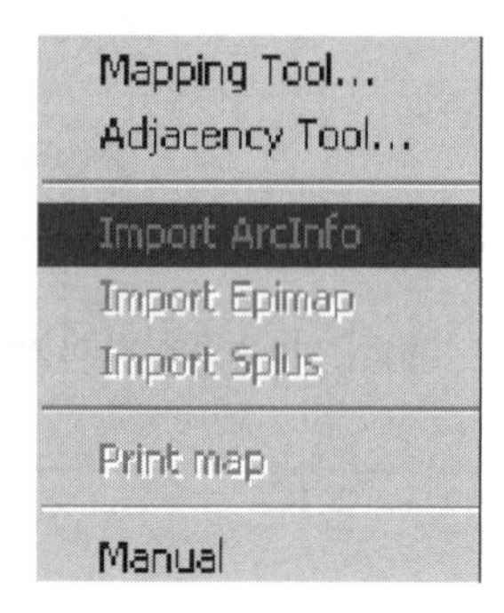

Figure 4.20 The Map menu.

the polygon files exported from those packages is necessary before they can be read into GeoBUGS. See the GeoBUGS manual (Thomas *et al.*, 2002) for details. See also appendix 2 for an S-Plus macro.

4.7.1 Producing maps using GeoBUGS 1.1

In order to produce a map of a summary statistic of the posterior distribution of a stochastic node, you must have already set a samples or summary monitor for that parameter and have carried out some updates. Select the **mapping tool** option (Figure 4.21) from the **map menu** in WinBUGS (if you are using version 1.4 of WinBUGS, GeoBUGS is already loaded; in other versions, you should install GeoBUGS first).

From the pull-down menu labelled **map**, select the name of the map you want to produce. By default, only the maps mentioned above are available. In the box labelled **variable**, type the name of the quantity to be mapped.

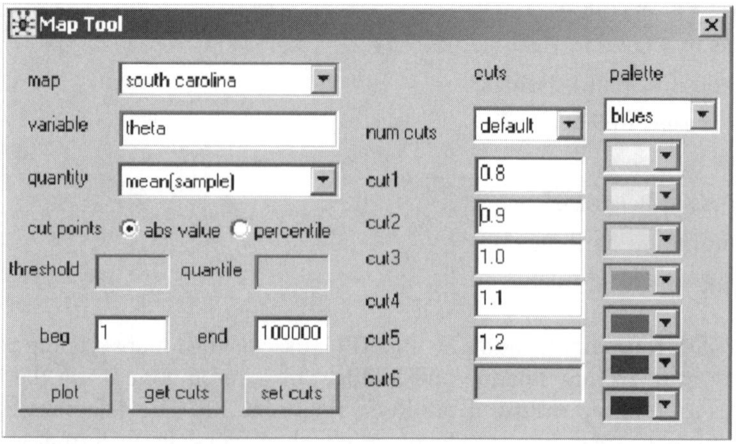

Figure 4.21 The Mapping tool.

If the quantity is data pick the value option in the menu labelled **quantity**.

If the quantity is a stochastic variable there are various options which you can select from the quantity menu:

- If you have monitored the variable by setting a summary monitor, then you must select the **mean(summary)** option from the menu.
- If you have monitored the variable by setting a samples monitor, you can select
  - **mean(sample)**: map the posterior means of the variable.
  - **percentile**: plot the posterior quantiles of the variable.
  - **prob greater** (or **prob less**): map the posterior probability that the value of the variable is greater than (less than) or equal to a specified threshold, which should be typed in the box labelled **threshold**.

GeoBUGS can work with *absolute cut-points* that are chosen to give equally spaced intervals or *percentile cut-points*; GeoBUGS gives, by default, the 10th, 50th and 90th percentiles of the empirical distribution of the variable being mapped but it allows the user to edit the values of the cut-points.

To obtain the map shaded according to the values of the variable click the **plot** button. Figure 4.22 shows the relative risks for cancer mortality in the state of South Carolina during the year 1998.

The index, label and value of an individual area on the map can be found by placing the cursor over the area of interest on the map and clicking with the left

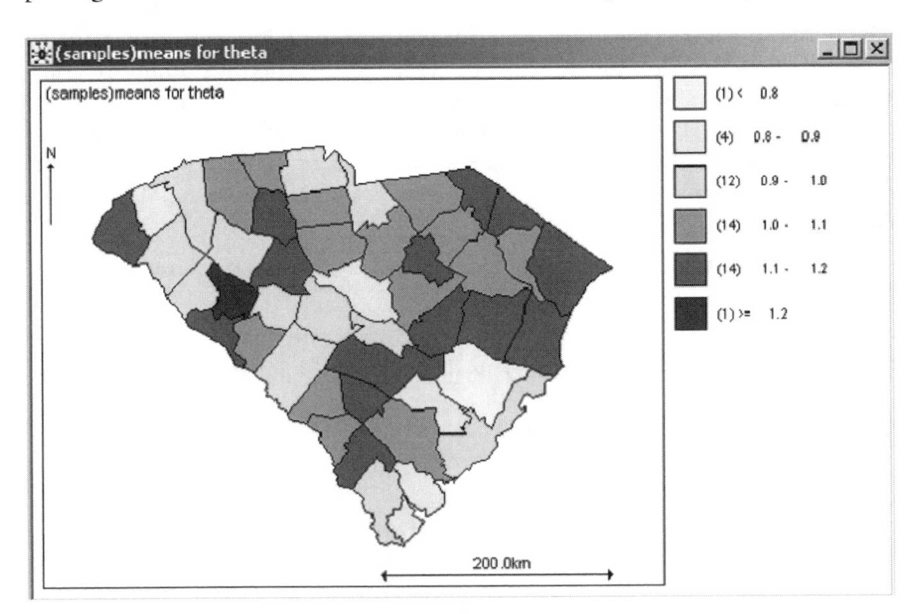

Figure 4.22 GeoBUGS map of relative risks for cancer mortality in the state of South Carolina in 1998.

mouse button. They are then displayed in the status bar of the WinBUGS window.

4.7.2 Spatial distributions in GeoBUGS 1.1

In WinBUGS 1.4, spatial models have been moved to GeoBUGS and new spatial distributions have been implemented. Table 4.5 shows the spatial distributions that are available in GeoBUGS.

The intrinsic Gaussian CAR prior distribution (or simply CAR) is specified using the distribution **car.normal** described in Table 4.5 where:

- *adj*[] is a vector listing the adjacent areas for each area;
- *w*[] is a vector of the same length as *adj*[] giving weights associated with each pair of areas;
- *n*[] is a vector of length m (total number of areas) giving the number of neighbours for each area;
- *tau* is a scalar representing the inverse-variance of the random effect.

The first three arguments are data and must be specified in the data files; *tau* is treated as unknown and is assigned a prior distribution.

A robust version of this model, **car.l1**, is available.

Since the CAR model is improper, it is necessary to impose a constraint to ensure that the model is identifiable. This means that an intercept term must be included in the model and it must be assigned an improper uniform prior. WinBUGS includes a distribution called *dflat*() which corresponds to an improper prior on the whole real line. GeoBUGS 1.1 includes a proper Gaussian CAR prior, **car.proper**, whose syntax is described in Table 4.5 and its arguments are as follows:

- *mu*[] is a vector giving the means for each area (this can be entered as data or assigned a prior distribution).
- *w*[]is a vector of normalized weights associated with each pair of areas and which must be entered as data. This differs from the intrinsic Gaussian CAR

Table 4.5 Spatial distributions available in GeoBUGS 1.1.

Name	Usage
CAR normal	*b*[]~car.normal(*adj*[], *w*[], *n*[], *tau*)
CAR l1	*b*[]~car.l1(*adj*[], *w*[], *n*[], *tau*)
car proper	*b*[]~car.proper(*mu*[], *w*[], *adj*[], *n*[], *M*[], *tau*, *gamma*)
spatial exp	*b*[]~spatial.exp(*mu*[], *x*[], *y*[], *tau*, *phi*, *kappa*)
spatial pred	*b*[]~spatial.pred(*mu*[], *x*[], *y*[], *s*[])

distribution where unnormalized weights should be specified. For disease mapping, Stern and Cressie (1999) suggest considering

$$w_{ij} = \begin{cases} w_{ij} & \text{for neighbouring areas} \\ 0 & \text{otherwise.} \end{cases}$$

- *M*[] is a vector of length m giving the diagonal elements of the conditional variance matrix. In the context of disease mapping Stern and Cressie (1999) chose the inverse of the expected count in area i:

$$M_{ii} = \frac{1}{e_i}, \; i = 1, \ldots, m.$$

- *gamma* is a scalar parameter representing the overall degree of spatial dependence.

The other arguments of the **car.proper** distribution are the same as the corresponding arguments to the **car.normal** distribution.

GeoBUGS 1.1 includes facilities for spatial prediction based on the known data at specific spatial locations. The distribution **spatial.exp** allows the user to fit Bayesian Gaussian kriging models. This is based on a full Bayesian kriging model with parameterized covariance. Estimation involves the inversion of a full covariance matrix. Its syntax is

$$b[] \sim spatial.exp(mu[], x[], y[], tau, phi, kappa)$$

where

- *mu*[] is a vector that gives the means for each area (this can be entered as data or assigned a prior distribution).
- *x*[] and *y*[] are vectors of length m that give the coordinates of each point or the centroid of each area.
- *tau* is a scalar representing the inverse-variance or precision in the model.
- *phi* is the rate of decline of correlation between points.
- *kappa* is a scalar parameter controlling the amount of spatial smoothing.

Finally, the distribution **spatial.pred** permits spatial interpolation or prediction. The syntax for this predictive distribution is

$$b[1:p] \sim spatial.pred(mu[], x[], y[], s[])$$

where

- p is a scalar that gives the number of prediction locations.
- *mu*[] is a vector of length p specifying the mean for each prediction location.

- $x[]$ and $y[]$ are vectors of length p that give the coordinates of the location of each prediction point.
- $s[]$ is a vector of observations to which the **spatial.exp** model has been fitted.

There is a version of this distribution, **spatial.unipred**, that carries out single-site prediction. The advantage of this distribution is that it is faster than the one that carries out the joint prediction at a set of target locations.

Examples of the use of these spatial distributions are given in Chapters 6, 7 and 8. For more information on the technical details of these spatial distributions and hyperprior specifications for them see the GeoBUGS 1.1 manual (Thomas *et al.*, 2002).

4.8 CONCLUSIONS

4.8.1 Practical considerations when using WinBUGS

In this chapter we have described some of the features of the WinBUGS package. In summary when using WinBUGS we would generally proceed through the following steps:

- Specify a Bayesian model. Learn as much as possible about the model before running the sampler (maximum likelihood, Bayesian approximations, ...).
- Construct a Markov chain whose target distribution is the joint posterior distribution of interest.
- Choose initial values.
- Run one or more chains (three parallel chains with overdispersed initial values is recommended).
- Monitor all unknown model quantities (not only parameters of interest) beginning from the first iteration. If it is not possible to monitor all parameters, then monitor at least representative samples of each kind of parameter.
- Assess convergence
  - retune model parameterization, priors, ...
  - increase number of iterations.
- Use samples for estimation and inference.

4.8.2 Additional information about WinBUGS

For problems or questions about WinBUGS, try the frequently asked questions (FAQ) section page of the BUGS website

(*http://www.mrc-bsu.cam.ac.uk/bugs/faqs/contents.shtml*).
The project team also has a user support service where users may email specific questions about the software, difficulties with installation, etc. to *bugs@mrc-bsu.cam.ac.uk*.

For sharing ideas and asking questions about modelling issues using BUGS you can email the WinBUGS discussion list (*bugs@jiscmail.ac.uk*).

5

MLwiN Basics

As discussed in Chapter 3, many multilevel modelling software packages, including MLwiN, were developed by statistical researchers in the field of education. It was then established that the same statistical model structures were applicable in many other application areas including public health and disease mapping. The MLwiN software package cannot fit as many of the models that we will consider in the later chapters of this book as the WinBUGS package, but it does offer both likelihood-based estimates and Bayesian estimates for the models it can fit. In this chapter we will describe some of the history and modelling features of MLwiN. We will particularly concentrate on the features in MLwiN that are important for disease mapping datasets.

5.1 ABOUT MLwiN

MLwiN is a computer program designed to fit multilevel statistical models. It is derived from a series of packages that began in the late 1980s with the software package ML2 that was designed to fit two-level statistical models. This package was extended to form the packages ML3 (for three levels) and MLN (for arbitrary numbers of levels). These three packages were command-driven packages that were all DOS-based. They used implementations of the IGLS and RIGLS algorithms (Goldstein, 1986) to calculate maximum likelihood (ML) and restricted maximum likelihood (REML) estimates for the parameters of the multilevel models.

The MLwiN software package was first released in 1998 (version 1.0) and took the DOS-based MLN package and incorporated it in a Windows environment. This allowed users to specify statistical models through an equations-based interface and also gave them much greater graphical facilities for plotting features of their model, for example the residuals. MLwiN also included the addition of a second estimation engine that produced MCMC estimates for the equivalent Bayesian models, and functionality to perform bootstrap estimation using the IGLS/RIGLS estimation engines.

Disease Mapping with WinBUGS and MLwiN A. Lawson, W. Browne and C. Vidal Rodeiro
 ISBN: 0-470-85604-1 (HB)

In 2000 a second release of MLwiN (version 1.1) was produced which included improvements in data handling functions and the ability to estimate some additional models. At the time of writing a third release (version 1.2) is available as a Beta test version and this version contains many additional model fitting functions, particularly using the MCMC engine, that extend MLwiN's scope even further.

In this book we will be using a further updated version of MLwiN which should be available to users in 2003 (version 2.0). Although not all of the estimation methods (IGLS, RIGLS, MCMC and bootstrapping) can be used for all of the models that MLwiN can fit, between them the methods can be used to fit a wide array of models. The package can handle many response types using the following distributions: normal, Poisson, binomial, negative binomial, multinomial, ordered multinomial and multivariate combinations of these distributions. It can handle both nested and crossed random effects, multiple membership models, missing data in multivariate responses, measurement errors in predictors, multilevel factor analysis modelling, autocorrelated (time series) residual structures and CAR models for spatial dependence between residuals. The package also has an underlying macro language (it was built on top of the Nanostat statistics package developed by Professor Michael Healy which pre-dates even ML2), which allows it to be easily used for simulation studies.

The package has many windows specifically designed for multilevel models. These include residual screens that produce plots and diagnostics for residuals and prediction and variance function screens that produce particular derived parameter estimates from the current model. The program has been developed since its inception by the multilevel modelling team headed by Professor Harvey Goldstein at the Institute of Education, London. The bulk of the program has been written by its main programmer, Jon Rasbash, although certain sections have been written by other members of the team, including the MCMC engine which has been developed by William Browne.

Unlike WinBUGS, MLwiN is not free but users pay a once-off fee for a licence, the money from which is used to fund some of the development of the program. Once you have bought a copy of the program, updates that include new features and bug fixes can be downloaded from the Multilevel modelling project website *http://multilevel.ioe.ac.uk/*. This website also contains lots of information about multilevel modelling including additional documentation on MLwiN and the latest working papers from the project team that can be downloaded.

MLwiN is designed primarily for researchers in the social and medical sciences who have varying statistical ability. So although it is used by many statisticians it also attempts to be accessible to researchers whose more limited statistical knowledge may mean that a program like WinBUGS is too complex for them. Consequently the program has a large volume of user documentation including step-by-step analyses of several datasets. Both the user's guide (Rasbash *et al.*, 2000) and the book, *MCMC Estimation in MLwiN* (Browne, 2003) are available for download from the Multilevel website.

Both books concentrate on a dataset from education that contains a continuous response variable to illustrate many of the features of the package and multilevel models that are common to many datasets. Then other chapters consider other datasets, concentrating on their differences from the first dataset, for example different response types and features of the dataset. With regard to disease mapping examples, both manuals consider a Poisson response dataset of malignant melanoma mortality in the European community from 1971 to 1980. The deaths were related to exposure to ultraviolet radiation, and in particular type B (UVB) radiation (between 290 and 320 nm) and the importance of different levels of geographic aggregation was considered. The dataset has been investigated more thoroughly in Langford *et al.* (1998).

Browne (2003) also considers the much-analysed Scottish lip cancer dataset which consists of observed counts of male lip cancer for the 56 regions of Scotland over the period 1975–1980. This dataset was analysed by amongst others, Clayton and Kaldor (1987) and is also one of the examples contained in the WinBUGS examples documentation. It is used to illustrate the additional multiple-membership and spatial modelling features that are available when using the MCMC estimation engine.

5.2 GETTING STARTED

MLwiN has its own file format known as a worksheet, in which is stored both the dataset and the last model and estimates run prior to saving. The data is stored in numbered columns (starting from column C1) which can be given labels, and so a worksheet looks very much like a spreadsheet, as we will see later.

If you are a new user then you may wish to try out the example chapters in the user manuals. For these examples the worksheets with the data required already exist and are supplied with the software. When, however, you are faced with performing your own analysis, you will need to create a worksheet from your dataset. We will now consider the melanoma mortality dataset and consider how to get the data into the package.

5.2.1 Inputting your own data

There are two ways to get your data into MLwiN; firstly you can input ASCII data files directly by specifying the desired columns into which you want the data input or alternatively (and perhaps more easily) you can simply copy and paste the data from another package, for example Excel. Of course, you also need to get the data into a format that MLwiN can use to estimate a model. In Table 5.1 are the first seven records for the melanoma mortality dataset.

Table 5.1 Melanoma mortality dataset.

Nation	Region	County	Observed	Expected	UVB
1	1	1	79	51.222	−2.9057
1	2	2	80	79.956	−3.2075
1	2	3	51	46.517	−2.8036
1	2	4	43	55.053	−3.0069
1	2	5	89	67.758	−3.0069
1	2	6	19	35.976	−3.4175
1	3	7	19	13.28	−2.6671
...	...	...	...	...	...

The data has three levels of nested geography, where counties are nested within regions and regions nested within nations. There are in total 354 counties, 78 regions and 9 nations in the dataset. Our response variable is the observed count that is collected at the county level. The first seven records all come from nation 1 (Belgium) and we see that the first county is in region 1 whilst the next five are in region 2 and the seventh county is in region 3. In fact our data has been sorted on regions within nations which, although not necessary prior to input into MLwiN, must be done before fitting multilevel models so that MLwiN can establish the data structure. The data is completed by an expected mortality count that is based on population size and number of years of counting (which varies between regions in this dataset) and an ultraviolet (type B) exposure predictor (UVB), which is calculated for each county and has been centred (hence the negative values for all these counties as Belgium is less sunny than southern Europe).

We will include the column names as a first row in the ASCII file and then read the data into Excel. To select all the data in Excel we simply click on the column headings and then choose **Copy** from the **Edit** menu. If we then start MLwiN we will have a blank worksheet and choosing the **Paste** option from the **Edit** menu will produce the following window.

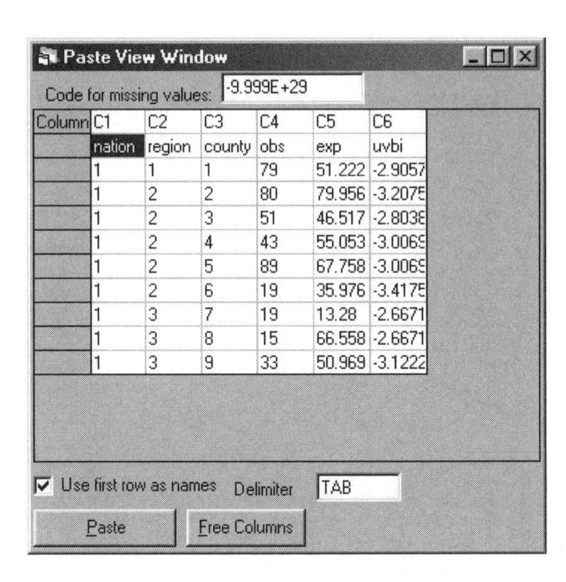

Column	C1	C2	C3	C4	C5	C6
	nation	region	county	obs	exp	uvbi
	1	1	1	79	51.222	-2.9057
	1	2	2	80	79.956	-3.2075
	1	2	3	51	46.517	-2.8038
	1	2	4	43	55.053	-3.0069
	1	2	5	89	67.758	-3.0069
	1	2	6	19	35.976	-3.4175
	1	3	7	19	13.28	-2.6671
	1	3	8	15	66.558	-2.6671
	1	3	9	33	50.969	-3.1222

Note that here we have selected the '**Use first row as names**' box and clicked on the **Free Columns** button. Clicking on the **Paste** button will now input the data into the worksheet. Two of MLwiN's many windows are useful for viewing the data. Firstly the **Names** window gives an overview of the data as shown below:

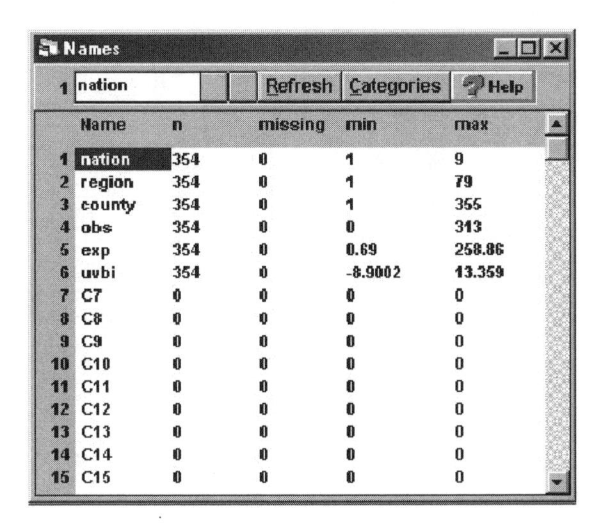

	Name	n	missing	min	max
1	nation	354	0	1	9
2	region	354	0	1	79
3	county	354	0	1	355
4	obs	354	0	0	313
5	exp	354	0	0.69	258.86
6	uvbi	354	0	-8.9002	13.359
7	C7	0	0	0	0
8	C8	0	0	0	0
9	C9	0	0	0	0
10	C10	0	0	0	0
11	C11	0	0	0	0
12	C12	0	0	0	0
13	C13	0	0	0	0
14	C14	0	0	0	0
15	C15	0	0	0	0

Secondly if we wish to look at the individual records or edit the data we can use the **View/Edit data** window:

Data

goto line 1 | view | Help | Font

	nation(354)	region(354)	county(354)	obs(354)	exp(354)	uvbi(354)
1	1	1	1	79	51.222	-2.9057
2	1	2	2	80	79.956	-3.2075
3	1	2	3	51	46.517	-2.8038
4	1	2	4	43	55.053	-3.0069
5	1	2	5	89	67.758	-3.0069
6	1	2	6	19	35.976	-3.4175
7	1	3	7	19	13.28	-2.6671
8	1	3	8	15	66.558	-2.6671
9	1	3	9	33	50.969	-3.1222
10	1	3	10	9	11.171	-2.4852
11	1	3	11	12	19.683	-2.5293
12	2	4	12	156	108.04	-1.1375
13	2	4	13	110	73.692	-1.3977
14	2	4	14	77	57.098	-.4386
15	2	4	15	56	46.622	-1.0249
16	2	5	16	220	112.61	-.5033
17	2	5	17	46	30.334	-1.4609
18	2	5	18	47	29.973	-1.8956
19	2	5	19	50	32.027	-2.5541
20	2	5	20	90	46.521	-1.9671

In the latest version of MLwiN, the **Paste** option has been improved so that it will accept alphanumeric data (which it will treat as a categorical variable). Alternatively we can set category names from the **Names** window as shown below:

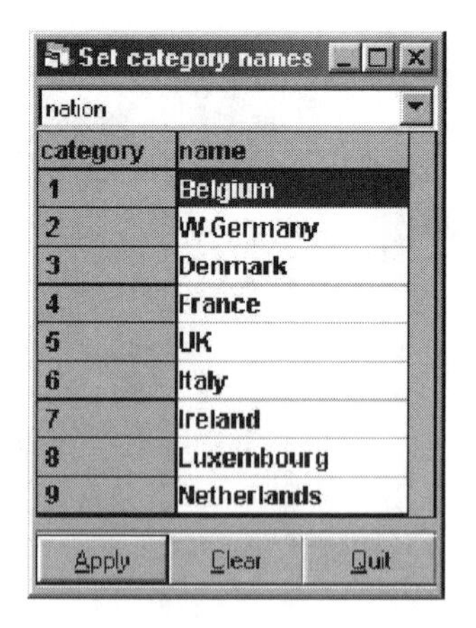

which will result in the data appearing as follows:

	nation(354)	region(354)	county(354)	obs(354)	exp(354)	uvbi(354)
1	Belgium	1	1	79	51.222	-2.9057
2	Belgium	2	2	80	79.956	-3.2075
3	Belgium	2	3	51	46.517	-2.8038
4	Belgium	2	4	43	55.053	-3.0069
5	Belgium	2	5	89	67.758	-3.0069
6	Belgium	2	6	19	35.976	-3.4175
7	Belgium	3	7	19	13.28	-2.6671
8	Belgium	3	8	15	66.558	-2.6671
9	Belgium	3	9	33	50.969	-3.1222
10	Belgium	3	10	9	11.171	-2.4852
11	Belgium	3	11	12	19.683	-2.5293
12	W.Germany	4	12	156	108.04	-1.1375
13	W.Germany	4	13	110	73.692	-1.3977
14	W.Germany	4	14	77	57.098	-.4386
15	W.Germany	4	15	56	46.622	-1.0249
16	W.Germany	5	16	220	112.61	-.5033
17	W.Germany	5	17	46	30.334	-1.4609
18	W.Germany	5	18	47	29.973	-1.8956
19	W.Germany	5	19	50	32.027	-2.5541
20	W.Germany	5	20	90	46.521	-1.9671

5.2.2 Creating variables within MLwiN

MLwiN also offers many data manipulation functions that can be used to create new variables. In our dataset we need to create firstly a column of '1's to represent an intercept and secondly a column of offset terms that are the logarithms of the expected counts. We may also want to create a column of SMRs (observed/expected). To create a constant term we can use the **Generate vector** window and put the constant in the first blank column (*c7*):

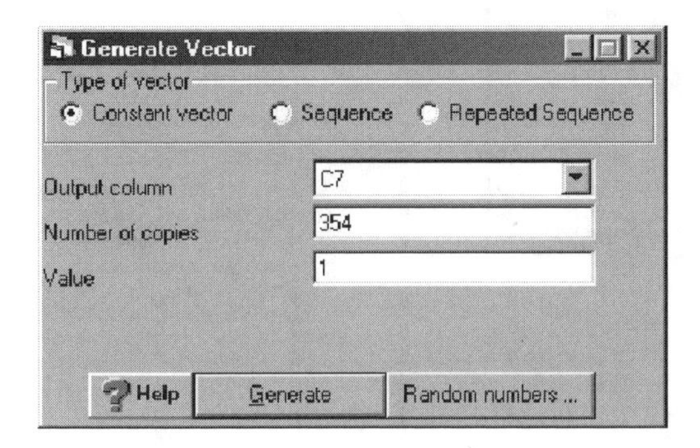

To create the offsets (and the SMR) column we can use the **Calculate** window which allows the user to input arbitrary formulae (we use the first empty column, *c8* here):

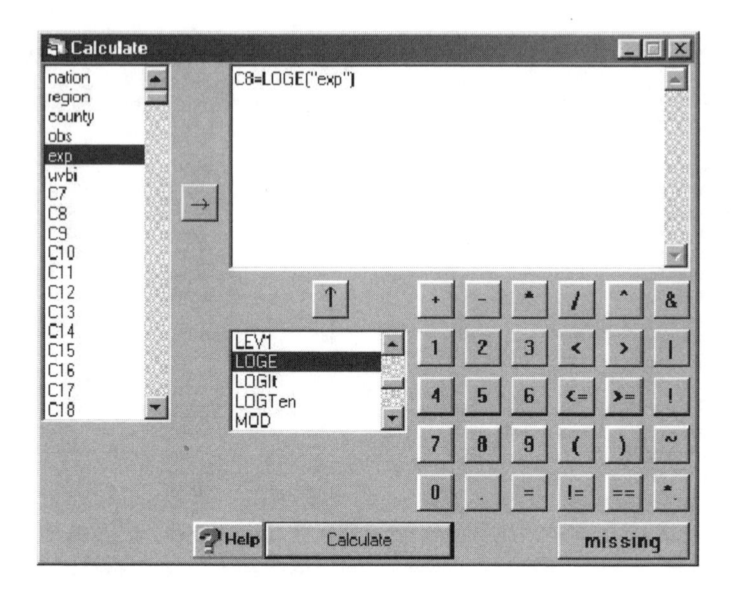

We can also now use the calculate window to create the SMR column ($c9$ = '*obs*' / '*exp*'). We can name our three new columns via the **Names** window. We will name the constant column '*cons*', the offsets '*logexp*' and the SMRs, '*smr*' as shown below:

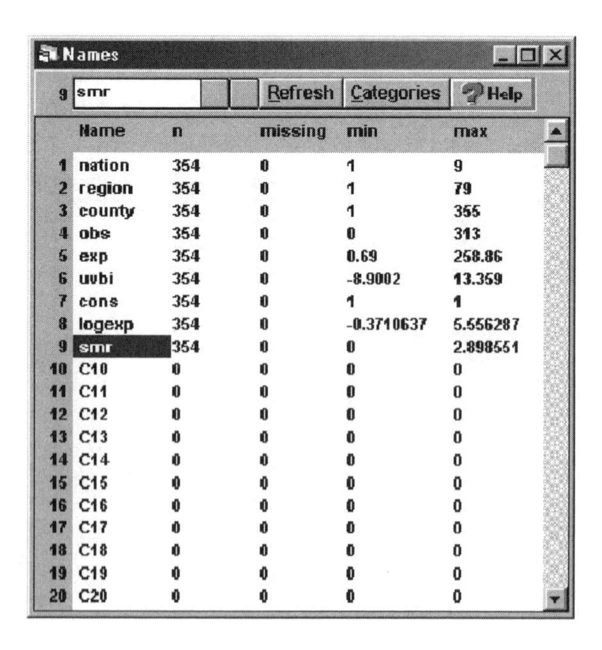

Names

9 smr | Refresh | Categories | Help

	Name	n	missing	min	max
1	nation	354	0	1	9
2	region	354	0	1	79
3	county	354	0	1	355
4	obs	354	0	0	313
5	exp	354	0	0.69	258.86
6	uvbi	354	0	-8.9002	13.359
7	cons	354	0	1	1
8	logexp	354	0	-0.3710637	5.556287
9	smr	354	0	0	2.898551
10	C10	0	0	0	0
11	C11	0	0	0	0
12	C12	0	0	0	0
13	C13	0	0	0	0
14	C14	0	0	0	0
15	C15	0	0	0	0
16	C16	0	0	0	0
17	C17	0	0	0	0
18	C18	0	0	0	0
19	C19	0	0	0	0
20	C20	0	0	0	0

5.2.3 Plotting the data

Before fitting statistical models to the data it may be useful to plot the data to look for any meaningful relationships. We could plot the observed counts against expected counts to look for unusual observations, or plot the SMRs against the ultraviolet exposure predictor. To set up graphs in MLwiN we need to specify the two responses in the **Customised graphs** window.

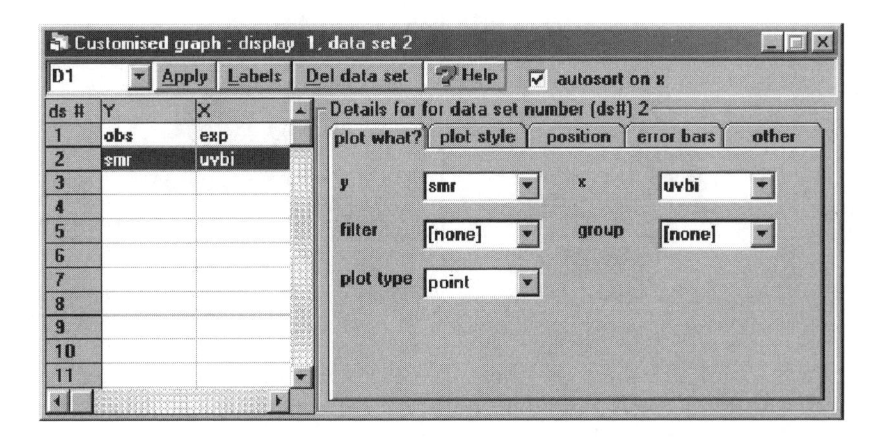

We can also add axis labels and highlight interesting points (although currently in MLwiN if you want to highlight points you need to set up a statistical model and so, to produce the highlighting seen here, we first need to run the first model in the next section):

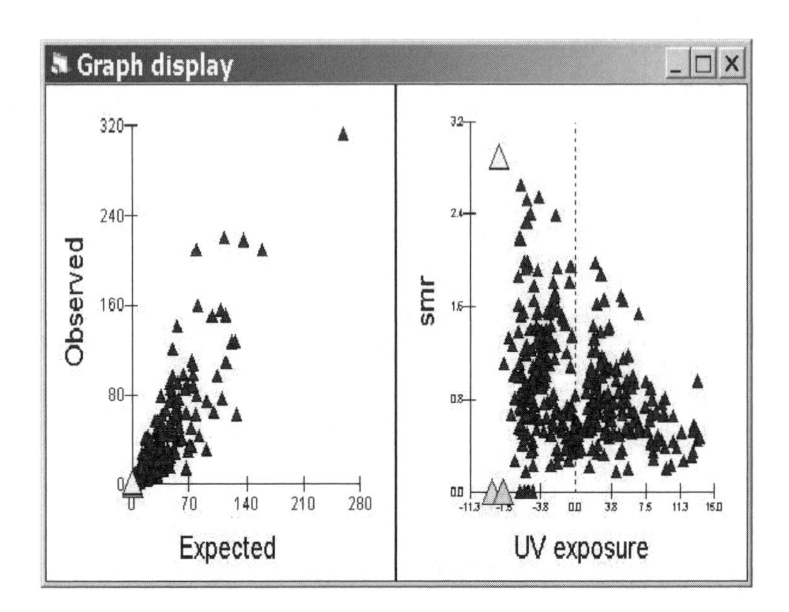

We have here highlighted the three counties with the smallest UV exposure in the right-hand graph (note that the equivalent points are highlighted in the left-hand graph but are all on top of each other). These counties are all in the UK and highlight how the variability in SMR is much higher for areas with small expected counts as the highest SMR corresponds to only two observed deaths. In fact the other four zero observed counts (that have not been highlighted) are all in Ireland and also have low expected counts.

5.3 FITTING STATISTICAL MODELS

Although we have an obvious three-level structure that we can fit in our model, perhaps a sensible place to start is to ignore this structure and fit a simple Poisson regression model. Although MLwiN is a multilevel modelling package it can of course also fit single-level models. To fit a model in MLwiN we need to firstly set it up in the **Equations** window.

5.3.1 The Equations window interface

If we select **Equations** from the **Model** menu we are faced with the following blank **Equations** window as we have not yet set up a model.

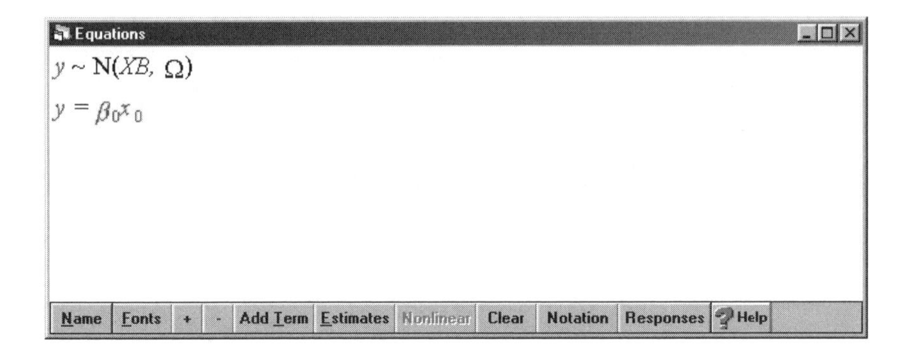

We need to set up our response variable and data structure. If we click on the red y we are faced with a screen in which we can input the response variable, the number of levels and the level identifiers. This we will fill in as follows:

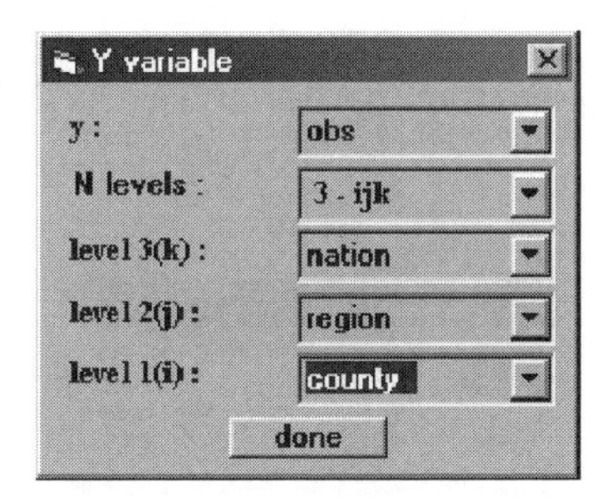

We now need to specify the response distribution, which by default is Normal and so we click on the N in the **Equations** window and pick *Poisson* from the response type window that appears:

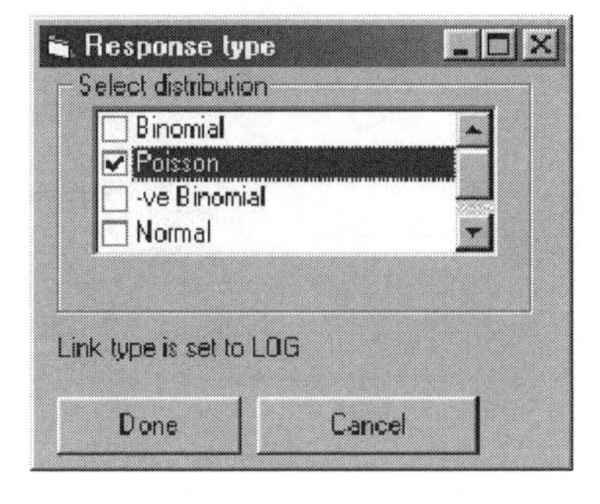

Having selected Poisson this window automatically chooses the log link, which is required for Poisson data. After clicking on the **Done** button, the **Equations** window will now look as follows:

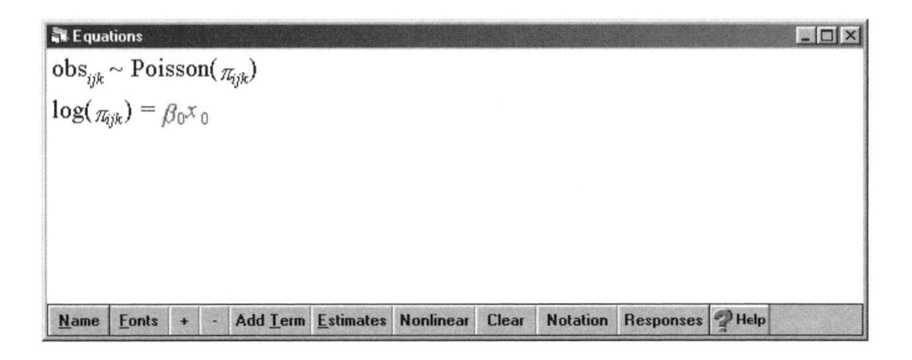

We now need to firstly specify the offset column. This is achieved by clicking on the π_{ijk} and choosing '*logexp*' from the list provided as shown and clicking on the **Done** button:

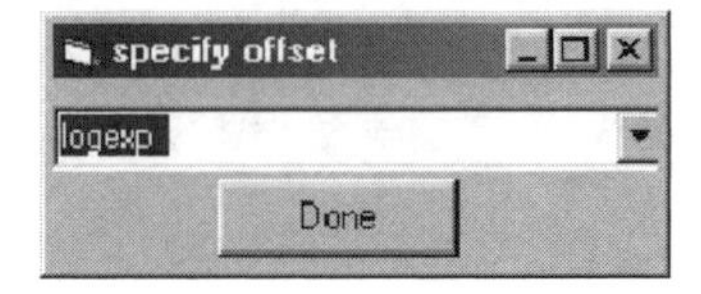

We now need to specify our intercept and predictor variables. We do this by clicking on the **Add Term** button and selecting first '*cons*' as shown below and secondly '*uvbi*'.

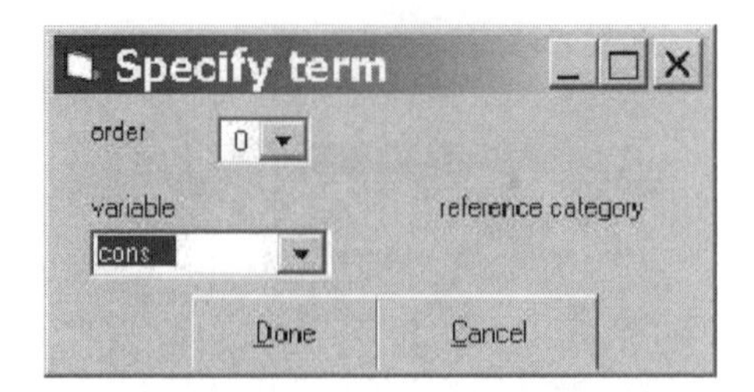

After adding the two terms and clicking on the **Estimates** button the **Equations** window will appear as follows:

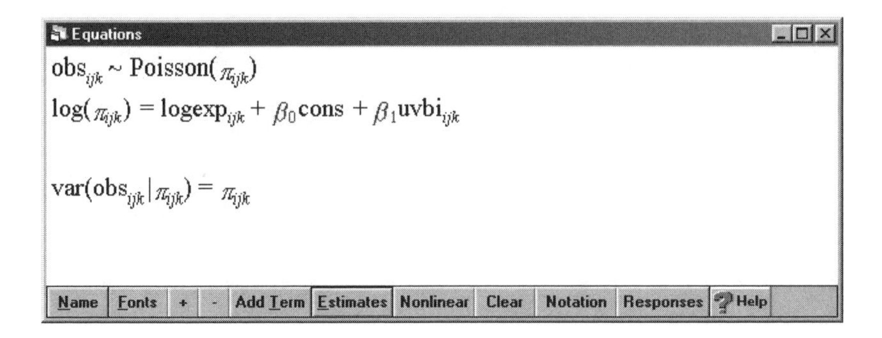

The estimates are currently shown in blue as we have yet to run the model. If we click on the **Nonlinear** button we can now select the type of quasi-likelihood method that we will use to fit the model. We will here choose to use the default method of first-order MQL estimation (see Chapter 3 for a description of the various methods).

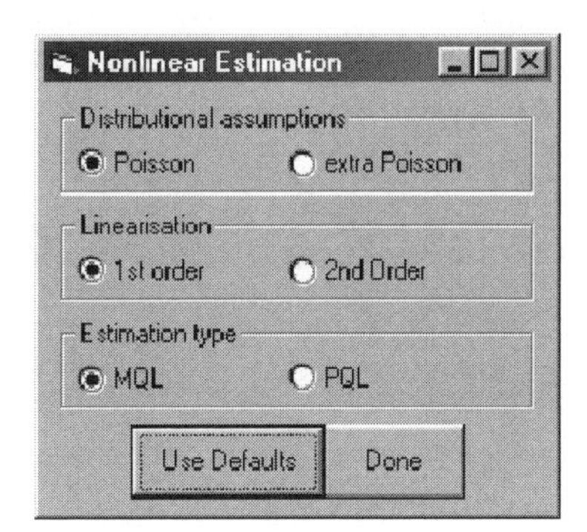

If we now click on the **Start** button the model will converge in seven iterations as the counter shows at the bottom of the screen. Clicking on the **Estimates** button again will give the actual numeric estimates of the parameters as shown in the **Equations** window below:

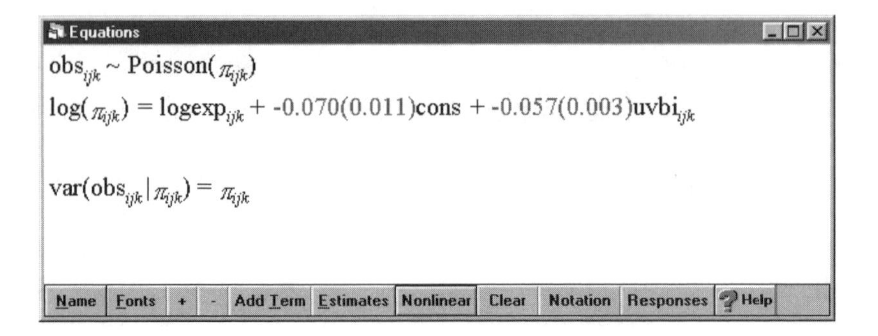

Here we see that an increase in UV exposure will result in a smaller risk of getting melanoma! Although this seems counter-intuitive what we are possibly seeing is that people in countries with higher UV exposure are genetically less likely to get skin cancer. Also the UV exposure is based on where the individuals live and does not take account of other factors, for example people going on holidays in hotter climes. All we can, however, establish from the model is that living in areas with higher UV exposure lowers the risk of mortality from melanoma.

The bottom line of the model describes the variance. Currently we have assumed Poisson variation and hence constrained this variance to equal π_{ijk}, we can, however, remove this constraint by choosing '*extra-Poisson*' variation on the **Nonlinear** window. If we do this we get the following:

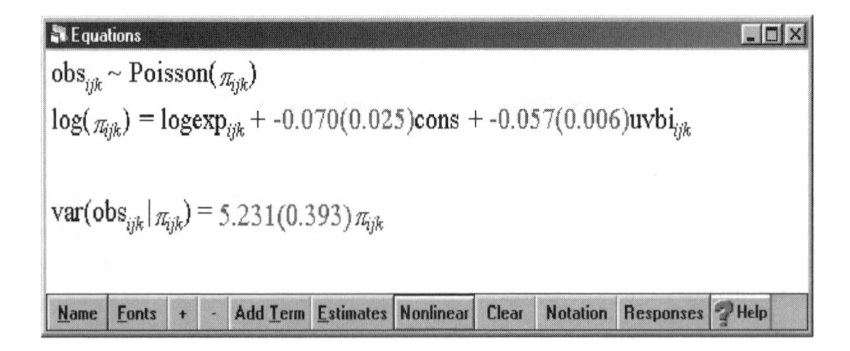

We can see here that the Poisson assumption is not a good fit to the data as we have far greater variability than we would expect from a Poisson distribution (5.231 versus 1.0). This would suggest that in fact the counties are not independent Poisson counts and that we should take account of some of the structure in the data by fitting random effects for the districts and nations and this will result in a Poisson response multilevel model. To do this we click on the constant term '*cons*' in the **Equations** window and, on the **X variable** window that appears, select the *nation* and *region* levels, implying that the intercept is random at these two levels.

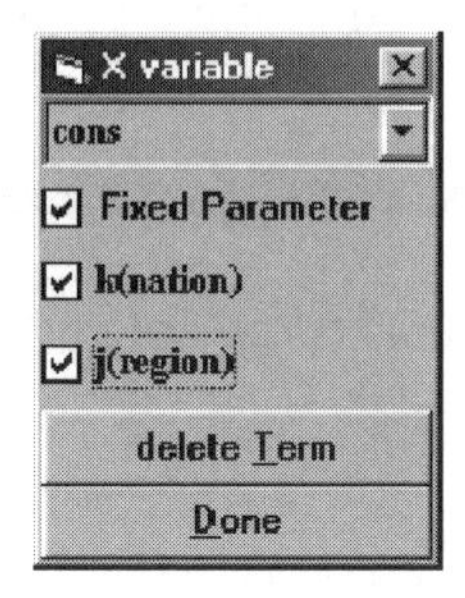

If we now click on the **Start** button without changing the nonlinear settings we see that the extra-Poisson variation virtually all vanishes as it has been explained by the variability between nations and districts:

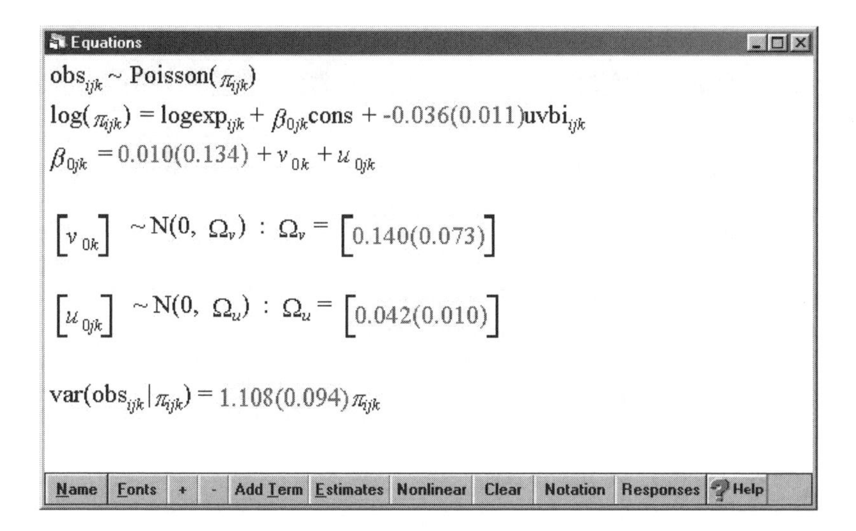

The Poisson assumption is therefore a reasonable fit, conditional on district and nation random effects. We also see that there is a larger variability between nations (0.140) than between districts within nations (0.042). The effect of UV exposure is reduced now that we have including effects for nations and districts giving some support to the earlier hypothesis of different genetic effects for different nations. When we fit random effects and hence a multilevel model to the data the choice of quasi-likelihood method matters. The default method in

MLwiN is first-order MQL estimation as it is more likely to converge and give estimates; however, the second-order PQL estimation method is generally a better approximation to maximum likelihood. If we now change the method to second-order PQL on the **Nonlinear** window and remove the extra-Poisson assumption we will get the following estimates upon pressing the **More** button.

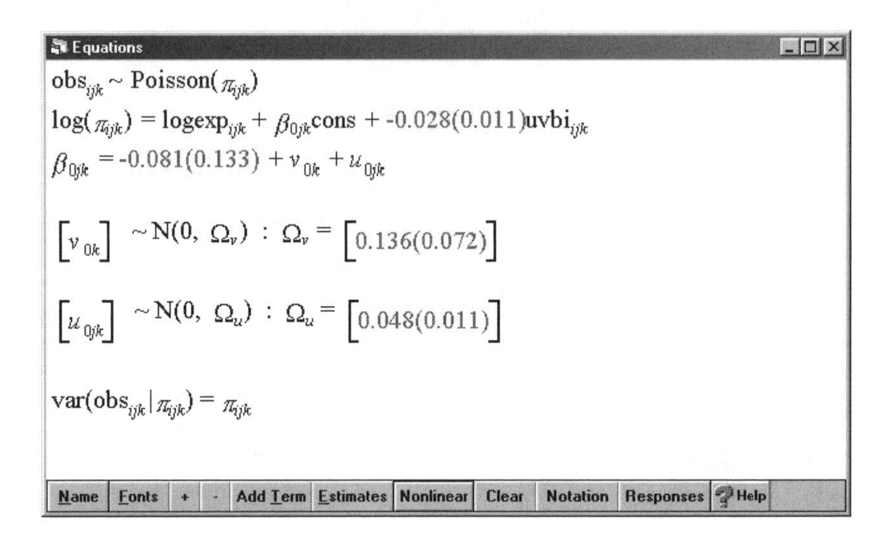

Here we see that in fact for this model the choice of quasi-likelihood method seems to make little impact on the variance components. We could now consider fitting further models and Rasbash *et al.* (2000) consider fitting nations as fixed instead of random effects (as there are only nine nations) and fitting different effects of UV exposure for each country. This chapter is, however, just an introduction to MLwiN so we refer the interested reader to Rasbash *et al.* (2000) and Browne (2003) for further models. We will now concentrate on some of the other useful functions available in MLwiN for our current model.

5.3.2 Residuals

In Section 2.6 the notion of residuals for a model was introduced. In our multilevel model we have two sets of higher-level residuals (random effects). The set u_{0jk} are the region-level residuals and represent the effects of region j in nation k whilst the v_{0k} are the nation-level residuals and represent the effect of nation k. MLwiN has a **Residuals** window which has options to calculate residuals and display specific graphs for residuals. Choosing **Residuals** from the **Model** menu gives the following window:

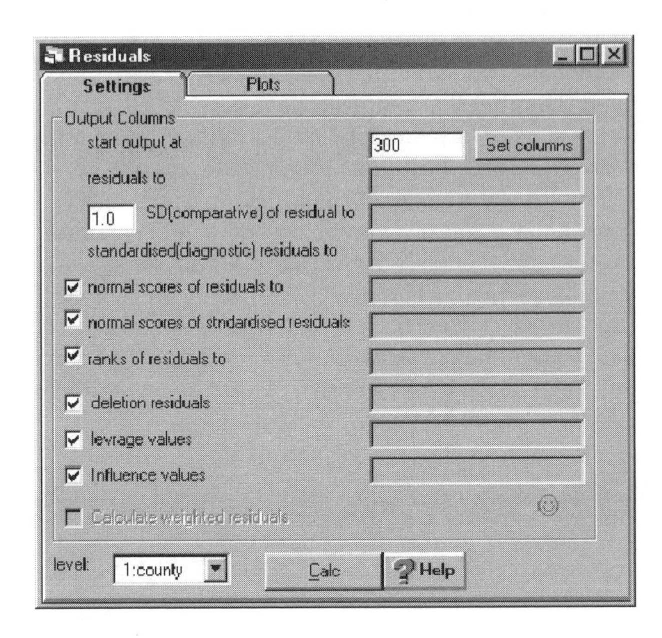

Here we see that as well as calculating residuals, this window also calculates functions of the residuals, such as their rank and normal scores, which we will use in the graphs that follow. The window also calculates diagnostic tools such as deletion residuals and leverage and influence values that are described elsewhere (Chapter 12 of Rasbash *et al*., 2000). If we now choose level 3 (*nation*) and change the SD multiplier to 1.96 then after calculating the residuals we can plot intervals for the nine nations against their ranks via the **plots tab** as shown below:

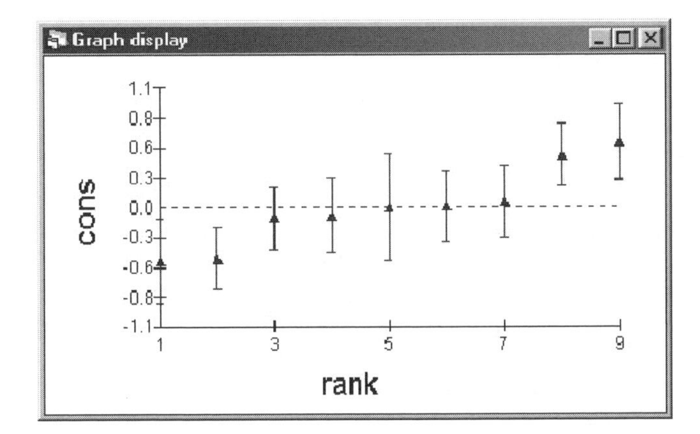

Here we see that two *nations* have significant positive residuals (Denmark and West Germany) whilst two *nations* have significant negative residuals (France and Ireland) and the other five countries have residuals that are close to zero. This pattern may suggest a heavier tailed distribution than a normal for the residuals, but given we only have nine nations perhaps a better approach is to assume a uniform distribution, i.e. fit the *nations* as fixed effects.

A plot of the residuals against normal scores does not help much at the *nation* level. However, the equivalent plot for the *region* level residuals suggests three or four (possible) outliers in the bottom left corner of the graph. Two of these residuals (including the lowest) are in neighbouring regions of Italy, which could be an interesting finding as Italy itself does not have a significant *nation* effect. In fact the area covered by these two regions had only 69 deaths when it was expected to have around 190.

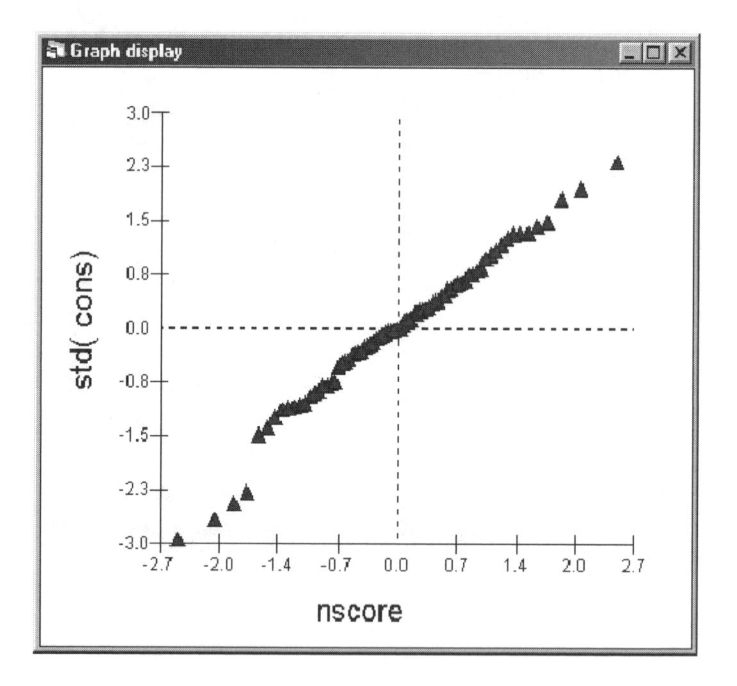

5.3.3 Predictions

One of the purposes of fitting a statistical model may be to predict future observations based on the current data. Alternatively we may wish to calculate the predicted estimates for our current dataset to evaluate the model fit. MLwiN has a **Predictions** window that can be used for this purpose. If we select the **Predictions** window from the **Model** menu and click on the variables '*cons*' and '*uvbi*' we will get the following:

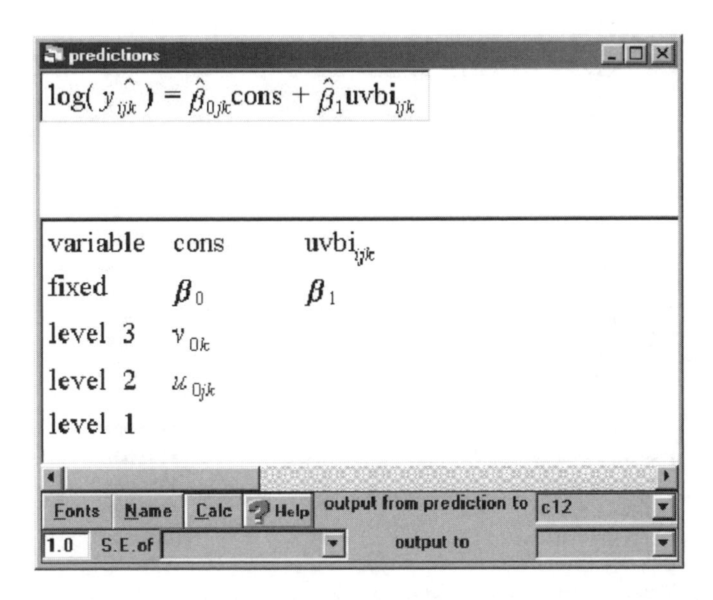

Here we have chosen column *c12* to output our predicted values into. It should be noted that the values output from the Predictions window do not include the offset term and so this must be added and then the values have to be exponentiated before we can relate them to the observed counts. This can be done (and the predictions output in column *c13*) by calculating c13 = expo(c12 + 'logexp') either in the **Calculate** or the **Command Interface** window.

We can now view the observed and predicted counts via the **View/Edit Data** window. Note that we have named column *c13 'pred'* and are also viewing the *nation* and *region* columns.

Data

goto line 1 | view | Help | Font

	nation(354)	region(354)	obs(354)	pred(354)
1	Belgium	1	79	69.22311
2	Belgium	2	80	77.23854
3	Belgium	2	51	44.42763
4	Belgium	2	43	52.8821
5	Belgium	2	89	65.08607
6	Belgium	2	19	34.9596
7	Belgium	3	19	7.782462
8	Belgium	3	15	39.0049
9	Belgium	3	33	30.25489
10	Belgium	3	9	6.513049
11	Belgium	3	12	11.49009
12	W.Germany	4	156	149.9807
13	W.Germany	4	110	103.0519
14	W.Germany	4	77	77.71716
15	W.Germany	4	56	64.5154

Here we see that for the first county our model predicts 69.2 deaths when in reality there were 79. We have fitted a term for each *region* and this is the only *county* in *region* 1 so you may wonder why we do not have a perfect fit for this *county*. This is because we are treating *regions* as random effects and so the effect for *region* 1 is shrunk towards the mean. If we believe our model is true then what this is saying is that, although we observed 79 deaths in this time period, then if we were to (hypothetically) take many equivalent populations (in *county* 1) and an equivalent time period and observe the number of deaths in each population, then we would expect on average to see 69.2 deaths.

5.4 MCMC ESTIMATION IN MLwiN

Although MLN was originally designed as a likelihood-based estimation package, since 1998, when MLwiN was first released, a second estimation engine that fits the equivalent Bayesian models using MCMC estimation was introduced. Let us consider the Poisson response model that we have just fitted that consists of a fixed intercept, a fixed effect for UV exposure and two sets of random effects for nations and regions. To convert this model into a Bayesian framework we need to add prior distributions for the two fixed effects and the two random variances. The model that is thus created is a form of Bayesian hierarchical model as described in Chapter 2.

In MLwiN if we want to use MCMC estimation we simply change estimation method by selecting **MCMC** from the **Estimation** menu. This will convert the model to a Bayesian model by selecting some 'default' priors as shown in the **Equations** window below:

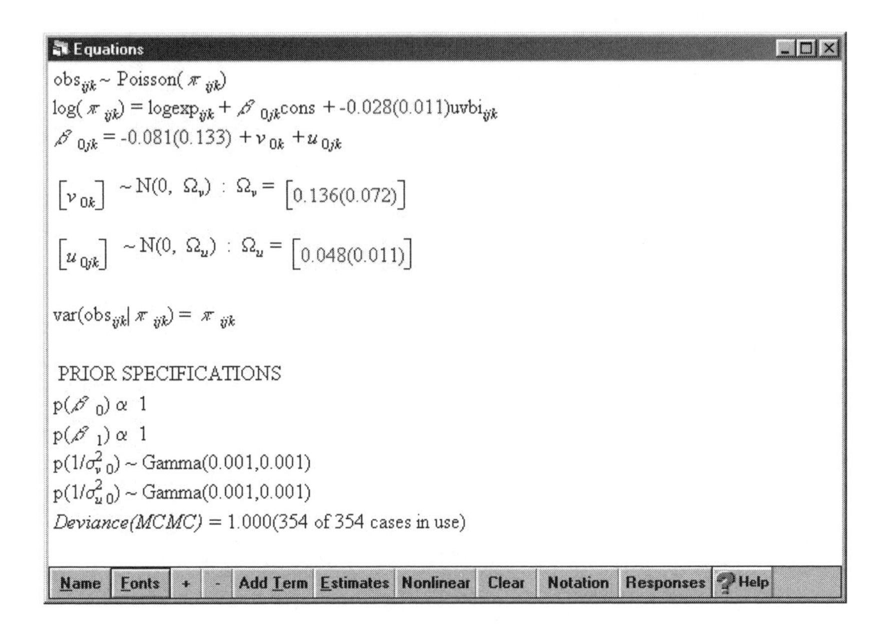

Note that we have pressed the + button to show the prior distributions. The default choice is to use a uniform prior for fixed effects and an inverse *Gamma*(e,e) prior for the variances (where $e = 0.001$). These priors can of course be changed and MLwiN will allow the use of any normal priors for the fixed effects and any gamma priors for variances (effectively any conjugate priors).

We have here firstly run the second-order PQL method prior to selecting MCMC and so the estimates that are shown in the window are from this method. MCMC will use these values as starting values for its estimation. The other settings, for example the length of the burn-in and the number of iterations to run for, are available on the **Estimation Control** window shown overleaf:

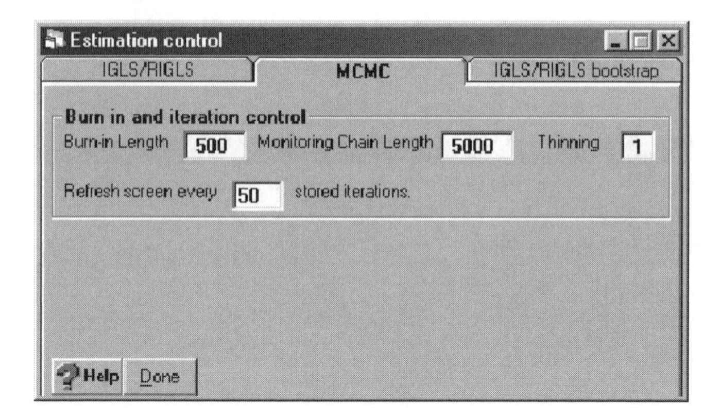

As with WinBUGS, MLwiN will by default also choose which MCMC methods to use for estimation, although (when possible) any other MCMC methods that can be used (in MLwiN) to fit the same model can be selected from the **MCMC methods** window available in the **MCMC** submenu of the **Model** menu.

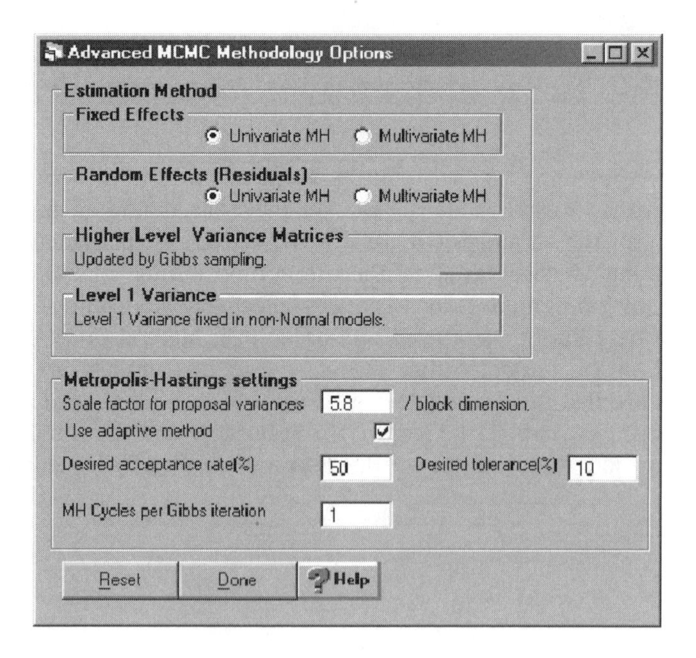

As shown, MLwiN by default uses univariate Metropolis sampling for fixed effects and residuals and a Gibbs sampling step to update the variance parameters from their inverse gamma conditional posterior distributions. This combination of methods is usually faster than the Adaptive Rejection (AR) sampler used by WinBUGS for the fixed effects and residuals in the same model although the AR sampler will provide less autocorrelated chains. Poisson response models often produce very autocorrelated chains in MLwiN and so we will increase the monitoring chain length to 50 000 and thin the chain by a factor of 10 (i.e. store only every tenth iteration). This we will do via the **Estimation Control** window. To run the sampler we simply press the **Start** button.

To choose proposal distributions MLwiN uses an adaptive method described in Browne and Draper (2000) that attempts to produce chains that have an acceptance rate of around 50%. This acceptance rate and some of the other settings that are used with the Metropolis method can be altered on the **MCMC methods** window. After running this adapting period and a burn-in, the 50 000 iterations are fairly rapid taking just over 30 seconds on a Pentium III 1.2 GHz machine. The point estimates (chain means) are given in the **Equations** window:

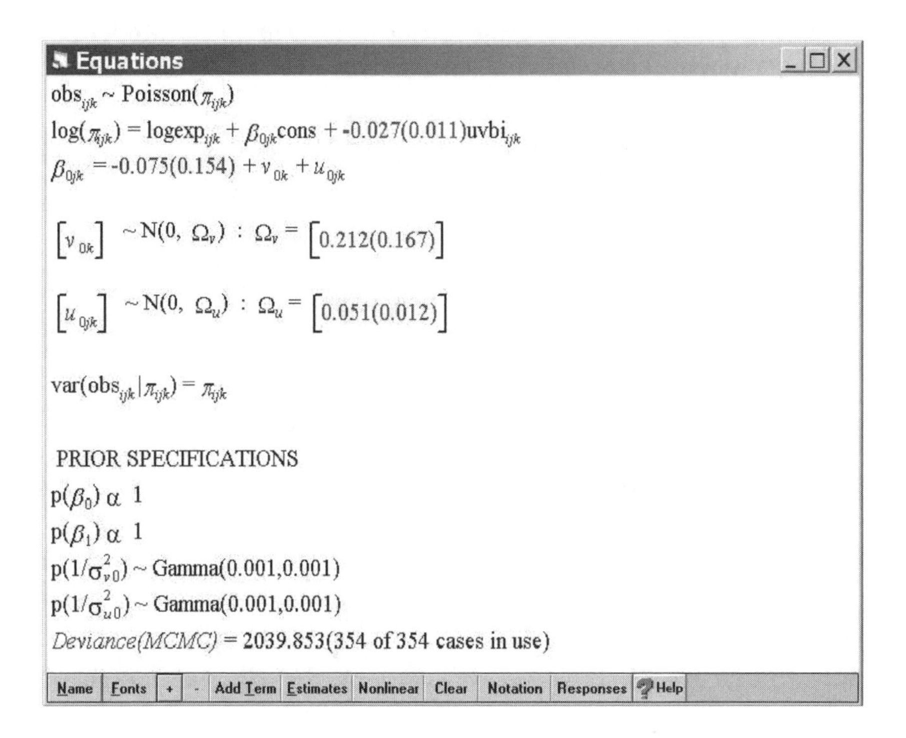

We can see here that the estimates are quite a bit different to those produced by second-order PQL even though the default priors are chosen to mimic the likelihood approach. We see that the fixed effects tell a similar story to the PQL estimates with the intercept not being significantly different from zero and the effect of UV exposure being significant and negative. The between-nations variance (σ^2_{v0}) is here a lot larger than the estimate produced by PQL. Perhaps the main reason for this is that there are only nine nations and hence getting an estimate of the variance between them is difficult There are several other potential reasons for the discrepancy. First the MCMC estimates are posterior means whilst the PQL method approximates a maximum likelihood estimate, which is equivalent to the mode of the posterior distribution. Second the quasi-likelihood methods involve approximations which mean that they may be inaccurate and finally as we will see when we look at the MCMC diagnostic plots we have not actually run the MCMC methods for long enough to be totally confident of our estimates.

5.4.1 MCMC estimates and chain diagnostics

The MCMC approach as described in earlier chapters involves producing chains of parameter estimates and the **Trajectories** window under the **Model** menu can be used to view the chains evolving over time.

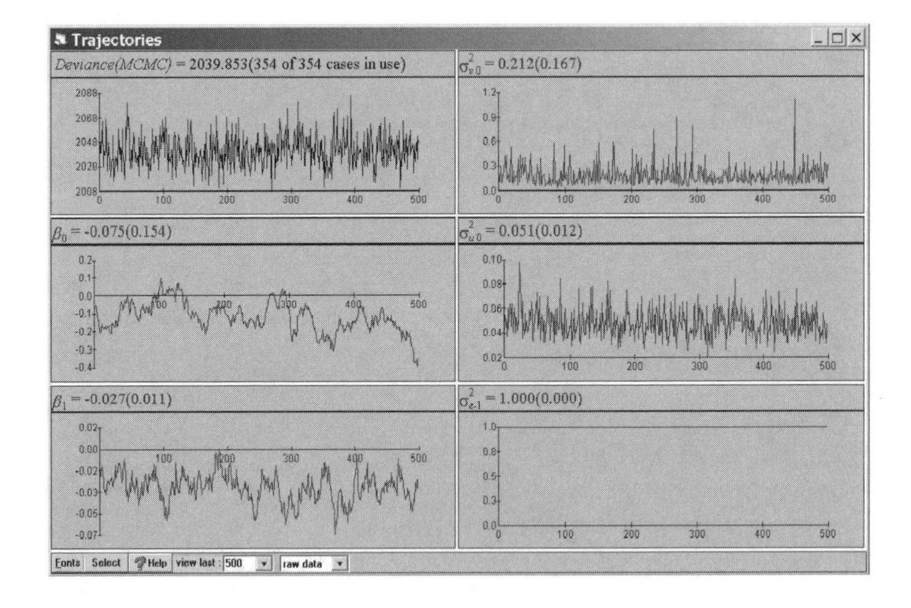

The above window shows the last 500 iterations of the (thinned) chains for the four main parameters of interest along with a chain for the deviance. We can see that even though we have thinned the chain there is still a large autocorrelation in the fixed effect chains. Also we see that the chain for σ^2_{v0} occasionally has jumps to large values suggesting a skewed distribution with a long right-hand tail. This would be a typical shape for a variance estimate based on a small number of observations. To investigate any parameters further we can click on their trace to bring up the **MCMC diagnostics** window. Firstly, if we click on the parameter β_1 (UV exposure):

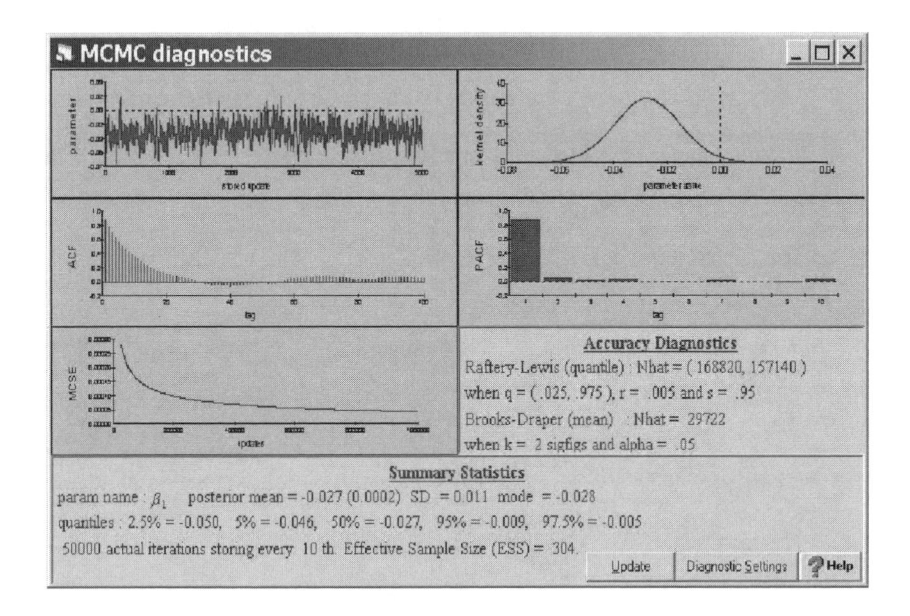

In this window we get more information about the parameter of interest, β_1. In the top left pane we see the whole chain of 5 000 stored values and next to it a kernel density (smoothed histogram) plot for the parameter. Here we see that this density looks fairly normal which is what we expect to see for a fixed effect. The next line contains two time series diagnostic plots, the autocorrelation function (ACF) and partial autocorrelation function (PACF). Here we see that even though we have thinned the chain we still have an autocorrelation at lag 1 of 0.85 suggesting this chain is very correlated. The third line contains a Monte Carlo standard error (MCSE) plot that indicates how many iterations are required to quote estimates with a particular MCSE.

We also have two MCMC diagnostics. Unlike WinBUGS, MLwiN does not offer the ability to fit multiple chains at once and so consequently the diagnostics used are both single-chain diagnostics. The Raftery–Lewis diagnostic is designed to calculate how many iterations are required to get accurate quantile estimates and hence accurate interval estimates. The Brooks–Draper diagnostic is designed to calculate how many iterations are required to quote the mean estimate to a given number of significant figures. For this parameter the diagnostics suggest that we can quote our estimate as −0.028 but that we need to run for a lot longer to have accurate quantiles. The final panel gives mean, median and mode estimates along with quantiles that can be used to give a

Bayesian credible interval and an effective sample size (ESS), which compares our chain with an independent chain. Here our 5 000 stored iterations have resulted due to autocorrelation to the equivalent of 304 independent draws.

If we now consider the chain for the between-nations variance, σ_{v0}^2 we get the following:

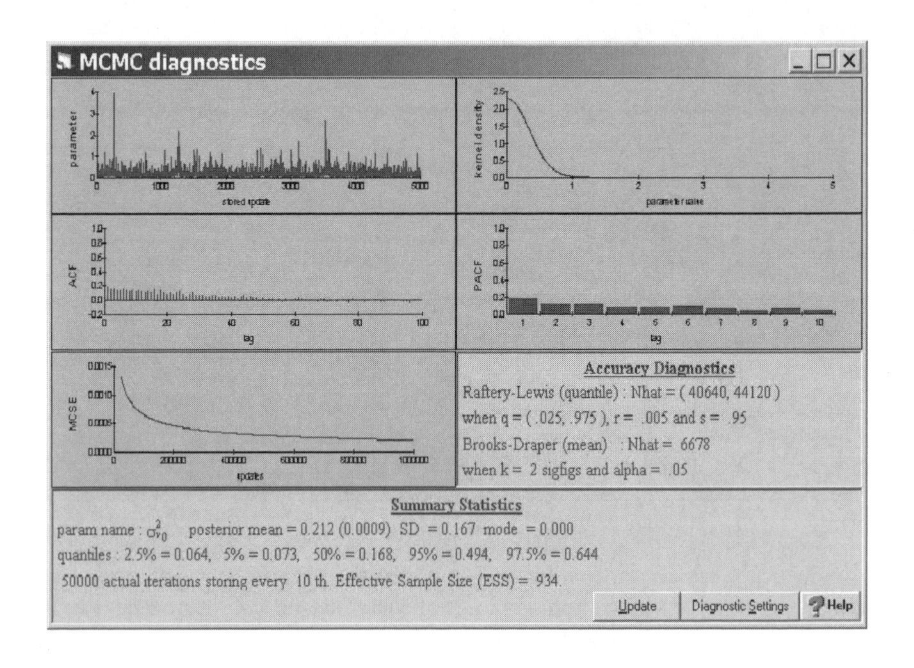

Here we see that the chain is less autocorrelated, and that the diagnostics all suggest that we have run the chain for long enough, although we only have effectively 934 independent values. We see that the parameter has the long right-hand tail that we expected and, interestingly, that the mode of the chain is 0.0 (based on the kernel density plot) which is smaller than the quasi-likelihood estimates. This zero modal estimate may explain why the different estimation procedures give conflicting estimates for this parameter.

5.4.2 Model diagnostics

When using MCMC estimation the user can still produce residual plots to check the data for outliers and to check for distributional assumptions etc., as they would when using the quasi-likelihood methods. MLwiN also offers the DIC diagnostic of Spiegelhalter *et al.* (2002a), which is calculated by selecting **DIC diagnostic** from the **MCMC** submenu. For our model this gives:

```
-> BDIC
Bayesian Deviance Information Criterion (DIC)
Dbar D(thetabar) pD DIC
2039.85 1978.83 61.02 2100.87
```

Browne (2003) fits several other models to this dataset and finds that the DIC suggests that fitting a model with effects for nations and regions is a great improvement on the simple Poisson regression. They also find that fitting the nation effects as fixed rather than random effects gives roughly the same DIC value, suggesting the additional random effect assumption for the nation effects is not necessary.

5.4.3 MLwiN to WinBUGS interface

In the last section we have seen how to use MCMC estimation in MLwiN and that the estimation procedures it uses can produce heavily autocorrelated chains. WinBUGS uses a different MCMC estimation procedure and MLwiN also offers a simple way into WinBUGS via an interfacing procedure. If we run again the three-level Poisson regression model using second-order PQL, then after switching to MCMC we can select from the **MCMC** submenu of the **Model** menu the **WinBUGS options** window as shown below:

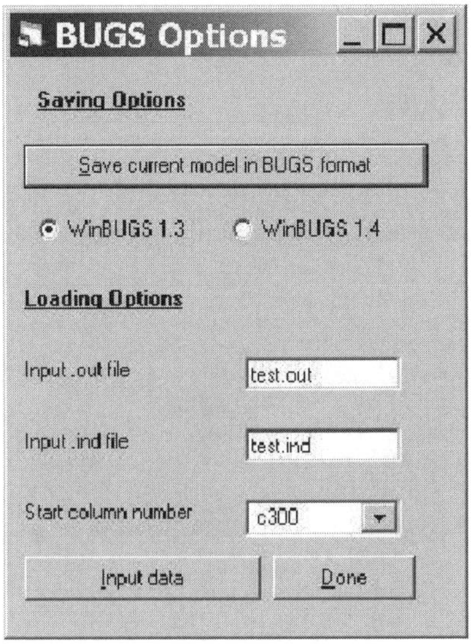

This window will create a text file containing the three parts of the WinBUGS code required to fit the same model using WinBUGS. The window will generate either WinBUGS 1.3 or WinBUGS 1.4 code and requires the user to give a filename for the code. We will here use WinBUGS 1.4.

If we load up the file in WinBUGS we will see the following three sections of code: First the model definition is given:

```
#--MODEL Definition-----
model
{
# Level 1 definition
for(i in 1:N) {
obs[i] ~ dpois(mu[i])
log(mu[i]) <-offs[i] + beta[1] * cons[i]
+ beta[2] * uvbi[i]
+ u2[region[i]] * cons[i]
+ u3[nation[i]] * cons[i]
}
# Higher level definitions
for (j in 1:n2) {
u2[j] ~ dnorm(0,tau.u2)
}
for (j in 1:n3) {
u3[j] ~ dnorm(0,tau.u3)
}
# Priors for fixed effects
for (k in 1:2) { beta[ k] ~ dflat() }
# Priors for random terms
tau.u2 ~ dgamma(0.001000,0.001000)
sigma2.u2 <-1/tau.u2
tau.u3 ~ dgamma(0.001000,0.001000)
sigma2.u3 <-1/tau.u3
}
```

Next we get the initial values (including residuals) that come from the second-order PQL fit of the model and are identical to those used by the MLwiN MCMC engine.

```
#--Initial values file--------
list(beta = c(-0.081039,-0.028178),
u2 = c( 0.381428,0.037167,-0.447428,
-0.105743,0.123108,0.353427,0.019626,0.275285,
-0.052795,-0.069364,-0.306009,
-0.007217,-0.166731,0.108048,0.204645,0.140898,
-0.131569,0.100346,0.071120,0.078510,
-0.183546,0.047904,0.197728,0.047583,
-0.062235,0.035894,0.047389,- 0.087250,0.119009,0.099712,
-0.159212,- 0.043417,-0.204221,-0.063588,
-0.412259,0.009563,0.067774,0.126635,0.086360,0.024277,
-0.197342,-0.177864,0.304836,0.265616,
-0.061356,-0.012419,0.058058,-0.091444,
-0.235664,0.019530,-0.104978,-0.370753,
```

```
-0.549701,0.191542,0.345851,0.148999,0.201420,
-0.021847,-0.031271,0.000691,0.264060,- 0.227158,-0.129567,
-0.205172,0.158149,0.060234,
-0.005343,0.110463,0.147333,-0.118915,
-0.019224,-0.028438,-0.008834,
-0.000797,-0.078599,-0.004529,0.231624,- 0.128155),
u3 = c( -0.081094,0.482732,0.601812,
-0.468458,-0.103901,0.006980,-0.493350,- 0.002242,0.057210),
tau.u2= 20.660154,
tau.u3 = 7.345692)
```

Finally we get the data section, which is the biggest chunk of code and so we only show the first few lines here:

```
#--Data File----------
list(N = 354, n2 = 78, n3 = 9,
region = c(1,2,2,2,2,2,3,3,3,3,3,4,4,4,4,5,5,5,5,5,
5,5,6,7,8,9,9,10,10,10,10,11,11,11,11,11,12,12,12,13,
14,15,15,15,16,16,16,17,17,17,17,17,17,17,17,18,18,19,19,19,
...
```

This code can be compiled and run as described in Chapter 4. For comparison we will run the code for 50 000 iterations (thinning with a factor of 10 as before). WinBUGS 1.4 took over 8 minutes to run this model which is 14 times as long as MLwiN but, as we mentioned previously, the methods it uses should be more efficient.

If we monitor the two fixed effects and the two variances then we get the following estimates which are similar to those from MLwiN:

node	mean	sd	MC error	2.5%	median	97.5%	start	sample
beta[1]	−0.08933	0.195	0.01781	−0.4833	−0.08178	0.2602	501	5000
beta[2]	−0.02737	0.01141	3.875E-4	−0.04876	−0.02755	−0.004825	501	5000
sigma2.u2	0.05106	0.01206	1.975E-4	0.03165	0.04963	0.07893	501	5000
sigma2.u3	0.2257	0.2009	0.01311	0.062	0.1746	0.7149	501	5000

We can read the chains back into MLwiN via the **CODA options** by saving the output files and reading them in using the **WinBUGS options** window in MLwiN. If we specify the correct filenames and press the **input data** button then the chains will be read into consecutive columns on the MLwiN worksheet as shown in the **Names** window below:

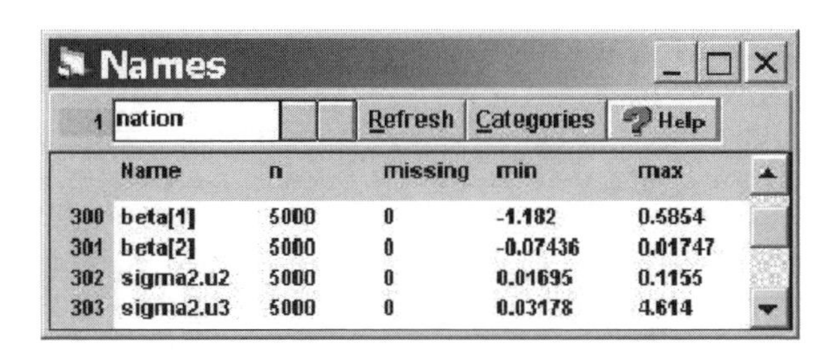

MLwiN will allow MCMC diagnostics to be carried out on any column via the **Column Diagnostics** window in the **Basic Statistics** menu. In this window we can select the fixed effect for UV exposure, *beta[2]* as shown. (Note WinBUGS counts from 1 and not 0 hence the numbering difference.)

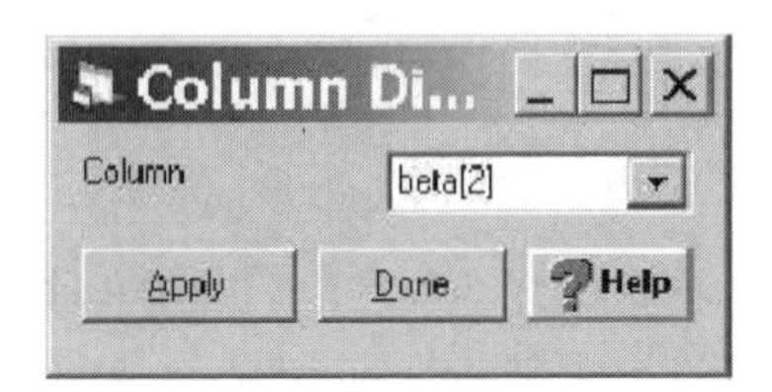

The diagnostics for this parameter are as follows and can be compared with the equivalent window we looked at earlier for the MLwiN engine:

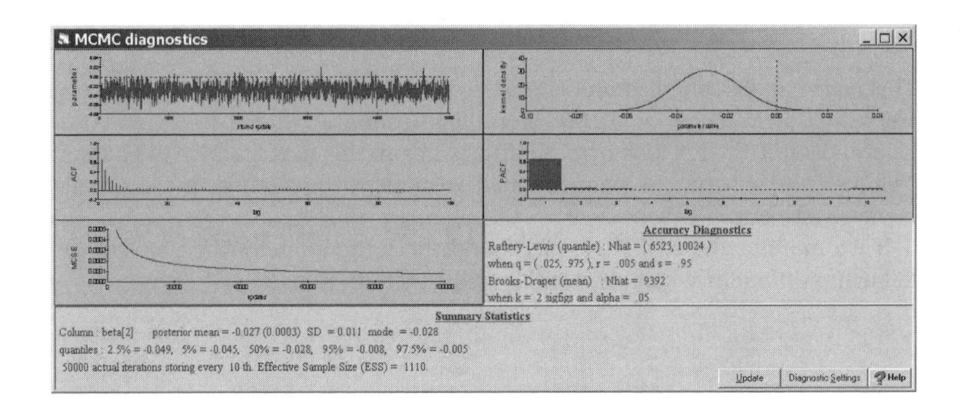

Here we see that the autocorrelation at lag 1 for this plot is about 0.65 as compared to 0.85 for the Metropolis method in MLwiN and that the effective sample size has increased from 304 to 1110, a three-fold increase, although this must be balanced against the 14-fold increase in time to run the model.

5.5 SPATIAL MODELLING

In this chapter we have so far only considered fitting standard hierarchical models to disease mapping data in MLwiN. As described in the earlier chapter on multilevel modelling it is possible to account for the spatial structure within a dataset by using an extension to multilevel models called multiple-membership models. The alternative (and perhaps more common) method to account for the fact that two areas are neighbouring is through CAR modelling. MLwiN allows the fitting of multiple membership modelling through both likelihood-based estimation methods and MCMC estimation, and also allows the fitting of some simple CAR models using MCMC estimation only.

The melanoma mortality dataset that we have considered does not have spatial information associated with it and so we will instead here consider the Scottish lip cancer dataset mentioned earlier. The dataset comes with the MLwiN software and is stored in the worksheet '*lips1.ws*'. It is analysed in great detail in Chapter 16 of Browne (2003) and already contains columns of neighbour identifiers and weights that are needed for spatial models. If we load up the worksheet and look at the **Names** window we see the following:

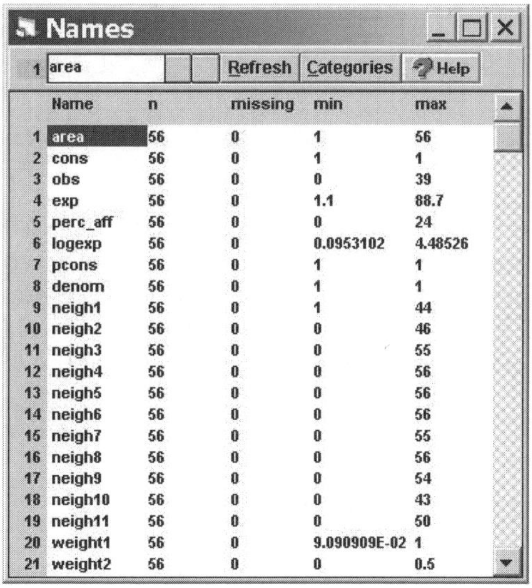

Names

1 area | Refresh | Categories | Help

	Name	n	missing	min	max
1	area	56	0	1	56
2	cons	56	0	1	1
3	obs	56	0	0	39
4	exp	56	0	1.1	88.7
5	perc_aff	56	0	0	24
6	logexp	56	0	0.0953102	4.48526
7	pcons	56	0	1	1
8	denom	56	0	1	1
9	neigh1	56	0	1	44
10	neigh2	56	0	0	46
11	neigh3	56	0	0	55
12	neigh4	56	0	0	56
13	neigh5	56	0	0	56
14	neigh6	56	0	0	56
15	neigh7	56	0	0	55
16	neigh8	56	0	0	56
17	neigh9	56	0	0	54
18	neigh10	56	0	0	43
19	neigh11	56	0	0	50
20	weight1	56	0	9.090909E-02	1
21	weight2	56	0	0	0.5

The Scottish lip cancer dataset contains counts (in the column labelled '*obs*') of lip cancer in males in the period 1975–1980 for the 56 regions in Scotland, and as with the melanoma dataset, interest is again on the effect of sun exposure. In this dataset we however use the surrogate measure, the percentage of the workforce that work in agriculture, fishing and farming ('*perc_aff*') as our main predictor variable. You will see that in the **Names** window above we have 11 consecutive columns named '*neighx*' where *x* runs from 1 to 11. If we look at these columns in the **View/Edit Data** window we will see the following:

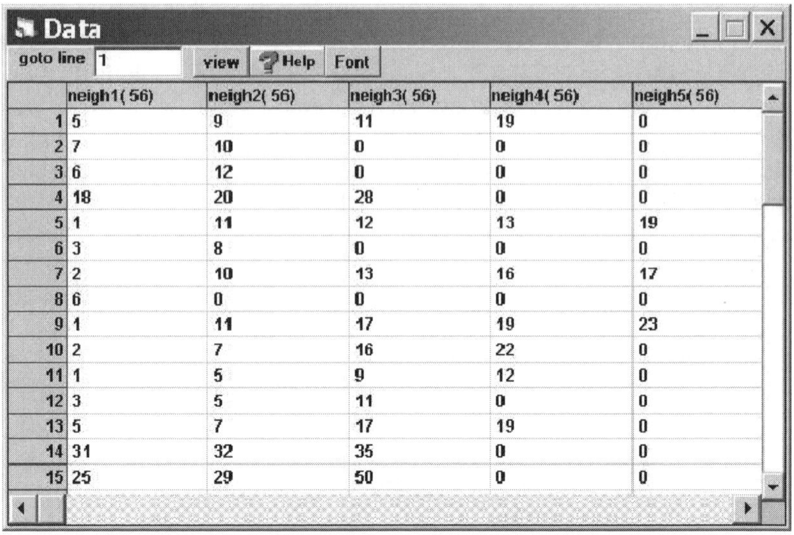

Data

goto line 1 | view | Help | Font

	neigh1(56)	neigh2(56)	neigh3(56)	neigh4(56)	neigh5(56)
1	5	9	11	19	0
2	7	10	0	0	0
3	6	12	0	0	0
4	18	20	28	0	0
5	1	11	12	13	19
6	3	8	0	0	0
7	2	10	13	16	17
8	6	0	0	0	0
9	1	11	17	19	23
10	2	7	16	22	0
11	1	5	9	12	0
12	3	5	11	0	0
13	5	7	17	19	0
14	31	32	35	0	0
15	25	29	50	0	0

Here we see that region 1, which is represented by the first row of the dataset has four neighbouring regions, regions 5, 9, 11 and 19, whilst region 2 only has two neighbouring regions, regions 7 and 10. We have 11 columns as 1 region (region 29) has 11 neighbours as can be seen from the **Data** window. Any regions that have less than 11 neighbours have the additional columns filled with zeros.

5.5.1 Multiple-membership models

These neighbour columns can be used to fit either multiple-membership models or CAR models to the dataset. In a multiple-membership model we also need a weight for each neighbouring region and hence following the 11 neighbour columns are 11 weight columns, the first five of which can be seen overleaf:

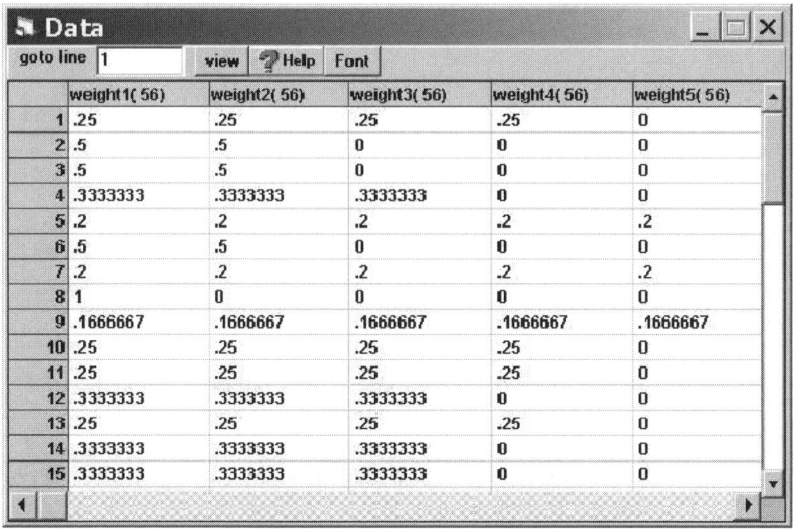

Data

goto line 1 | view | Help | Font

	weight1(56)	weight2(56)	weight3(56)	weight4(56)	weight5(56)
1	.25	.25	.25	.25	0
2	.5	.5	0	0	0
3	.5	.5	0	0	0
4	.3333333	.3333333	.3333333	0	0
5	.2	.2	.2	.2	.2
6	.5	.5	0	0	0
7	.2	.2	.2	.2	.2
8	1	0	0	0	0
9	.1666667	.1666667	.1666667	.1666667	.1666667
10	.25	.25	.25	.25	0
11	.25	.25	.25	.25	0
12	.3333333	.3333333	.3333333	0	0
13	.25	.25	.25	.25	0
14	.3333333	.3333333	.3333333	0	0
15	.3333333	.3333333	.3333333	0	0

The weights in a multiple membership model are often chosen to sum to 1 and so here we see that the first region which has four neighbours has four weights each of 0.25. MLwiN fits multiple-membership models in different ways depending on which estimation method is required. The method based on quasi-likelihood iterative least squares estimation devised by Hill and Goldstein (1998) actually involves expressing the multiple membership model as a constrained nested model with many sets of random effects constrained to have the same variance. We can firstly set up a three-level structure with '*area*' at levels 1 and 2 to account for the Poisson residuals and the unstructured normal residuals respectively. We will use '*cons*' as the level 3 identifier as we will here set the 56 constrained variances. We can firstly fit a two-level model by specifying '*obs*' as response, using '*logexp*' as the offset column and defining '*cons*' as both a fixed effect and random at level 2 with '*perc_aff*' as a fixed effect. This model can be set up in a similar way to those we fitted to the melanoma dataset and upon fitting using first-order MQL we should get the following estimates:

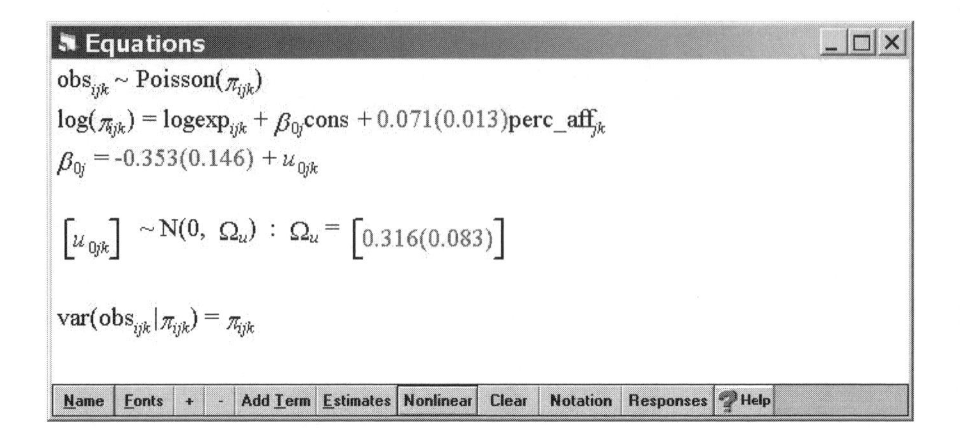

To construct the constrained model MLwiN has three commands that can be input in the **Command Interface** window. It is, however, sensible at this point to close the **Equations** window as the overhead of drawing the 56^*56 variance matrix at level 3 will really slow the program down. The first command *WTCOL* constructs what is effectively an N^*N matrix in N specified columns where N is the number of areas. In the lip cancer dataset this will involve the creation of 56 columns of length 56 and we can use the following command

```
wtcol 11 c9-c19 c20-c30 c301-c356
```

Here *c9–c19* are the 11 neighbour columns and *c20–c30* are the weight columns with the expanded weight matrix stored in columns *c301–c356*. We now need to add these 56 columns as random parameters at level 3 and constrain them to have equal variance. This is achieved by the following two commands:

```
addm 1 3 c301-c356 c357
rcon c357
```

We can now run the model by clicking on the **Start** button. After 16 iterations the model appears to converge and we can (as we are not looking at the **Equations** window) get access to the estimates by typing the commands RAND and FIXED in the **Command Interface** window. Interestingly the results that first-order MQL has produced are rather strange as it is estimating two of the variances as zero. If, however, we click on **More** the program will run for 15 more iterations and give more sensible estimates suggesting that it had not converged properly originally. In the table below we give estimates obtained by both first-order MQL and second-order PQL estimation along with, for comparison, MCMC estimation as described in the next section. Note to obtain second-order PQL estimates we changed estimation method via the **Command Interface** window by setting the boxes $b11 = 2$ and $b12 = 1$.

	1st MQL	2nd PQL	MCMC
β_0	−0.096 (0.184)	−0.276 (0.194)	−0.268 (0.211)
β_1	0.047 (0.013)	0.050 (0.014)	0.047 (0.015)
σ_v^2	1.039 (0.424)	0.972 (0.461)	1.230 (0.471)
σ_u^2	0.022 (0.048)	0.060 (0.067)	0.051 (0.051)

Here we see that the first MQL method underestimates both the intercept and the level 2 variance but that the second PQL method and MCMC have reasonable agreement on the parameters. The quasi-likelihood methods run quickly on the lip cancer dataset but as we can see there are some problems, for example the loss of the use of the **Equations** window, and as the size of the dataset gets bigger then the methods struggle to fit the constraints matrix and obtain convergence.

To fit a multiple-membership model using MCMC is a far simpler task as described in Chapter 16 of Browne (2003). Here we set up a three-level model with '*area*' as the level identifier for the first two levels and '*neigh1*' (the first of the 11 neighbour columns) as level 3. The first level will be the Poisson distributed residuals whilst the second level will contain random effects for each area. The third level will contain the neighbour multiple-membership residuals. Again we have '*obs*' as a response model and one predictor apart from the intercept, '*perc aff*'. The intercept '*cons*' is declared as a fixed effect and random at levels 2 and 3.

After running the IGLS method (which will think the model is nested and so may give strange estimates), we can then switch to MCMC and specify the correct structure to the data via the **Classifications** window that is available from the **MCMC** submenu of the **Model** menu. This window we set up as follows:

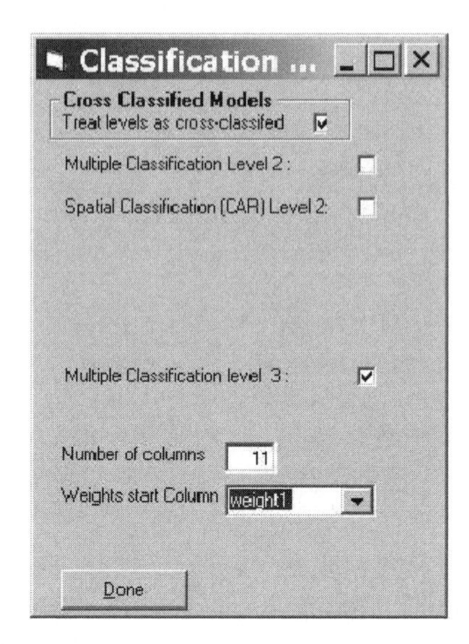

Here we are specifying that level 3 is a multiple-membership classification, and that there are 11 columns containing both the identifiers (neighbours) and weights. Note here that the columns for the identifiers must be consecutive as must the weights.

If we now click on the **Done** button and then run the model for 50 000 iterations by pressing **Start** we will get the following estimates in the **Equations** window.

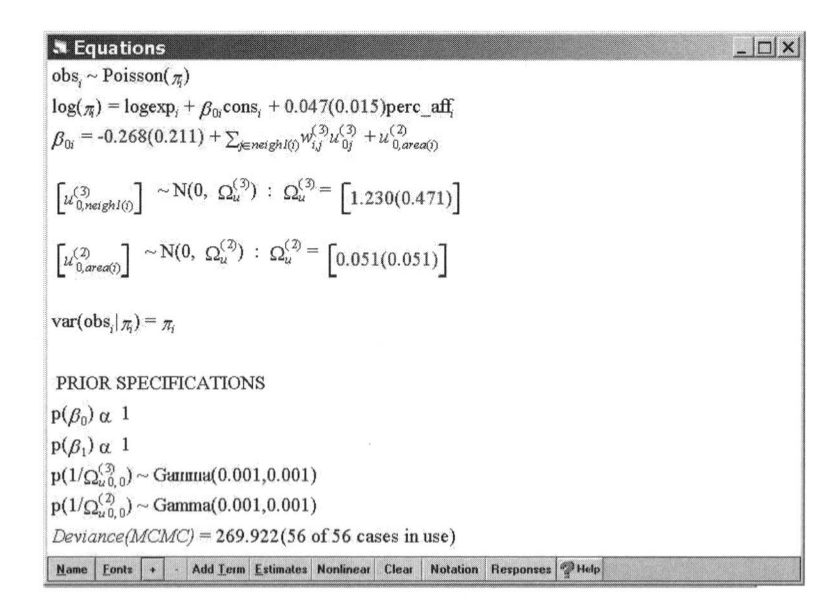

Note that in order to get the notation shown above that was introduced in Browne *et al.* (2001) we had to click on the **Notation** button and remove the tick under **multiple subscripts**. Here we see that the areas with larger percentages of outdoor workers have higher cancer rates than expected, as β_1 is significant and positive. Also we see that here the neighbour effects have a far greater variance than the area overdispersion effects (1.230 versus 0.051) so are more important in determining the rates.

5.5.2 CAR models

CAR models are a more common way of dealing with the spatial correlation between neighbouring areas than the multiple membership models. In MLwiN we can fit a limited set of CAR models using MCMC estimation only. The **Classifications** window will allow the user to specify one set of CAR residuals in a model, which means that we cannot fit spatially-correlated effects at differing levels of geography. Browne (2003) fits two models containing CAR residuals to the lip cancer dataset: a first model that only contains spatially-correlated residuals and a second model that is often called a *Convolution* model that also contains uncorrelated normally distributed effects.

When fitting a CAR model we again need a set of neighbours and a set of weights.

In a CAR model, however, it is common to specify a weight of 1 for each neighbour rather than constrain the weights to sum to 1. In the lip cancer dataset we have these weights stored in the columns labelled '*wcar1*' to '*wcar11*' as shown in the window below:

Data

goto line 1 view Help Font

	wcar1(56)	wcar2(56)	wcar3(56)	wcar4(56)	wcar5(56)
1	1	1	1	1	0
2	1	1	0	0	0
3	1	1	0	0	0
4	1	1	1	0	0
5	1	1	1	1	1
6	1	1	0	0	0
7	1	1	1	1	1
8	1	0	0	0	0
9	1	1	1	1	1
10	1	1	1	1	0
11	1	1	1	1	0
12	1	1	1	0	0
13	1	1	1	1	0
14	1	1	1	0	0
15	1	1	1	0	0

Unlike the multiple-membership model for the CAR model we need to specify the *areas* themselves as the level identifiers and then specify the first *neighbour* column in the **Classifications** window as shown below:

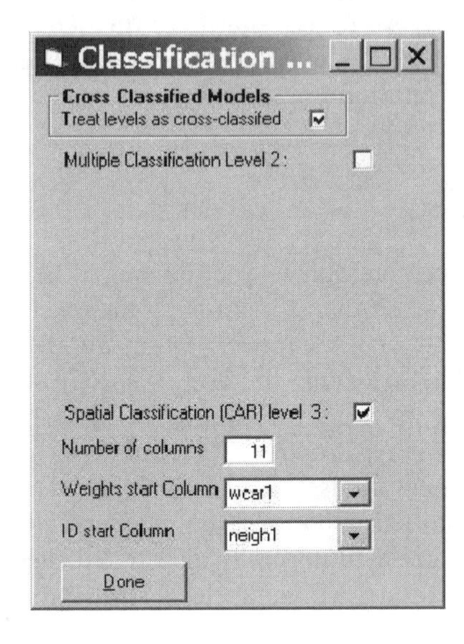

The CAR distribution is improper and in order to produce a model that has a unique solution we need to add a constraint of some kind to the model. In MLwiN the user must remove the intercept term to make the model identifiable. The results for the convolution model having been run for 50 000 iterations are shown below:

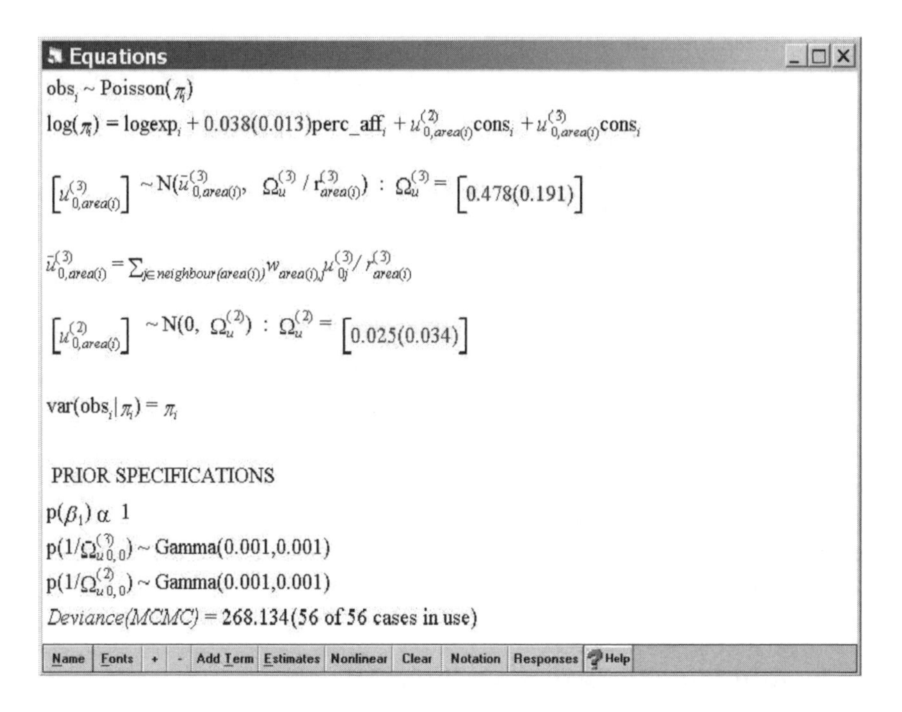

As with the multiple-membership model we see a similar significant positive effect for higher percentage of outdoor workers. We also see that the variance of the (structured) spatial effects is bigger than that of the unstructured effects although in this model they are not directly comparable due to the weights. Both the multiple-membership model and the CAR convolution model are attempting to fit a set of spatial/neighbour effects and a set of exchangeable unstructured normal random effects. In the lip cancer dataset if we compare the models using the DIC diagnostic we see that the convolution model has a slightly improved fit.

```
-> BDIC
   Bayesian Deviance Information Criterion (DIC)
   Dbar D (thetabar) pD DIC
   268.13 238.57 29.56 297.70 <Convolution model>
   269.92 237.32 32.60 302.53 <MM + exchangeable>
```

5.6 CONCLUSIONS

In this chapter we have given you a brief overview of some of the features of MLwiN, with a particular emphasis on disease mapping applications. MLwiN has the advantage over many competing packages that it encompasses both likelihood-based and Bayesian fitting of statistical models. It also offers many additional functions and plotting facilities on top of the estimation engines making it a fairly comprehensive package. It also has an underlying Macro language that allows the user (in a similar way to the WinBUGS script language) to write macros that will set up and run a statistical analysis. This language is fairly powerful and Browne (2003) shows how to write macros that will perform MCMC estimation of simple linear regression models independent of the MLwiN estimation engines.

Although MLwiN is not primarily designed to fit the spatial models that are often used in disease mapping or to produce the maps that GeoBUGS excels at, the package is continually being developed and there is likely to be more spatial functionality in the future. It is also useful to have an alternative program to WinBUGS that will fit some of the same (or similar) models to disease mapping data and to compare estimates from different estimation procedures in the two packages. In the case of very large datasets where WinBUGS becomes computationally unfeasible due to the time it takes to produce estimates, it may still be possible to get quasi-likelihood (or even MCMC) estimates from MLwiN in a much shorter time.

5.6.1 Additional information about MLwiN

For problems or questions about MLwiN, there is a frequently asked questions (FAQ) page on the MLwiN website (*http://multilevel.ioe.ac.uk/support/faq/index.html*). The project team also has a user support service run by Min Yang where users may e-mail their questions on MLwiN to *mlwin.support@ioe.ac.uk*. There is also an active multilevel modelling discussion list that is not solely about MLwiN but encapsulates multilevel modelling in other software packages and general statistical advice in such models. If you join the multilevel discussion list you can then e-mail your questions to *multilevel@jiscmail.ac.uk*.

6

Relative Risk Estimation

This chapter first focuses on the fitting of models for the estimation of relative risks in small areas using WinBUGS. In the latter part of the chapter parallel analyses are reported from MLwiN. In the study of geographical variation of disease risk in count data, there are basic models that, at least as a starting point, are usually applied. First, the basic model assumed for $(y_1, \ldots, y_m)$ is a Poisson likelihood with parameter $e_i\theta_i$. This is called the classical model (SMRs). The other models that will be fitted are extensions of the classical model when prior distributions for the relative risks are assumed. These models range from simple risk structures to more complex risk structures, including spatial correlation. Spatiotemporal or mixture models for relative risks will also be described.

6.1 RELATIVE RISK ESTIMATION USING WinBUGS

Step 1: Create a text file containing the model specification (graphically or in the BUGS language).
Step 2: Create a file (or files) containing the data.
Step 3: Create a file of starting values for every unknown parameter. These are used to initialize the simulation and can be arbitrary, although it helps to choose realistic values.

Once model and data are specified, it is possible to proceed to run the model in WinBUGS. Using the menus in WinBUGS the following steps can be followed:

4: Check the model syntax.
5: Load data.
6: Compile.
7: Load initial values.
8: Set sample to store sampled values and start simulation (update).
9: Check convergence.
10: Carry out further updates if necessary.

Disease Mapping with WinBUGS and MLwiN A. Lawson, W. Browne and C. Vidal Rodeiro
 ISBN: 0-470-85604-1 (HB)

11: Summarize samples of parameters.
12: Produce maps.

In the next sections we provide some examples to illustrate the use of WinBUGS in disease map modelling. The methods mentioned above are applied to disease data in the form of counts of cases within small areas (counties). In particular, we are going to analyse the spatial distribution of congenital abnormalities deaths in South Carolina in 1990, the spatial distribution of malignant neoplasms in South Carolina in 1999, and the spatial distribution of respiratory cancer in Ohio in the period 1979–1988. The database for the South Carolina data records the number of deaths from congenital abnormalities and for malignant neoplasms for the period 1990–1999 in each of the 46 counties in the state. For Ohio, the database records the number of deaths and the population at risk from 1968 to 1988 in each of its 88 counties.

6.1.1 Standardized mortality ratios

As a first step, it is often proposed to compute and map the Standardized Mortality Ratio (SMR) which is defined as

$$\hat{\theta}_i = SMR_i = \frac{y_i}{e_i},$$

where $(y_1, \ldots, y_m)$ and $(e_1, \ldots, e_m)$ denote the number of deaths and the expected number of deaths, respectively, from the disease during the study period.

Figure 6.1 shows the standardized mortality ratio for congenital abnormalities in the counties of South Carolina for the year 1990. The SMRs vary widely around their mean, 0.9282, (standard deviation: 0.7675). Although there is some suggestion of lower mortality in the north-east and south-west of South Carolina and higher mortality in the central counties, no clear spatial pattern emerges from this map.

Figure 6.2 shows the SMRs for respiratory cancer in the counties of Ohio for the year 1988. No clear spatial pattern emerges from this map although there are areas in the south with high mortality and areas in the north-west with lower mortality.

This estimator suffers from certain drawbacks (see Section 1.2.1.2 in this volume for details); statistical smoothing can be used to overcome them. The idea is that the smoothed estimate for each area ‘borrows strength’ (precision) from the data in other areas.

Hierarchical Bayesian models (described in Chapter 2) have an important role in modelling complexity of data structures in spatial epidemiology and

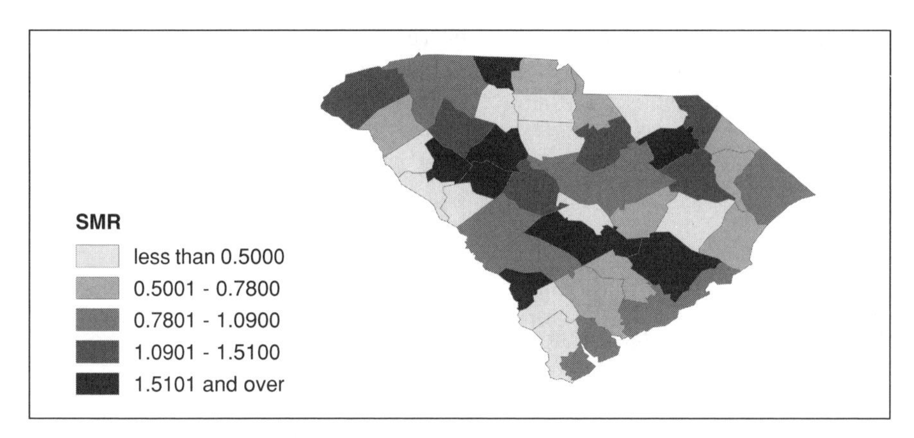

Figure 6.1 Congenital anomalies deaths, SMR. South Carolina, 1990.

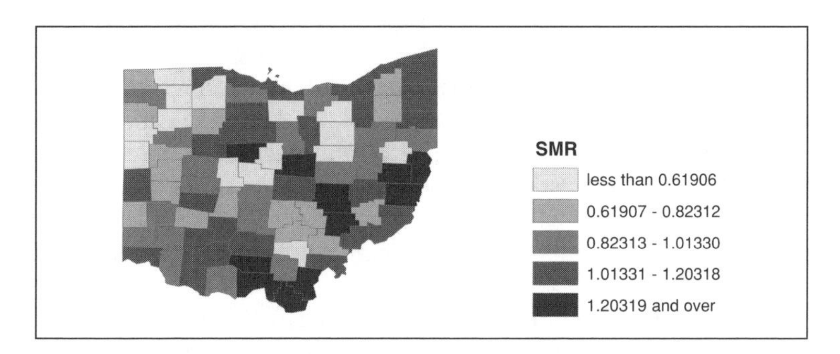

Figure 6.2 Respiratory cancer mortality, SMR. Ohio, 1988.

overcome the drawbacks of the SMRs. The Bayesian approach to disease mapping consist of considering, in addition to the observed events in each area, prior information on the variability of disease rates in the overall map; it can also take into account a spatial pattern in disease (i.e. the tendency of geographically close areas to have similar disease rates).

6.1.2 Poisson–gamma model

One of the earliest examples of Bayesian mapping is the Poisson–gamma model. Although it was introduced in Chapter 2, it is described here in more detail.

When the disease is non-contagious and rare, the numbers of deaths in each area are assumed to be mutually independent and to follow Poisson distributions

$$y_i \sim Poisson(e_i\theta_i) \quad \forall i.$$

A gamma prior distribution for the relative risks combines conveniently with the Poisson likelihood to give a gamma posterior distribution. If the prior distribution is a *Gamma(a, b)* then, the relative risk has the following posterior distribution

$$Gamma(a + y_i, b + e_i),$$

with mean given by

$$E[\theta_i|y_i, a, b] = \frac{a + y_i}{b + e_i} = \omega_i \text{SMR}_i + (1 - \omega_i)\frac{a}{b},$$

where

$$\omega_i = \frac{e_i}{b + e_i}.$$

The posterior mean of the relative risk for the *i*th area is a weighted average of the SMR for the *i*th area and the relative risk in the overall map, the weight being inversely related to the variance of the SMR. For rare diseases and small areas, this variance is large so the weight, ω_i, is small and the posterior mean tends towards a global mean a/b, thereby producing a smoothed map. In areas with abundant data the posterior mean of the relative risk is close to y_i/e_i.

One disadvantage of the Poisson–gamma model is its inability to cope with spatial correlation. In Section 6.1.5 a model which copes with it will be fitted.

Prior distributions for a and b are also specified. In this example, exponential distributions with mean 0.1 are considered for both parameters.

The WinBUGS code to fit a Poisson–gamma model to the South Carolina data is given in Figure 6.3.

To proceed with the estimation, check and compile the model following the steps described in Section 4.4. Two chains are computed using two different sets of initial values, since running multiple chains is one way to check the convergence of the MCMC simulations.

Before starting updating, use the sample tool to monitor

- the parameters a and b of the population distribution,
- the mean and the variance of the population,
- the relative risks.

After running 2000 iterations (this is an orientative number) it is appropriate to look at the convergence of each parameter that has been monitored. Figure 6.4 shows the sample trace plots for parameters a and b. From these plots we

```
model
{
for (i in 1:m)
{
   # Poisson likelihood for observed counts
   y[i]~dpois(mu[i])
   mu[i]<-e[i]*theta[i]
   # Relative Risk
   theta[i]~dgamma(a,b)
}

# Prior distributions for "population" parameters
a~dexp(0.1)
b~dexp(0.1)

# Population mean and population variance
mean<-a/b
var<-a/pow(b,2)
}
```

Figure 6.3 Poisson–gamma model in WinBUGS.

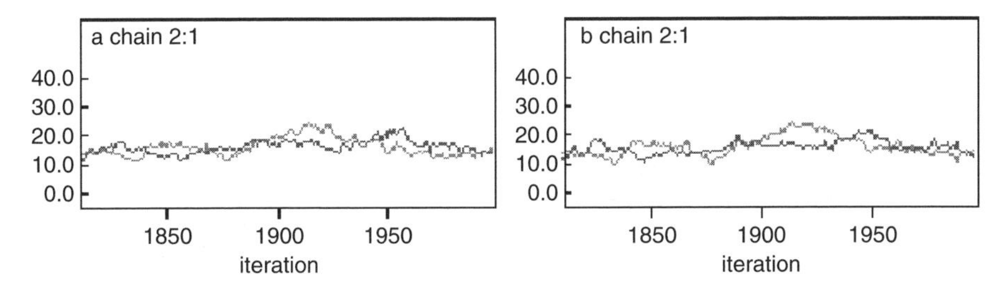

Figure 6.4 Sample trace plots for parameters a and b in the Poisson–gamma model after 2000 iterations.

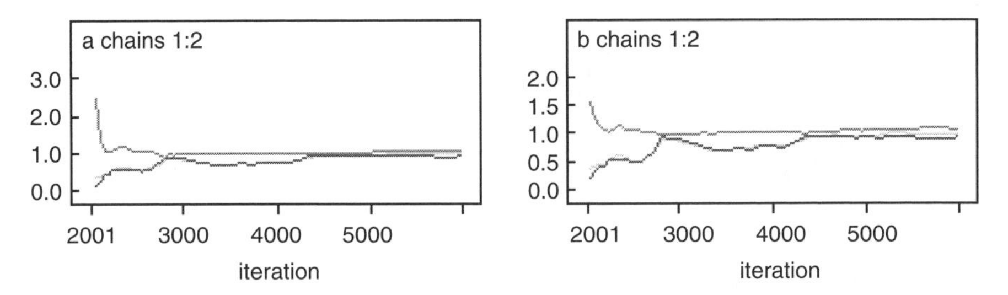

Figure 6.5 Gelman and Rubin convergence diagnostic for parameters a and b in the Poisson–gamma model after 4000 more iterations.

cannot conclude that the chains have converged, so it is advisable to carry out some more updates and check again.

The results of the Gelman and Rubin convergence diagnostic after 4000 more iterations are shown in Figure 6.5.

Table 6.1 Posterior statistics for the parameters in the Poisson–gamma model.

Parameter	mean	sd	MC error	2.5%	97.5%
a	16.27	7.397	0.3726	5.64	34.32
b	16.21	7.424	0.3734	5.59	34.05
mean	1.01	0.088	0.0026	0.85	1.19
var	0.07	0.035	0.0025	0.03	0.16

Table 6.2 Posterior means for the relative risk in five counties of South Carolina obtained using the Poisson–gamma model.

County	mean	sd	MC error	2.5%	97.5%
Abbeville	0.931	0.2673	0.002207	0.458	1.514
Aiken	1.020	0.2288	0.001349	0.618	1.521
Allendale	1.032	0.2874	0.001953	0.549	1.677
Anderson	0.910	0.2176	0.001955	0.521	0.521
Bamberg	1.011	0.2813	0.001772	0.532	1.637

Once one is happy with convergence, further iterations are carried out to obtain samples from the posterior distributions. Tables 6.1 and 6.2 show the posterior values for the parameters of the model after 5000 iterations. From Table 6.1, we can see that the mean of the relative risk is 1.0100, bigger than the mean for the SMRs, and their standard deviation has been reduced by almost 85%.

Figure 6.6 shows the map of the posterior expected relative risks for each county in South Carolina based on the Poisson–gamma model. The range of the posterior relative risks is reduced. The lowest estimated risk is now 0.877 and the highest 1.333. Thus, we removed from the data random variability due to the small counts. We are now dealing with a smoother map with less extremes in the relative risk estimates. Again, we observed high risk in the central counties and lower risks in the south-west.

6.1.3 Log-normal model

Although a gamma prior distribution for the relative risk is mathematically convenient, as shown in Section 6.1.2, it may be restrictive because covariate adjustment is difficult and there is no possibility for allowing spatial correlation between risks in nearby areas (appropriate if geographic trends in risk are anticipated).

A log-normal model for the relative risk is more flexible:

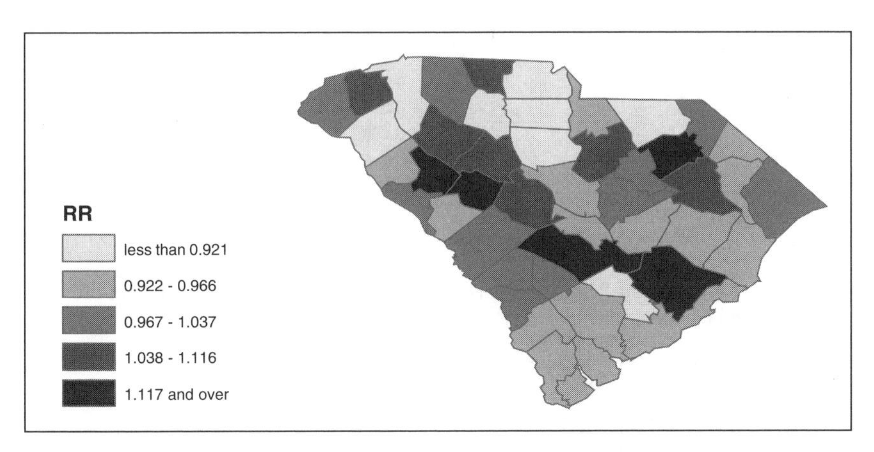

Figure 6.6 Posterior expected relative risk (Poisson–gamma model) for congenital anomalies: South Carolina, 1990.

```
model
{
for (i in 1:m)
{
# Poisson likelihood for observed counts
   y[i]~dpois(mu[i])
   log(mu[i])<-log(E[i])+alpha+v[i]
# Relative Risk
   theta[i]<-exp(alpha+v[i])
# Prior distribution for v
   v[i]~dnorm(0,tau)
}
# Hyperprior distribution on inverse variance parameter of the random effects
tau~dgamma(0.5,0.0005)
# Vague prior distribution for intercept (mean relative risk in study region)
alpha~norm(0,1.0E-5)
mean<-exp(alpha)
}
```

Figure 6.7 Log-normal model in WinBUGS.

$$
\begin{aligned}
y_i &\sim Poisson(e_i\theta_i),\\
\log\theta_i &= \alpha + v_i,\\
v_i &\sim N(0, \tau_v^2).
\end{aligned}
$$

We are going to repeat the analysis in Section 6.1.2 using the log-normal model (code in Figure 6.7).

Convergence has not been reached with only 2000 iterations. The Gelman and Rubin diagnostic plot for the parameter τ_v is shown in Figure 6.8 which also displays the autocorrelations for the parameter τ_v.

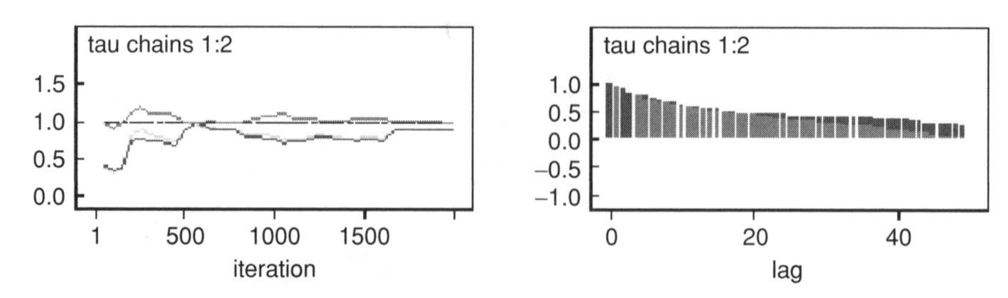

Figure 6.8 Gelman and Rubin convergence diagnostic and autocorrelation plot for τ_v in the log-normal model.

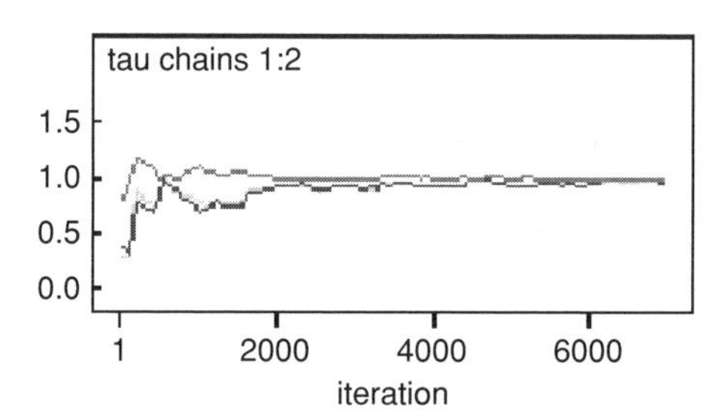

Figure 6.9 Gelman and Rubin convergence diagnostic for parameter τ in the log-normal model after 4000 more iterations.

Figure 6.9 shows the Gelman and Rubin convergence diagnostic after 4000 more iterations. From this plot we can see that convergence has improved.

Figure 6.10 displays the relative risk estimates for all the counties in South Carolina after convergence was achieved. The spatial pattern in relative risk is very similar to the one obtained using the Poisson–gamma model. Again, we are dealing with a smoother map with less extremes in the relative risk estimates.

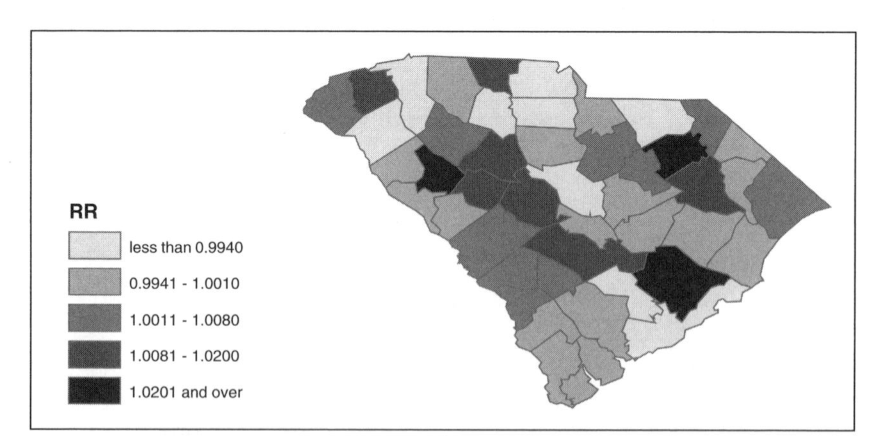

Figure 6.10 Posterior expected relative risk (log-normal model) for congenital anomalies: South Carolina, 1990.

6.1.4 Choice of the hyperprior distributions

In the fully Bayesian approach to disease mapping, the choice of the hyperprior distribution of the parameters is a key issue. To investigate the influence of the choice of hyperprior on estimates of relative risk, it is necessary to carry out sensitivity analyses to different choices of priors or parameters. If the data are numerous, the data dominates the prior and hence it matter less which values of the hyperparameters are chosen. If the data are scarce, the choice of a suitable combination of hyperparameters is important. For studies concerning sensitivity of the estimates to the choice of the hyperpriors see Bernardinelli *et al.* (1995b) and Mollié (2000).

6.1.5 Besag, York and Mollié (BYM) model

In this model for relative risks, area-specific random effects are decomposed into a component that takes into account the effects that vary in a structured manner in space (clustering or correlated heterogeneity) and a component that models the effects that vary in an unstructured way between areas (uncorrelated heterogeneity).

The model, introduced by Clayton and Kaldor (1987) and developed by Besag *et al.* (1991), is formulated as follows

$$y_i \sim Poisson(e_i \theta_i),$$
$$\log \theta_i = \alpha + u_i + v_i,$$

where α is an overall level of the relative risk, u_i is the correlated heterogeneity and v_i is the uncorrelated heterogeneity.

Bayesian modelling requires specification of prior distributions for random effects. The distribution model for the uncorrelated heterogeneity is

$$v_i \sim N(0, \tau_v^2).$$

For the clustering component, a spatial correlation structure is used, where estimation of the risk in any area depends on neighbouring areas. The conditional autoregressive (CAR) model proposed by Besag *et al.* (1991) is used

$$[u_i|u_j, i \neq j, \tau_u^2] \sim N\left(\bar{u}_i, \tau_i^2\right)$$

where

$$\bar{u}_i = \frac{1}{\sum_j \omega_{ij}} \sum_j u_j \omega_{ij},$$

$$\tau_i^2 = \frac{\tau_u^2}{\sum_j \omega_{ij}},$$

$$\omega_{ij} = 1 \text{ if } i, j \text{ are adjacent (or 0 if they are not).}$$

Parameters τ_v^2 and τ_u^2 control the variability of v and u. In a full Bayesian analysis, prior distributions must be specified for those parameters. We considered gamma distributions for both, as suggested by Bernardinelli *et al.* (1995b).

To fit the model using WinBUGS, in addition to the files with the model specification and the observed and expected counts, it is necessary to include a file containing the data on the adjacency matrix for South Carolina. This matrix can be generated using the **Adjacency Tool** from the **Map menu** in GeoBUGS 1.1.

The weights must also be entered as data. The easiest way to define them is to create a loop in the WinBUGS code (lines 18–22 in Figure 6.11; *sumNumNeigh* is the length of the adjacency matrix and is also an output if the matrix is generated by the adjacency tool mentioned before).

Model fitting was carried out using two separate chains starting from different initial values. Convergence was checked by visual examination of time series plots of samples for each chain and by computing the Gelman and Rubin diagnostic. The first 5000 samples were discarded as a burn-in; each chain was run for a further 10 000 iterations.

The Bayesian estimates of the relative risk (Figure 6.12) show less variation than the observed SMR (Figure 6.1). The posterior relative risk vary from 0.1150 to 3.7070 with an overall mean of 0.9520. Some extreme SMR estimates have disappeared and much of the map has been smoothed. Figure 6.13 shows the posterior probability of the relative risk being bigger than 1.

Table 6.3 shows the posterior mean and 95% credible intervals of the variance components for the BYM model. The variability of the relative risk is attributed more to the uncorrelated heterogeneity than to the spatially structured effects. However, as τ_u^2 is small, the risk in any given area is similar to that in neighbouring areas.

As mentioned in Section 4.7.2, the CAR model is improper and it is necessary to impose a constraint to ensure that the model is identifiable (we included α in the model and it was assigned an improper prior). We are going to use now a proper Gaussian CAR distribution, available only in GeoBUGS 1.1, rather than the intrinsic CAR distribution for the area-specific random effects. It should be noted that the **car.proper** distribution as defined by Stern and Cressie (1999) has been criticized for the fact that it conditions on the neighbourhood totals of

```
model
{
for (i in 1:m)
{
# Poisson likelihood for observed counts
  y[i]~dpois(mu[i])
  log(mu[i])<-log(e[i])+alpha+u[i]+v[i]
# Relative Risk
  theta<-exp(alpha+u[i]+v[i])
# Posterior probability of θ_{i}>1
  PP[i]<-step(theta[i]-1+eps)
# Prior distribution for the uncorrelated heterogeneity
  v[i]~dnorm(0,tau.v)
}
eps<-1.0E-6
# CAR distribution for the spatial correlated heterogeneity
u[1:m]~car.normal(adj[],weights[],num[],tau.u)
# Weights
for (k in 1:sumNumNeigh)
{
  weights[k]<-1
}
# Improper distribution for the mean relative risk in the study region
alpha~dflat()
mean<-exp(alpha)
# Hyperprior distributions on inverse variance parameters of random effects
tau.u~dgamma(0.5,0.0005)
tau.v~dgamma(0.5,0.0005)
}
```

Figure 6.11 BYM model in WinBUGS.

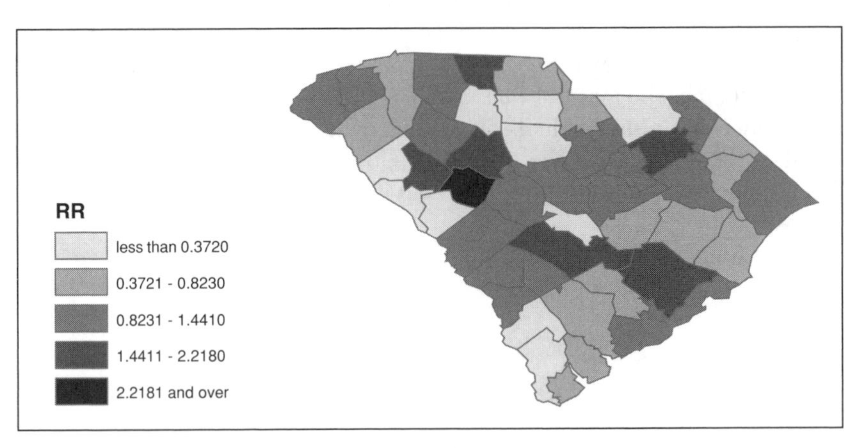

Figure 6.12 Posterior expected relative risk (BYM model) for congenital anomalies: South Carolina, 1990.

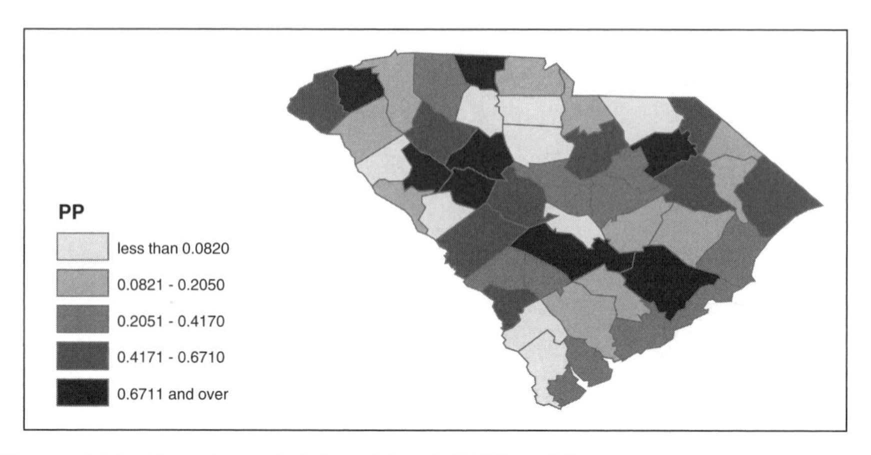

Figure 6.13 Posterior probability of $\theta_i > 1$ (BYM model).

Table 6.3 Posterior statistics of the variance components.

	Mean	SD	Credible interval
τ_u^2	0.0666	0.0223	(0.0316,0.1170)
τ_v^2	0.1156	0.0307	(0.0639,0.1843)

the random effect and not on the means and that there is no dependence on the number of neighbours of each region in the model. Figure 6.14 displays the results after fitting the model with WinBUGS.

Maps in Figures 6.12 and 6.14 present a similar geographical pattern for the relative risks but under the model with proper Gaussian CAR distribution, they range from 0.9800 to 1.0350, which shows a bigger degree of smoothness.

The models fitted previously can be compared using overall goodness of fit measures, such as the *Deviance Information Criterion* (Spiegelhalter *et al.*, 2002a), and also by examining residuals plots (the residuals are easily obtainable from the output of WinBUGS). The overall goodness of fit measures are useful for helping model selection but give little help in assessing how well the model fits the data.

In order to investigate the local goodness-of-fit of the models described in this section, we can compare their Bayesian residuals (described in Section 2.6)

$$r_i = y_i - e_i\hat{\theta}_i \quad \forall i.$$

Figures 6.15 and 6.16 display maps of these residuals. Figure 6.16 displays a reasonable pattern of high and low residuals with a small overall range in the residual values. On the other hand, the residuals displayed in Figure 6.15 show a greater range in values (for the BYM model with proper CAR distribution for

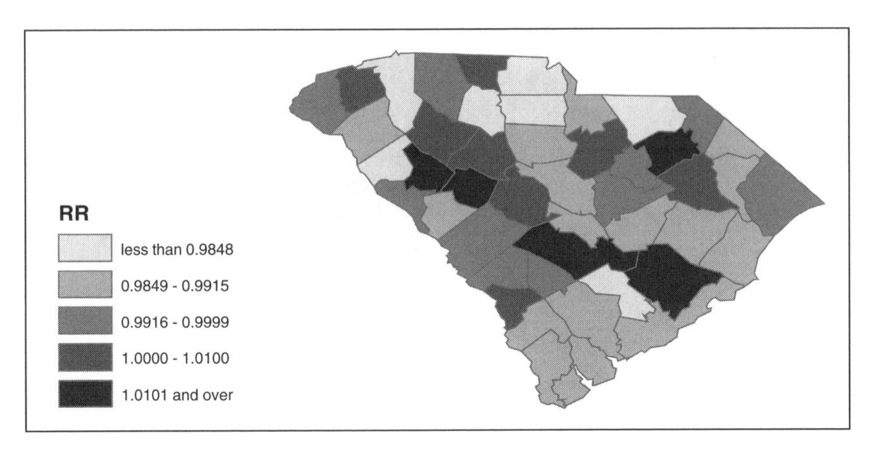

Figure 6.14 Posterior expected relative risk (BYM model with proper Gaussian CAR distribution) for congenital anomalies: South Carolina, 1990.

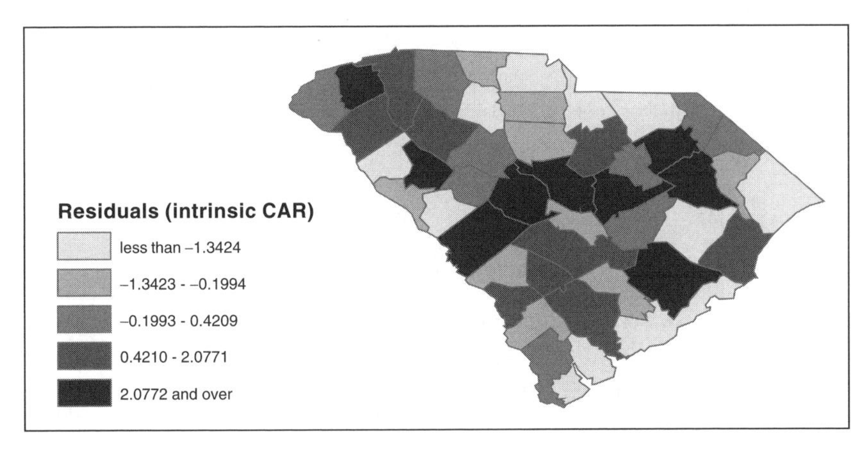

Figure 6.15 Residuals for the BYM model with **car.normal** prior for area-specific random effects.

area specific random effects, the range was reduced by almost 50%). The areas where the largest deviations occur are in the central and north-east counties in both maps, which means that both models have problems in estimating the true relative risk in the same areas.

Comparison of observed and fitted counts can also be carried out using the Pearson's chi-square goodness-of-fit measure (*RSS*). For the BYM model with proper CAR distribution for area-specific random effects we have lower values of the *RSS*. Also, in terms of the *Deviance Information Criterion (DIC)*, the **car.proper** distribution yields better estimations for the relative risk (Table 6.4).

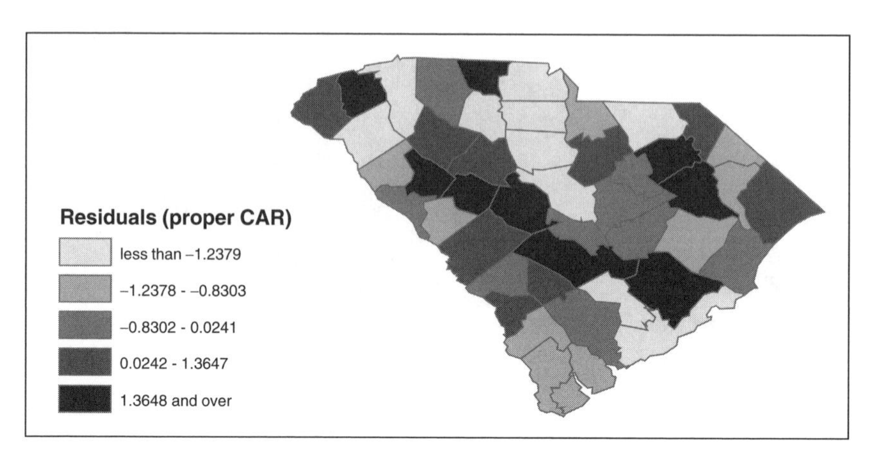

Figure 6.16 Residuals for the BYM model with **car.proper** prior for area-specific random effects.

Table 6.4 Goodness of fit for the BYM models.

Model	RSS	DIC	ΔDIC
BYM (intrinsic CAR)	14.133	192.041	
BYM (proper CAR)	3.731	168.974	−23.067

6.1.6 Space × time models

Many uses of disease mapping, such as the identification of spatial heterogeneity of disease risk or cluster investigation, are frequently constrained to a single time period, but data in public health are often available for time windows of several years. It is possible to consider the analysis of disease maps which have a temporal dimension. The most common format for observations is counts of cases of disease within small areas that are available for a sequence of T time periods.

There has been much recent interest in the analysis of disease rates over space and time. Most Bayesian methods (Bernardinelli *et al.*, 1995; Waller *et al.*, 1997; Knorr-Held and Besag, 1998) propose extensions of the purely spatial models by Clayton and Kaldor (1987) and Besag *et al.* (1991) to space × time data.

In this section we are going to analyse the space × time distribution of respiratory cancer mortality in Ohio over a period of ten years (1979–1988) using two different models for space × time variation of disease risk.

Model 1 Bernardinelli *et al.* (1995) suggest a model in which both area-specific intercept and temporal trend are modelled as random effects. This formulation allows for spatiotemporal interactions where temporal trend in risk may be different for different spatial locations and may even have spatial structure. However, all temporal trends are assumed to be linear, which is a restrictive assumption. Define y_{ik} as the count of disease and e_{ik} the expected count in the ith region and kth time period.

The model for the relative risk is of the form

$$\log \theta_{ik} = \alpha + u_i + v_i + \beta \cdot t_k + \delta_i \cdot t_k,$$

where α is an intercept (overall rate), u_i and v_i are area random effects (as defined in the BYM model), $\beta \cdot t_k$ is a linear trend term in time t_k, and δ_i is an interaction random effect between space and time. Prior distributions must be assumed for the parameters in this model. In this formulation there is no spatial trend, only a simple linear time trend and no simple temporal random effects (Figure 6.17).

```
model
{
for (i in 1:m)
{
   for (k in 1:T)
   {
   # Poisson likelihood for observed counts
   y[i,k]~dpois(mu[i,k])
   log(mu[i,k])<-log(e[i,k])+alpha+u[i]+v[i]+beta*t[k]+delta[i]*t[k]
   # Relative Risk in each area and period of time
   theta[i,k]<-exp(alpha+u[i]+v[i]+beta*t[k]+delta[i]*t[k])
   }
   theta_area[i]<-exp(u[i]+v[i])
   TT[i]<-exp(beta+delta[i])
}

# CAR prior distribution for spatial correlated heterogeneity
u[1:m]~car.normal(adj[],weights[],num[],tau.u)
delta[1:m]~car.normal(adj[],weights[],num[],tau.delta)

# Prior distributions for the Uncorrelated Heterogeneity
for(i in 1:m)
{
v[i]~dnorm(0,tau.v)
}

# Weights
for(k in 1:sumNumNeig)
{
   weights[k]<-1
}

# Improper distribution for the mean relative risk in the study region
alpha~dflat()
mean<-exp(alpha)

# Hyperprior distributions on inverse variance parameter of random effects
beta~dnorm(0,1.0E-5)
tau.v~dgamma(0.5,0.0005)
tau.u~dgamma(0.5,0.0005)
tau.delta~dgamma(0.5,0.0005)
}
```

Figure 6.17 Space × time model (Bernardinelli *et al.* 1995) in WinBUGS.

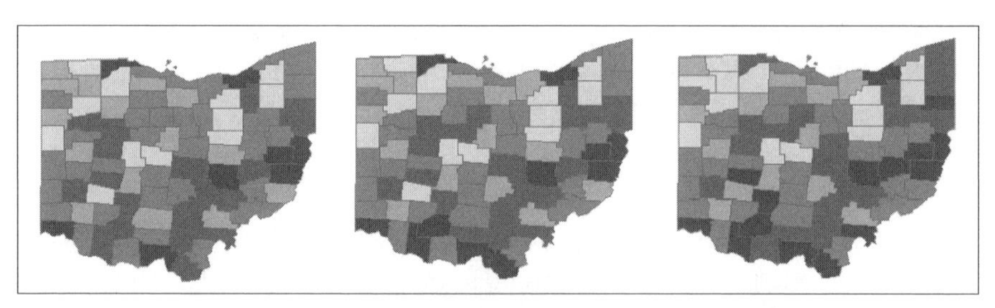

Figure 6.18 Posterior expected relative risk (space × time model (Bernardinelli *et al.*, 1995)) for respiratory cancer mortality: Ohio, 1979, 1983, 1988.

Model fitting was carried out using two separate chains starting from different initial values. Convergence was checked by visual examination of time series plots of samples for each chain and by computing the Gelman and Rubin diagnostic. The first 5000 samples were discarded as a burn-in; each chain was run for a further 10 000 iterations.

A value of 0.0026 was obtained for β, the coefficient of the trend term in time t_k. As $\exp(\beta)$ is the rate ratio between two consecutive years, the risk was multiplied by approximately 1.0026 every year. This finding shows that although there is an increasing trend in the mortality from respiratory cancer in Ohio, it is very smooth. Figure 6.18 maps the estimates of the relative risks in the first (1979), middle (1983) and last (1988) year of the study period. Respiratory cancer deaths rates are increasing over time, as indicated by a gradual darkening of the counties.

Figure 6.19 shows the temporal trend estimates in the counties of Ohio for the period 1979–1988. This is computed as $\exp(\beta + \delta_i)$ from the temporal model terms. In general, in the south-west and in the east of the state the mortality relative risk increases.

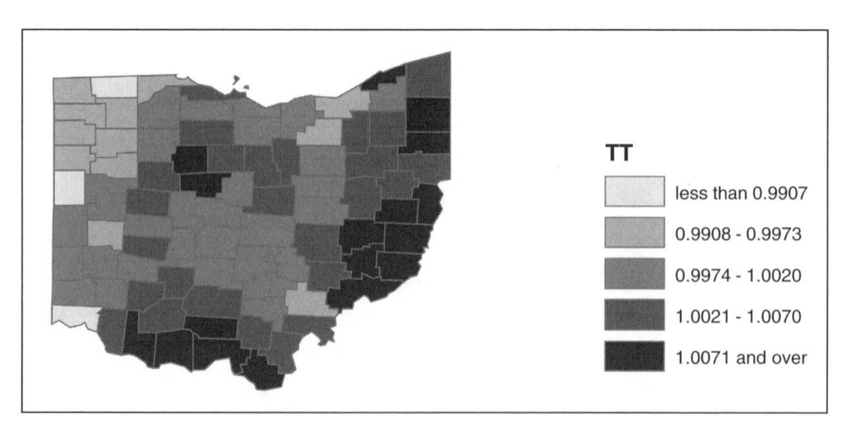

Figure 6.19 Posterior expected temporal trend (space × time model (Bernardinelli *et al.*, 1995)) for respiratory cancer mortality: Ohio, 1979–1988.

Model 2 Waller *et al.* (1997) use a nested model, where the hierarchical specification by Besag *et al.* (1991) is applied to each time point separately. The model does not have a single spatial main effect and allows the spatial patterns at each time point to be completely different. The log of the relative risk is parameterized as

$$\log \theta_{ik} = \alpha + u_i^{(k)} + v_i^{(k)},$$

where $u_i^{(k)}$ and $v_i^{(k)}$ are correlated and uncorrelated heterogeneity terms that can vary in time (effects for the region i in time k) (Figure 6.20).

```
model
{
for (t in 1:T)
{
  for (i in 1:m)
  {
#     Poisson likelihood for observed counts
      y[i,t]~dpois(mu[i,t])
      log(mu[i,t])<-log(e[i,t])+alpha+u[i,t]+v[i,t]
#     Relative risk
      theta[t,i]<-exp(u[i,t]+v[i,t])
  }
}
# CAR prior distribution for spatial correlated
  heterogeneity
u1[1:m]~car.normal(adj[],weights[],num[],tau.u[1])
u2[1:m]~car.normal(adj[],weights[],num[],tau.u[2])
u3[1:m]~car.normal(adj[],weights[],num[],tau.u[3])
u4[1:m]~car.normal(adj[],weights[],num[],tau.u[4])
u5[1:m]~car.normal(adj[],weights[],num[],tau.u[5])
u6[1:m]~car.normal(adj[],weights[],num[],tau.u[6])
u7[1:m]~car.normal(adj[],weights[],num[],tau.u[7])
u8[1:m]~car.normal(adj[],weights[],num[],tau.u[8])
u9[1:m]~car.normal(adj[],weights[],num[],tau.u[9])
u10[1:m]~car.normal(adj[],weights[],num[],tau.u[10])
for (i in 1:m)
{
  u[i,1]<-u1[i]
  u[i,2]<-u2[i]
  u[i,3]<-u3[i]
  u[i,4]<-u4[i]
  u[i,5]<-u5[i]
  u[i,6]<-u6[i]
  u[i,7]<-u7[i]
  u[i,8]<-u8[i]
  u[i,9]<-u9[i]
  u[i,10]<-u10[i]
}
# Prior distributions for the uncorrelated
  heterogeneity
for(i in 1:m)
{
  v1[i]~dnorm(0,tau.v[1])
  v2[i]~dnorm(0,tau.v[2])
  v3[i]~dnorm(0,tau.v[3])
  v4[i]~dnorm(0,tau.v[4])
  v5[i]~dnorm(0,tau.v[5])
  v6[i]~dnorm(0,tau.v[6])
  v7[i]~dnorm(0,tau.v[7])
  v8[i]~dnorm(0,tau.v[8])
  v9[i]~dnorm(0,tau.v[9])
  v10[i]~dnorm(0,tau.v[10])
}
for (i in 1:m)
{
  v[i,1]<-v1[i]
  v[i,2]<-v2[i]
  v[i,3]<-v3[i]
  v[i,4]<-v4[i]
  v[i,5]<-v5[i]
  v[i,6]<-v6[i]
  v[i,7]<-v7[i]
  v[i,8]<-v8[i]
  v[i,9]<-v9[i]
  v[i,10]<-v10[i]
}
# Weights
for (k in 1:sumNumNeigh)
{
  weight[k]<-1
}
# Improper prior distribution for the mean relative
risk in the study region
alpha~dflat()
mean<-exp(alpha)
# Hyperprior distributions on inverse variance
parameter of random effects
for (i in 1:T)
{
  tau.v[i]~dgamma(0.5,0.0005)
  tau.u[i]~dgamma(0.5,0.0005)
}
}
```

Figure 6.20 Space × time model (Waller *et al.*, 1997) in WinBUGS.

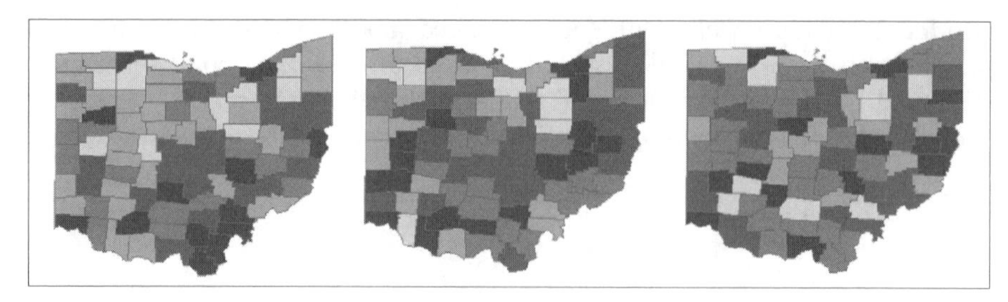

Figure 6.21 Posterior expected relative risk (space × time model (Waller *et al.*, 1997)) for respiratory cancer mortality: Ohio, 1979, 1983, 1988.

Model fitting was carried out using two separate chains starting from different initial values. The first 5000 samples were discarded as a burn-in; each chain was run for a further 10 000 iterations. Graphical monitoring of chains for a representative subset of the parameters, along with sample autocorrelations and Gelman and Rubin diagnostics indicate an acceptable degree of convergence.

Figure 6.21 maps the estimates of the relative risks in the first (1979), middle (1983) and last (1988) year of the study period. Respiratory cancer death rates are increasing over time, as indicated by a gradual darkening of the counties. We can also see increasing evidence of clustering among the high-rate counties.

If we investigate the clustering effect (τ_u) over time k, an increase is observed suggesting that the similarly of the respiratory cancer cases is increasing over the 10-year period (Figure 6.22). The posterior mean of the uncorrelated

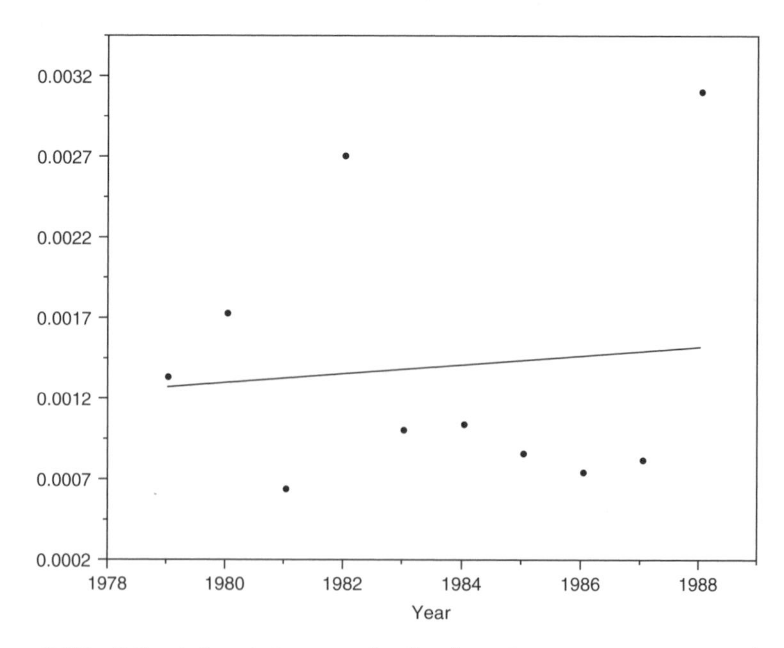

Figure 6.22 Estimated posterior mean for the clustering parameter τ_u versus k.

heterogeneity effect (τ_v) plotted versus k provides less indication of a trend suggesting that there is no additional heterogeneity in the data beyond that explained by the CAR prior.

Other examples of spatiotemporal modelling include models that combine the spatial model by Besag *et al.* (1991) with dynamic models which allow for a nonparametric estimation of temporal trends in disease (Knorr-Held and Besag, 1998), mixture models (Boehning *et al.*, 2000) and generalized additive models (MacNab and Dean, 2002).

6.1.7 Mixture models

All previous methods for the statistical analysis of risks in small areas are mostly based on smoothing techniques. However, these methods often smooth over large discontinuities in the risk surface which can be important to maintain. Methods which simply detect discontinuities in surfaces have been proposed in the literature (Schlattman and Boehning, 1993; Knorr-Held and Rasser, 2000). While those models address the issue of discontinuities, they do not admit the possibility that certain areas of a map have smooth rate transitions and it may be appropriate to allow both smoothness and discontinuities within the same map. To do this, it is possible to employ a special type of spatial mixture which admits different forms of spatial variation.

The mixture model, proposed by Lawson and Clark (2002), assumes that the log-relative risk can be decomposed into three additive components, one fixed component representing the unstructured heterogeneity, (v), which measures the overdispersion in a individual region, and two mixing components, (u, φ), representing different aspects of spatial correlation. The two mixing components are a combination of a spatial correlation component and a component which models discrete jumps (Figure 6.23).

The mixture model is specified as

$$\log \theta_i = \alpha + v_i + p_i u_i + (1 - p_i)\varphi_i.$$

Special cases of this formulation arise depending on the value of p_i. If all the $p_i = 1$, then the standard BYM model arises, and for $p_i = 0\ \forall i$, a pure jump model arises.

For the autocorrelation component, the usual intrinsic normal specification is adopted (see Section 6.1.5 for details). For the jump component a prior which examines the total absolute difference between neighbours, with neighbours defined as for the BYM model, was chosen

$$\pi(\varphi_1, \ldots, \varphi_m) \propto \frac{1}{\sqrt{\lambda}} \exp\left(-\frac{1}{\lambda} \sum_{i \sim j} |\varphi_i - \varphi_j| \right).$$

```
model
{
for (i in 1:m)
{
   # Poisson likelihood for observed counts
   y[i]~dpois(mu[i])
   log(mu[i])<-log(e[i])+alpha+v[i]+p[i]*u[i]+(1-p[i])*fi[i]
   # Relative Risk
   theta[i]<-exp(v[i]+p[i]*u[i]+(1-p[i])*fi[i])
   # Prior distribution for the uncorrelated heterogeneity
   v[i]~dnorm(0,tau.v)
   # Prior distribution for the p[i]
   p[i]~dbeta(0.5,0.5)
}

# CAR prior distribution for spatial correlated heterogeneity
u[1:m]~car.normal(adj[],weights[],num[],tau.u)

# CAR-L1 prior for the term that models discrete jumps
fi[1:m]~car.l1(adj[],weights[],num[],tau.fi)

# Weights
for(k in 1:sumNumNeig)
{
   weights[k]<-1
}

# Improper prior distribution for the mean relative risk in the study region
alpha~dflat()
mean<-exp(alpha)

# Hyperprior distributions on inverse variance parameter of random effects
tau.u~dgamma(0.5,0.0005)
tau.v~dgamma(0.5,0.0005)
tau.fi~dgamma(0.5,0.0005)
}
```

Figure 6.23 Mixture relative risk model in WinBUGS.

The parameter λ acts as a constrained term and it has a gamma prior distribution.
The priors for the other components in the mixture model are:

$$v_i \sim N(0, \tau_v^2)\ \forall i,$$
$$p_i \sim Beta(0.5, 0.5)\ \forall i.$$

The posterior expected relative risks for the mixture and the BYM models are given in Figures 6.24 and 6.25, respectively. The mixture model has maintained the main areas with jumps while displaying smoothing in the others. The BYM model relative risks are smoother and appear to have lost the jump discontinuities which were apparent in the SMR map (Figure 6.2). Overall, the BYM map has a lower relative risk range. In fact, much of the map is

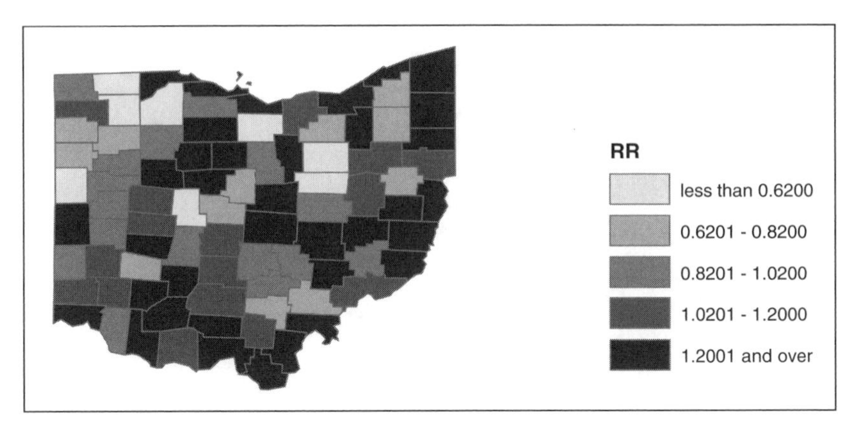

Figure 6.24 Posterior expected relative risk (mixture model) for respiratory cancer mortality: Ohio, 1988.

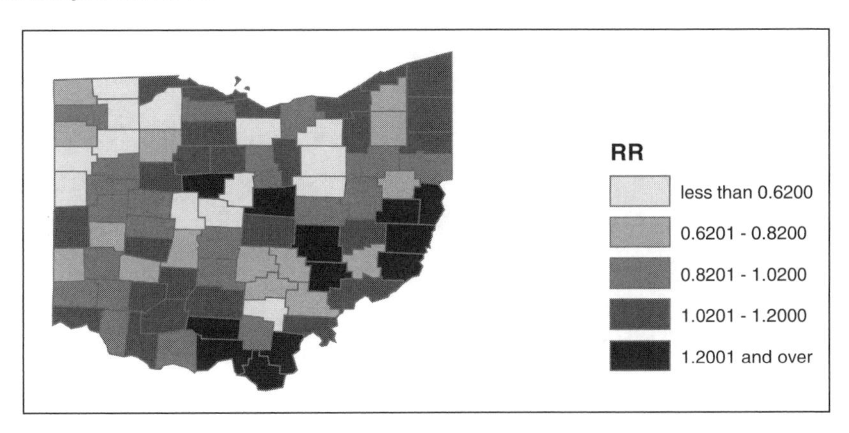

Figure 6.25 Posterior expected relative risk (BYM model) for respiratory cancer mortality: Ohio, 1988.

categorized by two levels of risk: a lower level in the west and a higher level in the south-east. The main difference between these maps is that the BYM model has smoothed the relative risk considerably more than the mixture model.

Figures 6.26 to 6.28 display the posterior expectations for the mixture model of u_i, v_i and φ_i.

As may be expected, the uncorrelated heterogeneity (Figure 6.27) displays a relatively random variation across the map, while the correlated heterogeneity (Figure 6.26) displays blocks of values.

The φ_i component (Figure 6.28) has patches of large jumps, particularly in the western counties.

Overall, there is evidence in this data example that the two-component mixture model provides an enhanced model for the relative risk compared to the BYM model. The improvement in the DIC (Table 6.5) supports this contention.

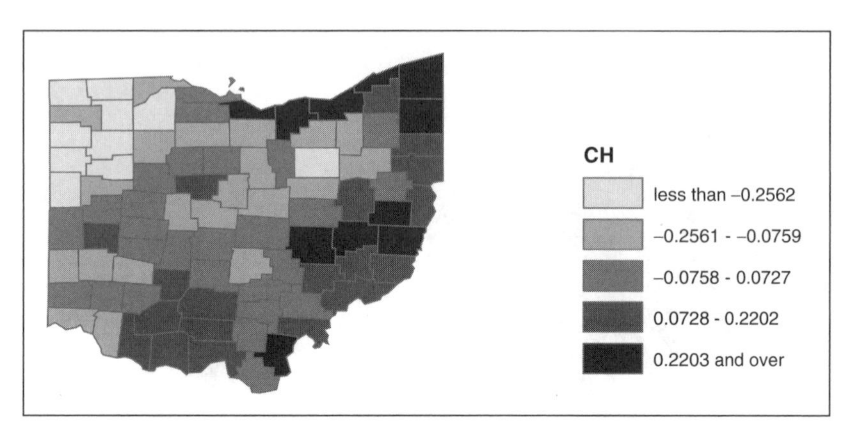

Figure 6.26 Posterior expectation of u for the mixture model.

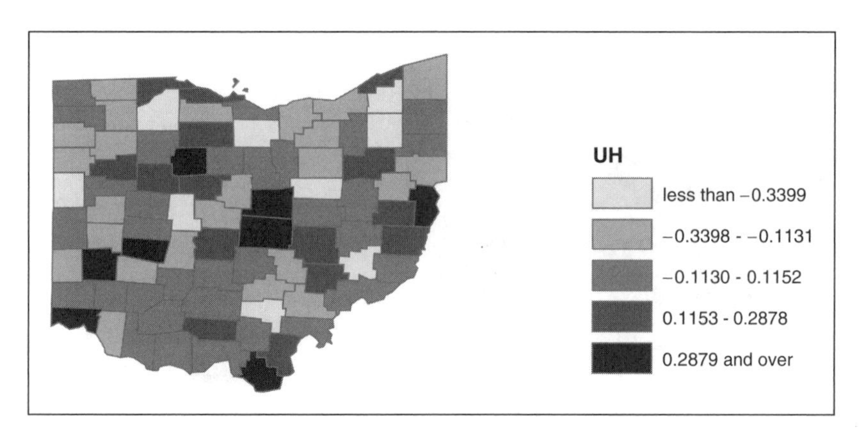

Figure 6.27 Posterior expectation of v for the mixture model.

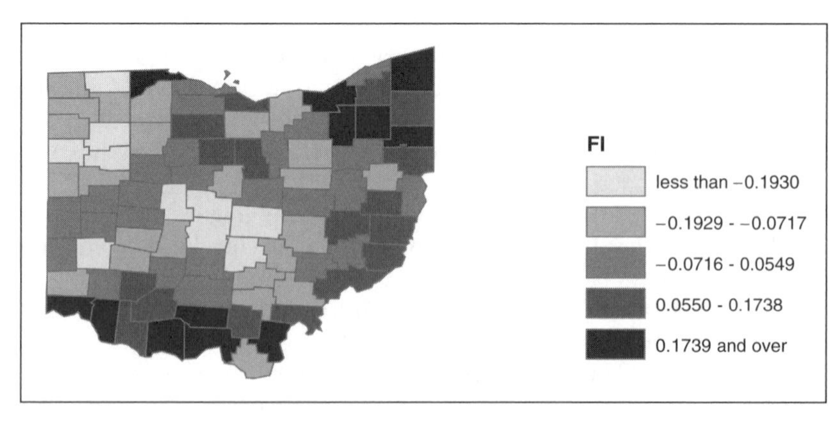

Figure 6.28 Posterior expectation of φ for the mixture model.

Table 6.5 Goodness of fit for the mixture and BYM models.

Model	Deviance	DIC	ΔDIC
Mixture	563.1	638.25	−11.53
BYM	562.9	649.78	

6.2 SPATIAL PREDICTION

In previous sections we have illustrated different models for the estimation of relative risk in a set of counties. However, what happens if we know the value of the relative risk in certain locations and based on that knowledge we want to predict the value in a location where it is unknown? The distribution **spatial.pred** implemented in GeoBUGS 1.1 (syntax in Section 4.7.2) permits spatial interpolation and prediction. In this section we illustrate the use of that distribution.

The data for this example consist of values of three variables: log SMR, **x** and **z**. The SMR is the standardized mortality ratio for cancer (malignant neoplasms) during the year 1999 in 36 counties of South Carolina, $\{(x_i, z_i), i = 1, \ldots, m\}(m = 36)$ are the coordinates of the centroids of each county and log SMR is the vector of the logarithm of the SMRs. The data file also contains a set of N locations **x.pred** and **z.pred** representing the centroids of counties at which we wish to predict the SMRs (in this example $N = 10$). The **spatial.exp** function allows the fitting of a fully parameterized covariance function within a multivariate normal distributional model. The parameters ϕ and κ define this covariance function based on a function of interpoint distance d_{ij} as: $\exp[-(\phi\, d_{ij})^{\kappa}]$ with a Gaussian covariance model for $\kappa = 2$ and a purely exponential covariance with $\kappa = 1$. Parameter ϕ controls the rate of distance decay of correlation between sites while κ controls the form of the decline. **Spatial.unipred** provides a method of predicting values of the fitted surface at unsampled locations. The code in WinBUGS for this analysis is given in Figure 6.29.

Table 6.6 shows the true and predicted values of the SMR in the 10 counties of South Carolina where we have assumed they are unknown.

Figures 6.30 and 6.31 display, respectively, the map of the true SMRs and the map with the predicted values. From the pictures we can see that when we predict some values, there are smoother areas that were not present before. Notice that this example does not use any information concerning expected rates within the predicted areas and so there is no adjustment for these local conditions.

```
model
{
for (i in 1:m)
{
   mu[i]<-beta
}
logSMR[1:m]~spatial.exp(mu[],x[],z[],tau,phi,kappa)
for (j in 1:N)
{
   logSMR.pred[j]~spatial.unipred(beta,x.pred[j],z.pred[j],logSMR[])
}

# Priors
beta~dflat()
tau~dgamma(0.1,0.001)
phi~dunif(0.001,0.8)
kappa~dunif(0.05,1.95)
}
```

Figure 6.29 WinBUGS code for spatial prediction.

Table 6.6 Predicted values and errors in the prediction for the SMR in ten counties of South Carolina.

County	SMR	Prediction	Error
45001	1.2739	1.0210	0.2529
45017	1.2499	1.0180	0.2319
45019	0.9305	1.0146	−0.0841
45027	1.0894	1.0152	0.0742
45033	0.9548	1.0248	−0.0701
45049	0.9564	1.0216	−0.0652
45051	1.1824	1.0248	0.1576
45061	1.1600	1.0215	0.1384
45069	1.1414	1.0241	0.1174
45077	0.9429	1.0241	−0.0812

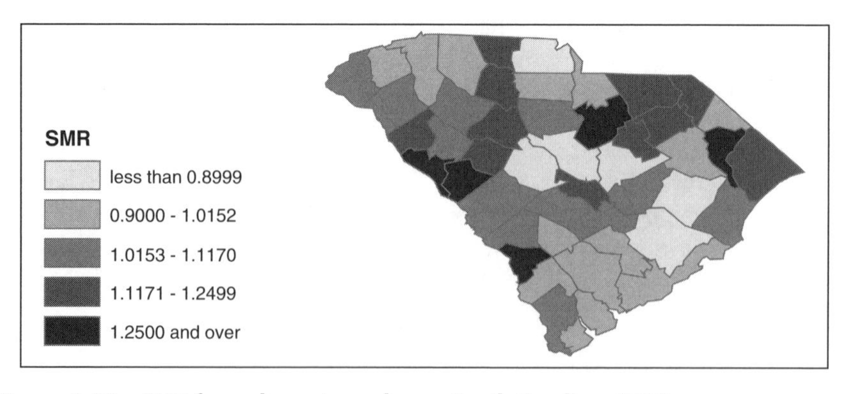

Figure 6.30 SMR for malignant neoplasms. South Carolina, 1999.

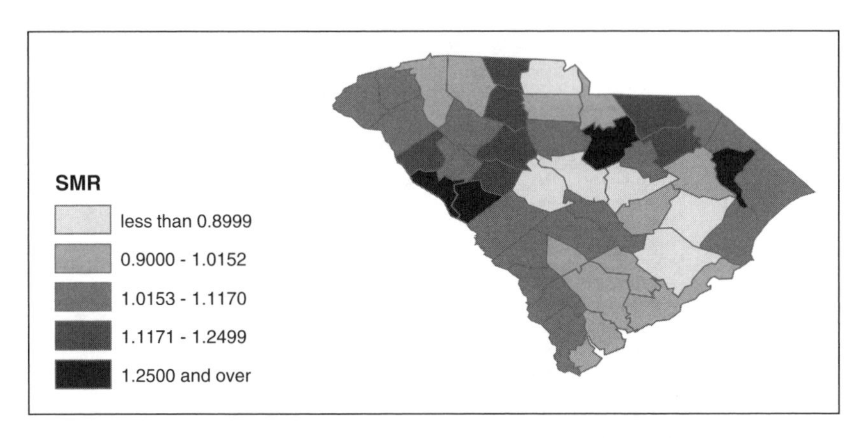

Figure 6.31 Predicted SMR for malignant neoplasms. South Carolina, 1999.

6.3 AN ANALYSIS OF THE OHIO DATASET USING MLwiN

In this section we are going to analyse a dataset of respiratory cancer deaths in the US state of Ohio from 1979 to 1988. This dataset differs from the Scottish lip cancer dataset discussed in the introductory MLwiN chapter as we not only have a spatial element to the dataset but also a time element. In this section we will consider some simple approaches to incorporate both space and time effects in a statistical model and leave the more complex modelling to the WinBUGS software sections described elsewhere. Firstly we will consider how to deal with repeated measures data in MLwiN.

6.3.1 Setting up a repeated measures dataset in MLwiN

The Ohio dataset that we are analysing has three parts to the data. Firstly we have the counts of the observed respiratory cancer deaths in each of the 88 counties within Ohio for each of the 10 years (1979–1988). Secondly we have the corresponding expected deaths that are calculated based on population size in each county and finally we have spatial information in the form of a list of neighbouring counties for each county. If we load up the worksheet *'ohio1.ws'* we will see the following 28 columns of data in the **Names** window.

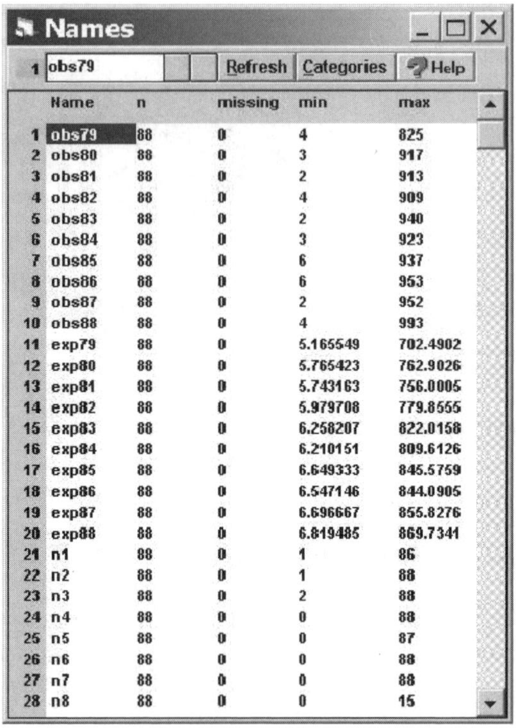

Names

1 obs79 | Refresh | Categories | Help

	Name	n	missing	min	max
1	obs79	88	0	4	825
2	obs80	88	0	3	917
3	obs81	88	0	2	913
4	obs82	88	0	4	909
5	obs83	88	0	2	940
6	obs84	88	0	3	923
7	obs85	88	0	6	937
8	obs86	88	0	6	953
9	obs87	88	0	2	952
10	obs88	88	0	4	993
11	exp79	88	0	5.165549	702.4902
12	exp80	88	0	5.765423	762.9026
13	exp81	88	0	5.743163	756.0005
14	exp82	88	0	5.979708	779.8555
15	exp83	88	0	6.258207	822.0158
16	exp84	88	0	6.210151	809.6126
17	exp85	88	0	6.649333	845.5759
18	exp86	88	0	6.547146	844.0905
19	exp87	88	0	6.696667	855.8276
20	exp88	88	0	6.819485	869.7341
21	n1	88	0	1	86
22	n2	88	0	1	88
23	n3	88	0	2	88
24	n4	88	0	0	88
25	n5	88	0	0	87
26	n6	88	0	0	88
27	n7	88	0	0	88
28	n8	88	0	0	15

Here we see that columns *c1–c10* contain the observed counts, *c11–c20* the expected counts and *c21–c28* contain the neighbour identifiers. Note that there are at least three and at most eight neighbours to each county. We could now consider the observed and expected counts for a particular year and fit any of the models that we fitted to the Scottish lip cancer dataset previously. Alternatively we could sum up both the observed and expected counts for the 10 years and analyse the 10-year period as a whole. We are here, however, looking to include time-based effects, for example trend effects for individual counties, to spot unusual patterns. Consequently we would like to analyse all the 880 observed and expected counts simultaneously and to do this requires reformatting the data so that we have one column of observed and one column of expected counts. Before formatting the data we will, however, add in two additional variables to represent a constant and a county indicator. These we will add in columns *c29* and *c30* respectively. We will use the **Generate vector** window that is found under the **Data manipulation** menu. We saw how to construct a constant vector in the earlier MLwiN chapter, and for the county indicator we use the option to generate a sequence by setting up the window as shown below and then pressing the **Generate** button. Here the counties will be numbered 1 to 88.

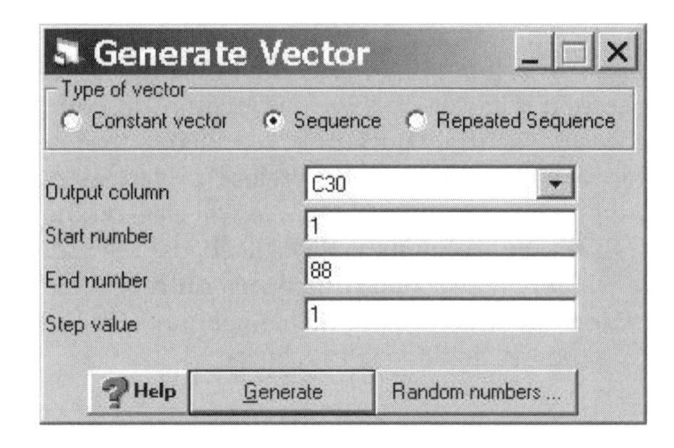

We will also name the two columns, *c29* and *c30* as '*cons*' and '*county*' respectively via the **Names** window. Reformatting the data by stacking counts into single columns is commonly required for all types of repeated measures data (see Rasbash *et al.*, 2000, Chapter 10). To transform the data we can use the **Split records** window that is available from the **Data Manipulation** menu and looks as follows:

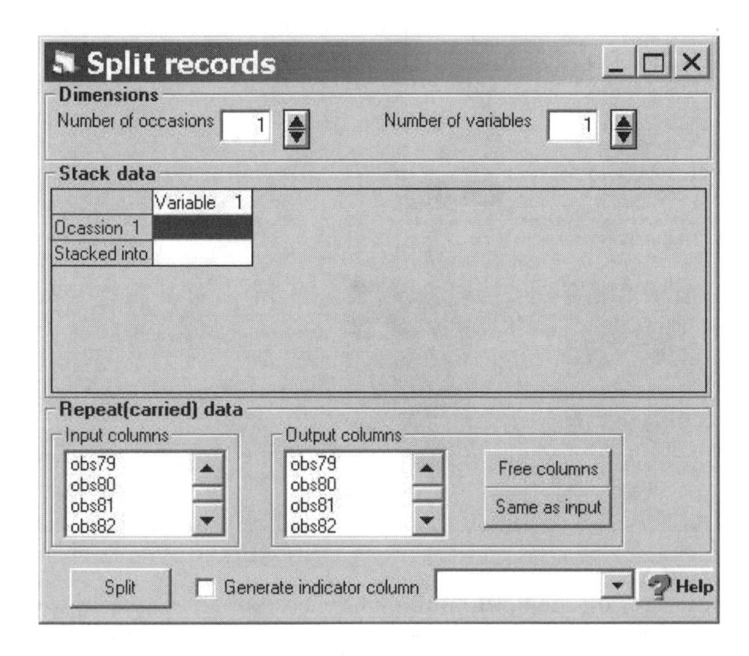

For the Ohio dataset we need to set the number of occasions (years) to 10 and the number of variables to 2 (observed and expected counts). We will then fill in the stacked data section with the 10 observed and expected counts. We will stack these two sets of columns into two empty columns, *c41* and *c42*. The **Repeated data** section is important as we also require any other variables, e.g. the county number and its neighbours, to be repeated so that we have values of these variables for each observation. We therefore need to select all the columns from '*n1*' to '*county*' as input columns here. We will click on the **Free columns** button to choose the output columns as this will choose the first empty columns (*c31* to *c40*). We will finally select an **indicator column** (which will represent the year of each observation) and store this in column *c43*. Having selected all these options our window should look as follows:

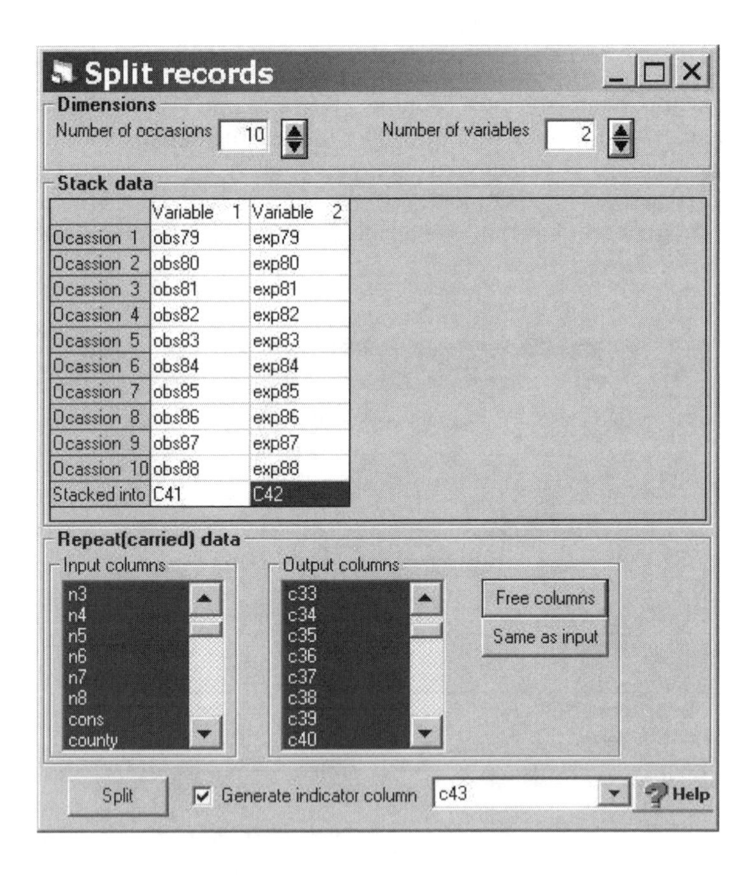

If we now click on the **Split** button and select **No** when asked if we want to save our worksheet then we will get 13 new variables of length 880 stored in columns *c31–c43*. We now no longer need the data in columns *c1–c30* and so we can use the following two commands in the **Command Interface** window: ERASE c1–c30 erases the columns whilst MOVE moves the columns to the top of the worksheet, starting at column *c1* once again. We will now have to rename our 13 columns and this can be done in the **Names** window as shown below:

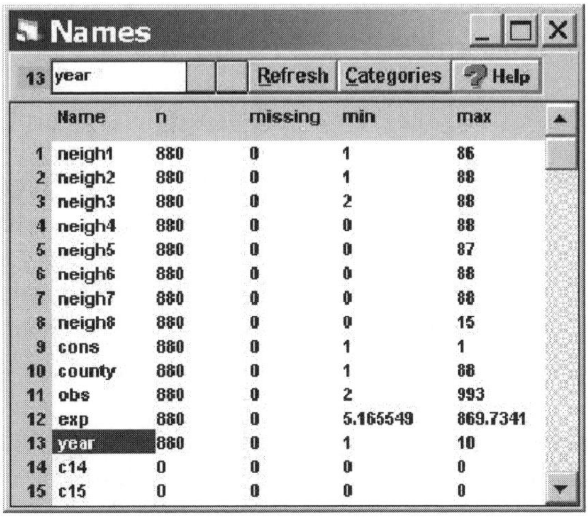

Names

13 year | Refresh | Categories | Help

	Name	n	missing	min	max
1	neigh1	880	0	1	86
2	neigh2	880	0	1	88
3	neigh3	880	0	2	88
4	neigh4	880	0	0	88
5	neigh5	880	0	0	87
6	neigh6	880	0	0	88
7	neigh7	880	0	0	88
8	neigh8	880	0	0	15
9	cons	880	0	1	1
10	county	880	0	1	88
11	obs	880	0	2	993
12	exp	880	0	5.165549	869.7341
13	year	880	0	1	10
14	c14	0	0	0	0
15	c15	0	0	0	0

It is probably now a good idea to save your new MLwiN worksheet as '*ohio2.ws*' at this point so that you will not have to repeat this procedure again if you want to return to the worksheet later. We can look more closely at the data structure by selecting columns *c10–c13* in the **View/Edit data** window that is available under the **Data Manipulation** menu.

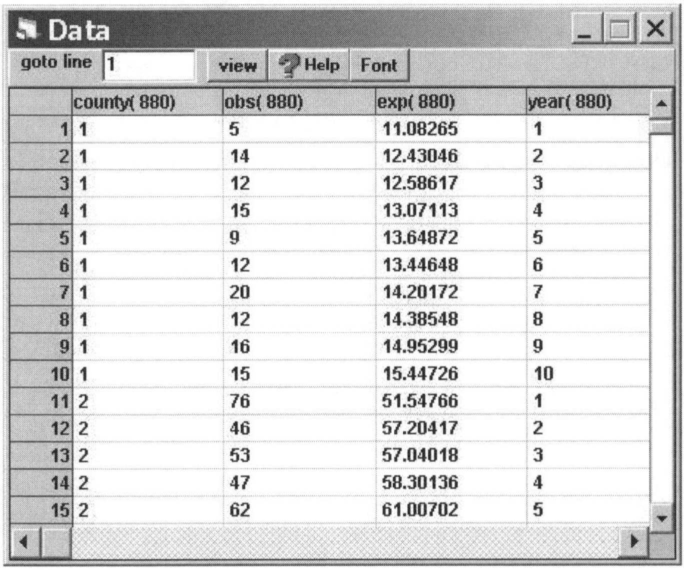

Data

goto line 1 | view | Help | Font

	county(880)	obs(880)	exp(880)	year(880)
1	1	5	11.08265	1
2	1	14	12.43046	2
3	1	12	12.58617	3
4	1	15	13.07113	4
5	1	9	13.64872	5
6	1	12	13.44648	6
7	1	20	14.20172	7
8	1	12	14.38548	8
9	1	16	14.95299	9
10	1	15	15.44726	10
11	2	76	51.54766	1
12	2	46	57.20417	2
13	2	53	57.04018	3
14	2	47	58.30136	4
15	2	62	61.00702	5

Here we see the data for county 1 is stacked on top of the data for county 2 and the years 1979–1988 are coded 1–10. We are now ready to start analysing the data.

6.3.2 Preliminary analysis of the Ohio dataset

As this chapter is titled relative risk estimation perhaps a good place to start is to look at some simple plots of the standardized mortality ratio (SMR) for each county. To do this we need to construct the SMRs and we can do this by typing the following commands in the **Command Interface** window: CALC C14 = 'obs'/'exp' to calculate the SMR and NAME C14 'SMR' to name the column. We can now plot the SMRs of each county against time via the **Customised Graph** window to look for any interesting patterns. To do this we set up the window as follows:

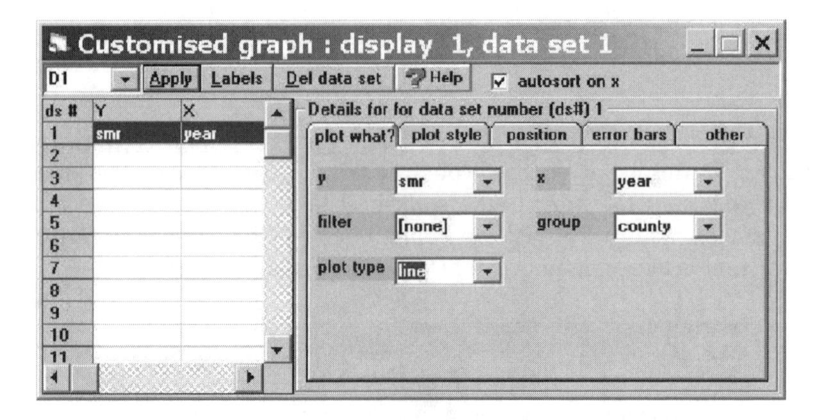

Upon clicking on the **Apply** button we will get 88 lines, one for each county. As we can see from the following graph there is no particular pattern to be observed. One county has the two highest SMRs in two consecutive years which may be worth investigating further but generally no immediately obvious patterns emerge. We will of course be unlikely to spot trends here as we are plotting SMRs which have a mean of 1.0 at each time point, but we are hoping to identify counties which have SMRs that are consistently above or below 1.

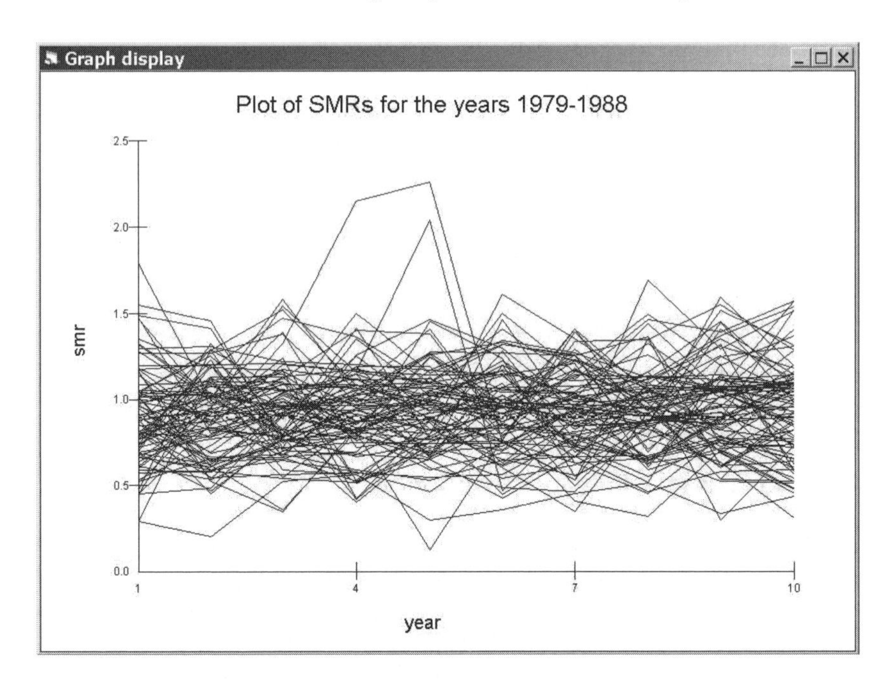

Another two interesting graphs are shown below. The first is a histogram of all the SMRs and the second is a histogram of the number of years a county's SMR is greater than 1. This second graph can be produced by firstly constructing a column that contains whether each SMR is greater than 1 and then using both the **Multilevel Data Manipulation** and **Unreplicate** windows to convert this information into a column of 88 counts.

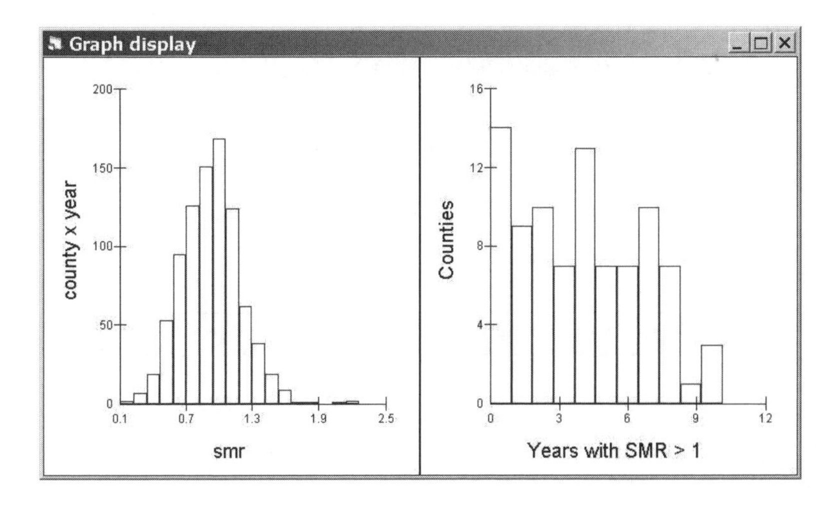

Here we see that there are several outlying large SMRs with the largest observed SMR equalling 2.26 although the county in question had a low expected count (9.7) and it is well known that counties with low expected counts tend to be more variable in terms of their SMRs. There are differences between counties that cannot be explained as simple random variation as the second graph shows that three counties (counties 18, 31 and 41) exhibited SMRs that were greater than 1 for the whole of the period and there are 14 counties that had SMRs of less than 1 for the whole period. The three counties with the consistently high SMRs are randomly spread in the dataset which does not suggest there is a focused clustering type effect from a point source near these counties but there may still be some underlying spatial effects that are not picked up by these simple summaries.

6.3.3 Poisson models for the Ohio dataset

We will now consider fitting some statistical models to the dataset. Perhaps the simplest model to fit is a Poisson distribution to all the counts fitting the logarithm of the expected counts as an offset. We can construct a variable that is the logarithm of the expected counts in column *c15*. This is done by typing the following commands in the **Command Interface** window: `CALC C15 = LOGE('EXP')` to calculate the offset and `NAME C15 'LOGEXP'` to name the column.

We will next set up a simple model with '*obs*' as a Poisson response variable, three levels (for use with later models) with level 1 being '*year*', level 2 being '*county*' and level 3 being '*neigh1*' (which will be used in later models for spatial effects). We will add '*cons*' as a fixed effect to fit simply an intercept in the model. Upon convergence (using first-order MQL estimation) the model should look as follows:

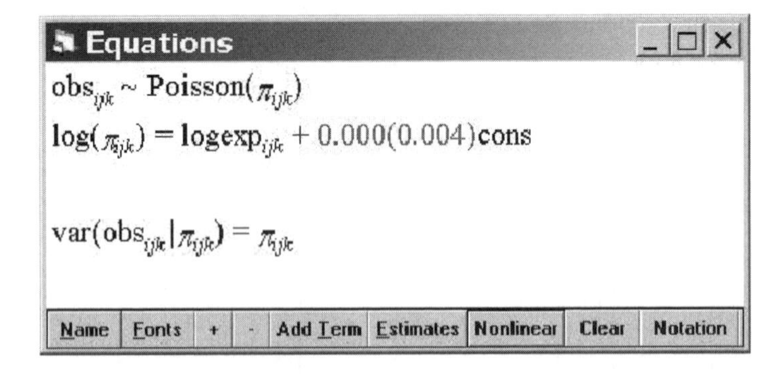

Here as expected we get an estimate of 0 for β_0 as this corresponds to an average SMR of 1.0. If we now remove the Poisson constraint and allow extra Poisson variation then we find the following:

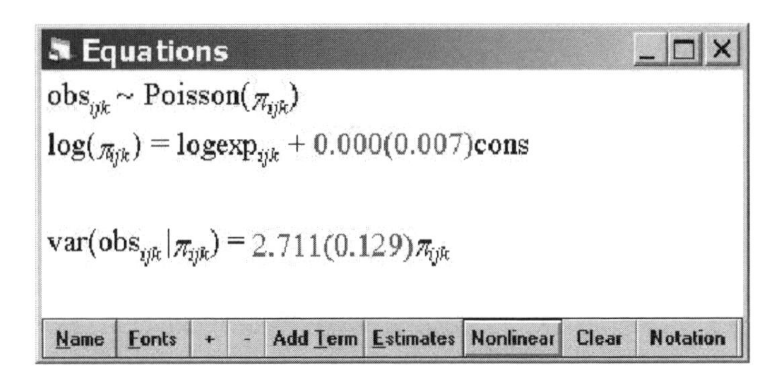

Here there is clearly the overdispersion that was suggested by the earlier histograms with nearly three times the variability expected for a truly Poisson distributed variable. The histograms suggested that this overdispersion was due to differences between the various counties and so to study this we can add an (unstructured) random effect for each county. To do this we click on the '*cons*' variable in the **Equations** window and click on the '*j(county)*' tickbox. If we fit this model with extra-Poisson variation still set we observe the following:

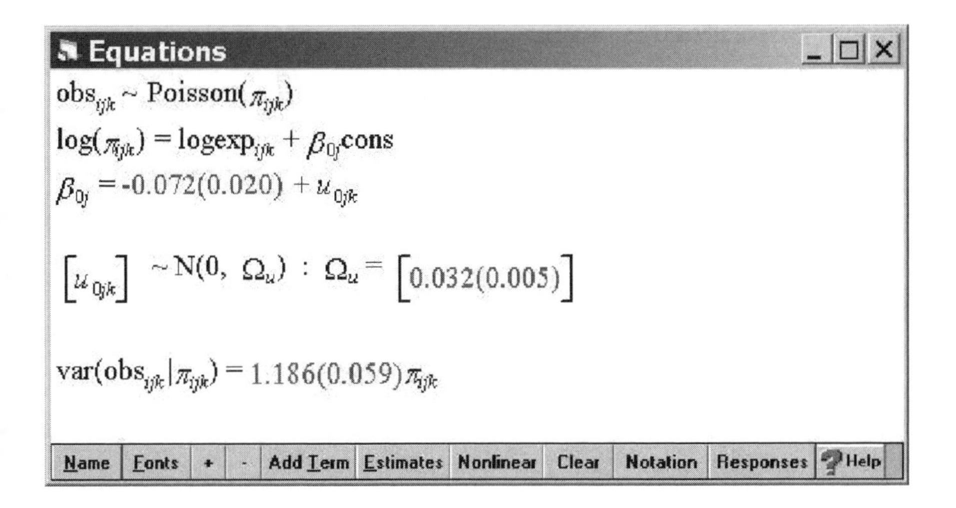

Here we see that there is still a little more variation than we would expect from a Poisson variable but the assumption of (conditional) Poisson counts is now much more plausible than for the model without county effects. If we look at a residual plot for the counties (available from the **Residuals** window) we see the following:

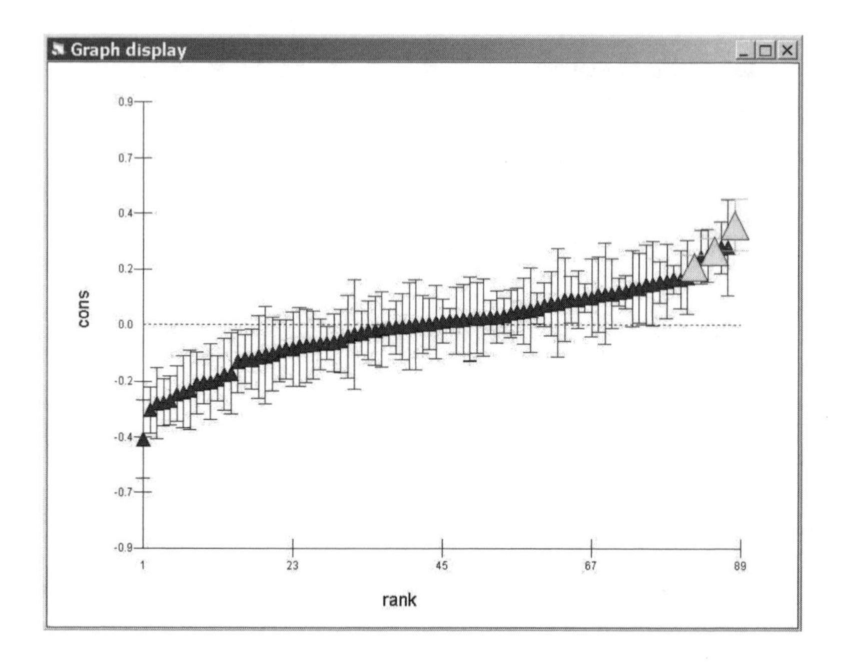

Here we have highlighted (towards the right of the graph) the three counties earlier identified as having SMRs consistently above 1. As we would expect all three of these counties have large positive residuals that are significantly bigger than zero. We have as yet neglected the time aspect of the data. We know that on average the effect of year in the data should be zero as each time point has mean SMR equalling 1 so we will add in not only a fixed effect for year but also a set of random effects for the effect of year in each county. This can be done by adding the variable '*year*' via the **Add Term** window and then clicking on it in the **Equations** window and selecting the '*j(county)*' tickbox. If we fit this model using first-order MQL Poisson estimation, we get the following:

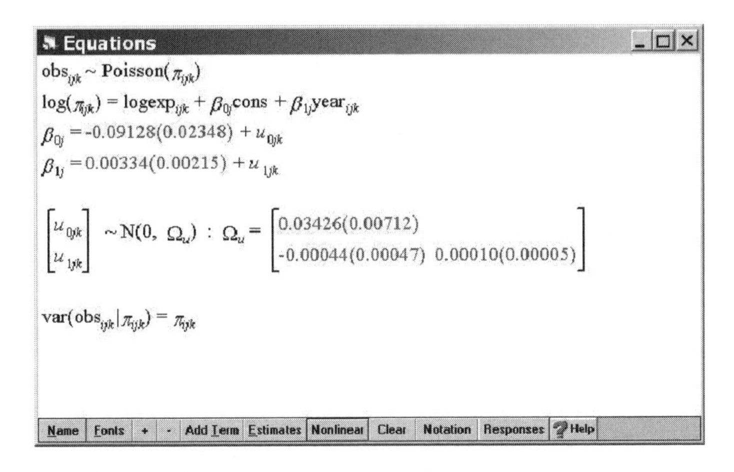

Note that here we have altered the number of decimal places that are given in the **Equations** window from 3 to 5. (This is achieved via the **Numbers** screen under the **Options** menu.) From this model we can plot the predicted relative risk graphs via the **Predictions** window. This will give predictions that need to be exponentiated to transform them to relative risks. A graph of the relative risks is shown in the following figure with once again the three extreme counties highlighted.

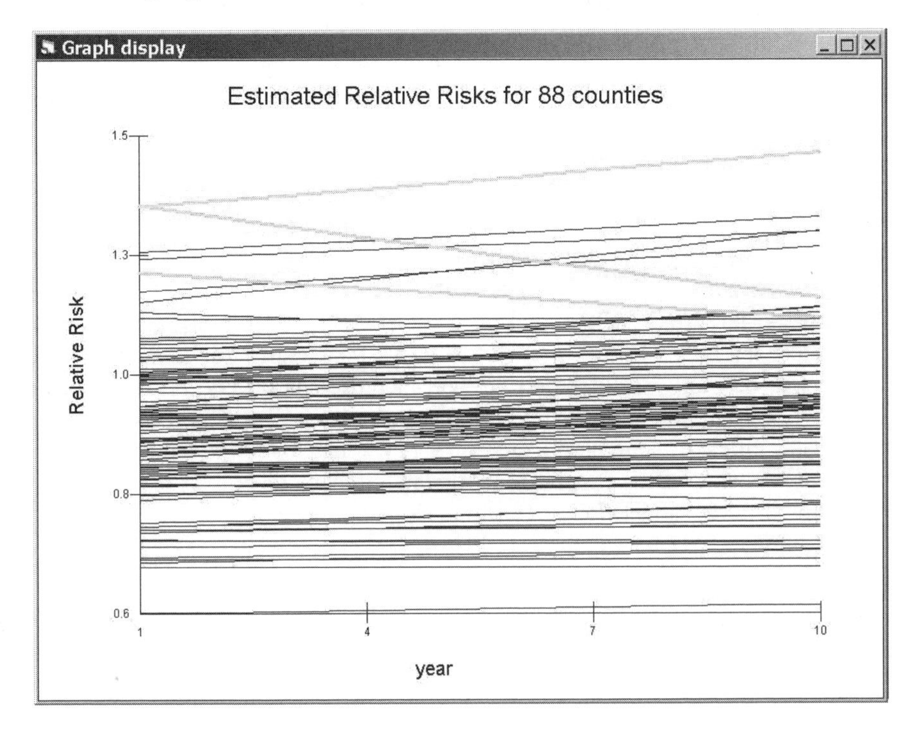

We can of course extend the model to fit higher-order polynomial effects and these can be created in the **Command Interface** or **Calculate** window, for example the command `CALC C19 = 'year' * 'year'` will create a squared term that can be added to the model. If we fit a quadratic relationship for each county we will get the following graph:

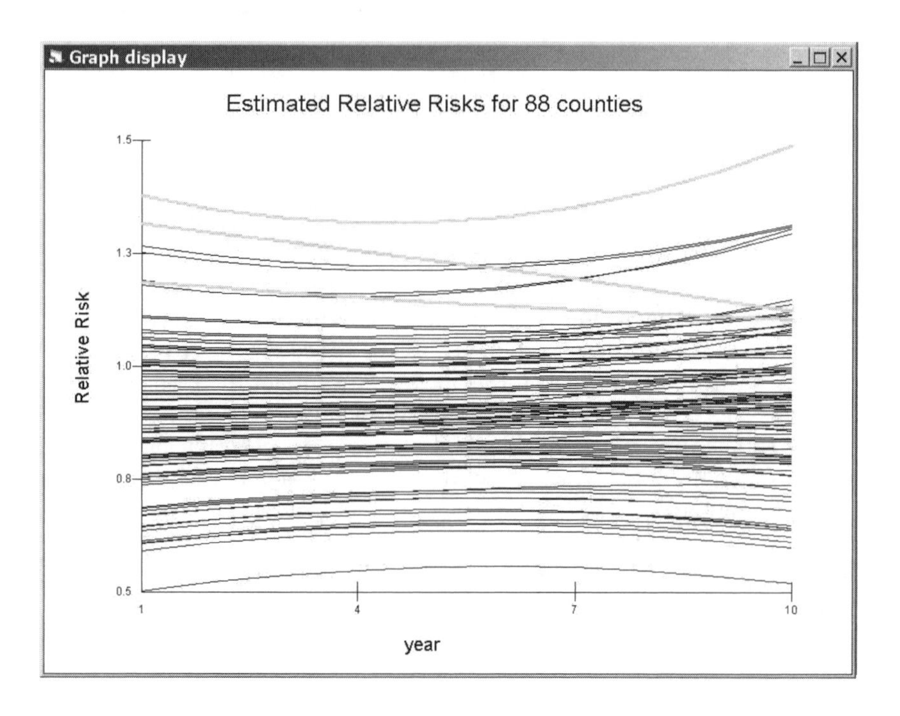

6.3.4 MCMC estimation

We have so far fitted each model with quasi-likelihood methods. We can of course also fit all the same models using MCMC estimation in MLwiN and MCMC will be more useful when we consider the spatial effects that are the next part of the modelling process. Rather than repeat details of the analysis of the models fitted thus far here using MCMC, we instead simply summarize the model fit statistics (DIC) for the four models fitted in the table below (results based on running models for 50 000 iterations):

Model	$\bar{D}$	$D(\bar{\theta})$	pD	DIC
Null	7185.23	7184.20	1.03	7186.26
County random effects	5679.21	5600.18	79.03	5758.23
Linear time effects	5644.87	5545.76	99.11	5743.98
Quadratic time effects	5643.60	5486.92	156.68	5800.28

Here we see that including both unstructured random effects and linear trend terms for each county improves the model fit but that adding quadratic terms for each county results in a larger DIC statistic and hence a worse model.

6.3.5 Adding spatial effects

When we analysed the Scottish lip cancer dataset we considered two methods of including spatial effects, multiple membership models and CAR spatial effects. To use either of these methods we need to create columns of weights for each neighbouring unit. We can create these columns in MLwiN by typing the required commands in the **Command Interface** window; however, here we need to type many commands and so it may be better to instead write the commands to a **macro** file which can then be executed. Below we have the macro commands in the **Macro** window which we have created via the **New Macro** option in the **File** menu. We have named this macro '*ohiomac.txt*' when we saved the file via the **Save Macro** option.

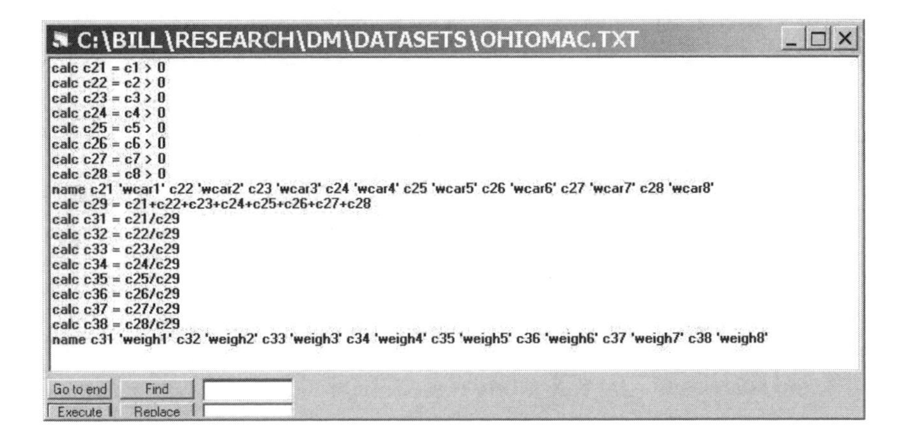

We will firstly set up a multiple membership model for the spatial random effects while keeping a linear trend term for each county in the model. This model is set up as for the lip cancer dataset example and when we have used MCMC estimation for 50 000 iterations after a 500 burn-in the results should be as shown below. It should be noted here that as the two effects at the county level are correlated we have to use an inverse Wishart prior distribution here.

Equations

$\text{obs}_i \sim \text{Poisson}(\pi_i)$

$\log(\pi_i) = \text{logexp}_i + \beta_{0i}\text{cons}_i + \beta_{1i}\text{year}_i$

$\beta_{0i} = -0.11138(0.03566) + \sum_{j\in neigh1(i)} w^{(3)}_{i,j} u^{(3)}_{0j} + u^{(2)}_{0,county(i)}$

$\beta_{1i} = 0.00326(0.00228) + u^{(2)}_{1,county(i)}$

$$\left[u^{(3)}_{0,neigh1(i)}\right] \sim \text{N}(0,\ \Omega^{(3)}_u)\ :\ \Omega^{(3)}_u = \left[0.06333(0.03743)\right]$$

$$\begin{bmatrix} u^{(2)}_{0,county(i)} \\ u^{(2)}_{1,county(i)} \end{bmatrix} \sim \text{N}(0,\ \Omega^{(2)}_u)\ :\ \Omega^{(2)}_u = \begin{bmatrix} 0.02818(0.00756) & \\ -0.00043(0.00043) & 0.00009(0.00004) \end{bmatrix}$$

$\text{var}(\text{obs}_i|\pi_i) = \pi_i$

PRIOR SPECIFICATIONS

$p(\beta_0)\ \alpha\ 1$

$p(\beta_1)\ \alpha\ 1$

$p(1/\Omega^{(3)}_{u0,0}) \sim \text{Gamma}(0.00100,0.00100)$

$$p(\Omega^{(2)}_u) \sim \text{inverse Wishart}_2[\,2{*}S_u,\ 2],\ S_u = \begin{bmatrix} 0.03400 & \\ -0.00040 & 0.00010 \end{bmatrix}$$

Deviance(MCMC) = 5644.37100(880 of 880 cases in use)

Name | Fonts | + | - | Add Term | Estimates | Nonlinear | Clear | Notation | Responses | Help

If we examine the DIC diagnostic here we see that the DIC equals 5743.01 which is slightly smaller than the DIC for the model without the spatial MM effects. We could here also consider fitting a neighbour by time interaction term which would give two (correlated) sets of neighbour effects, one set of neighbour effects as in the last model plus a second set of neighbour effects on the trends of the various counties. In this model we use two inverse Wishart priors and find that the DIC diagnostic is reduced slightly to 5742.08 (note that this will depend partly on the choice of parameter values for the inverse Wishart priors). The table below compares the DIC diagnostic for three possible models that incorporate spatial effects with the best-fitting non-spatial model.

Model	$\bar{D}$	$D(\bar{\theta})$	pD	DIC
Linear time effects	5644.87	5545.76	99.11	5743.98
Multiple-membership model	5644.37	5545.74	98.63	5743.01
MM plus temporal interaction	5641.66	5541.24	100.42	5742.08
CAR model	5644.07	5545.60	98.47	5742.54

The final model that we have compared in the table was produced by adding a CAR spatial effect to the model that fits linear time effects for each county. Note that to fit such a model in MLwiN we have to remove the global intercept term to make the model identifiable. Upon running the model for 50 000 iterations with a suitable Wishart prior distribution we get the following results:

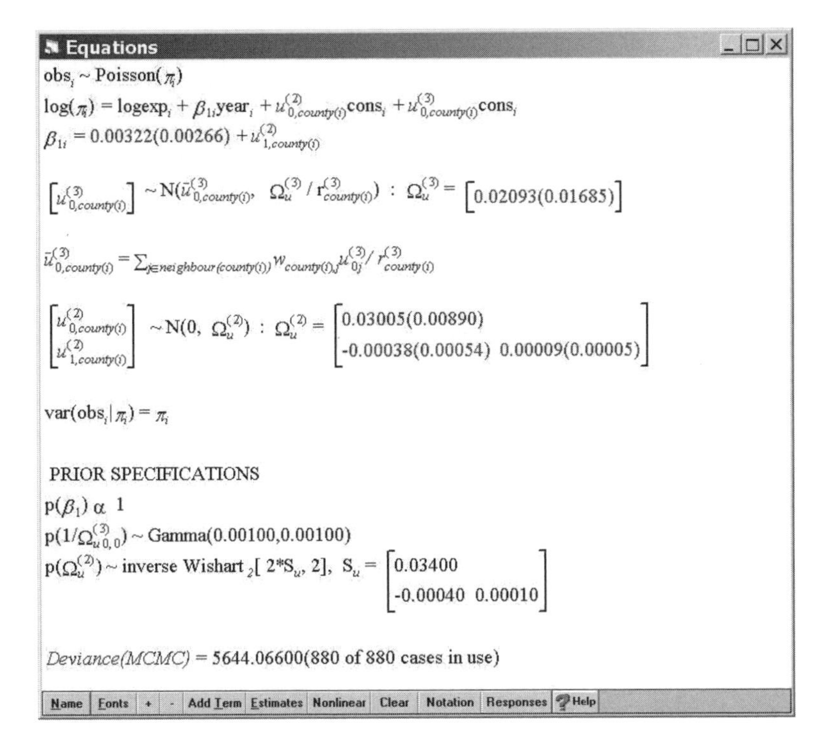

As the table above showed, the CAR model is a very slight improvement over the equivalent multiple-membership model and so our next step would be to fit a temporal–spatial interaction effect so that we have two sets of CAR distributed residuals. This requires the use of the WinBUGS package as we saw earlier in this chapter. We have in this section illustrated MLwiN's features (and limitations) for fitting disease mapping data with both spatial and temporal characteristics.

7

Focused Clustering: the Analysis of Putative Health Hazards

7.1 INTRODUCTION

The assessment of the impact of sources of pollution on the health status of communities is of considerable academic and public concern. The incidence of many respiratory, skin and genetic diseases is thought to be related to environmental pollution, and hence any localized source of such pollution could give rise to changes in the incidence of such diseases in the adjoining community.

In recent years, there has been growing interest in the development of statistical methods useful in the detection of patterns of health events associated with pollution sources. Here we concentrate primarily on the data analysis of observed point patterns of events rather than specific features of a particular disease or outcome.

A number of studies utilize data based on the spatial distribution of such diseases to assess the strength of association with exposure to a pollution source. Raised incidence near the source, or directional preference related to a dominant wind direction may provide evidence of such a link (see, for example, Lawson, 1993). Hence, the aim of the analysis of such data is usually to assess specific spatial variables rather than general spatial modelling. That is, the analyst is interested in detecting patterns of events near (or exposed to) the focus and less concerned about aggregation of events in other locations. The former type of analysis has been named *'focussed clustering'*. To date, most pollution source studies concentrate on incidence of a single disease (e.g. childhood leukaemia around nuclear power stations or respiratory cancers around waste product incinerators).

Disease Mapping with WinBUGS and MLwiN A. Lawson, W. Browne and C. Vidal Rodeiro
 ISBN: 0-470-85604-1 (HB)

The types of data observed can vary from disease event locations (usually residence addresses of cases) to counts of disease (mortality or morbidity) within census tracts or other arbitrary spatial regions. The two different data types lead to different modelling approaches. Spatial point process models are appropriate for event location data. In the case of count data, one may use properties of regionalized point processes and approximate inference is often made based on Poisson constant rate models. That is, conditional on knowledge of contributory factors, an independent Poisson model for regional counts is often assumed, resulting in the use of log-linear models. In the Bayesian context these models, at their simplest level, will have fixed effects (such as distance and directional repressors) but will have coefficients which have prior distributions.

The effects of pollution sources often are measured over large geographic areas containing heterogeneous population densities (usually both urban and rural areas). As a result, the underlying intensity of the point process model is heterogeneous and this population variation must be taken into consideration in any analysis.

7.2 STUDY DESIGN

In what follows, we consider a delimited geographical study area or window within which data concerning disease occurrence and exposure to the pollution source are collected. Issues concerning the strategic aims of the study must be considered prior to detailed consideration of the appropriate study region and data collection requirements.

7.2.1 Retrospective and prospective studies

During the 1980s, a number of studies of disease occurrence in geographical regions around putative sources of risk were carried out. Most of these were 'reactive', in that suspicion of a health risk, due to the past operation of a pollution source, instigated a review of the historical evidence for a link between disease incidence and exposure to the source. In essence, a *retrospective* study of disease occurrence was carried out. In some cases, continued monitoring of the source was also recommended or initiated. However, solely *prospective* studies of sources are seldom encountered. These two approaches and their respective strengths and weaknesses are well-known in the epidemiological literature.

Such studies of effects of pollution have a number of limitations, however. First, typically the emission characteristics of a source are not recorded for a suitable time period. Retrospective data on emissions may not be available and prospective monitoring data is expensive to collect over a long time period for a wide range of substances of interest. Often, no direct information is available on correlation between emission and disease occurrence. Furthermore, exposure and disease data are often collected by separate groups at different levels of resolution (even in prospective studies). Also, the nature of available data may be limited for particular diseases or health status indicators, or for particular time periods. Often, nationally-collected data rather than data from a specially designed study must be utilized. In some cases, the level of resolution in available data constrains the analysis considerably. For example, some diseases are reported only as counts from postal zones or census enumeration districts and not as exact addresses, due to confidentiality. In that case, methods based on analysis of counts rather than point events are appropriate. Inevitably, such regionalization leads to some loss of information. For example, very small clusters cannot be detected if they occur within a large census tract as the aggregate disease rate for the tract as a whole may not differ from the background disease rate. Only if the spatial pattern of events occurs at a larger scale than the measurement unit will it be detectable in regionalized data. Finally, for chronic outcomes like cancers, the temporal lag between exposure and an event of interest may be on the order of years or decades. Mobility of individuals over such a time period can confound exposure–outcome relationships and cause prohibitive costs in prospective studies over large areas.

7.2.2 Study region design

The design of a study region or window is of great practical importance. Usually, a study will concern the distribution of events (e.g. incident disease cases) within a fixed map area of given size and shape. The choice of size and shape can have considerable impact on study results and, while often it is not possible to choose the most appropriate region, some consideration should be given to these issues.

7.2.3 Replication and controls

Few studies examine replicated realizations of disease events around pollution sources. The main use of replication in such studies should be to provide estimates of variability not available from single realizations. An alternative use of replication is to study other areas where potential pollution sources exist but where no evidence has been demonstrated for adverse health links to the source or sources.

If substantial hypotheses concerning an individual source are to be examined, then control areas may be of some use. However, the use of replication to provide increased sample size by pooling, without examination of variability, only provides evidence for hypotheses concerning the sources in general, and not as individual sites. Local effects, which may be 'unusually' marked at an individual site, may be swamped in such a pooled sample.

In any study of disease incidence within a population, one must take some account of population structure. A standard epidemiological case-control design can be used where individuals are selected as controls and matched to cases with respect to confounding factors (e.g. age and occupation). Another standard approach in the conventional analysis of small area count data involves the use of strata-specific standardized rates to represent the 'background' population effect. The ratio of observed count to expected count, based on such a standardization, can be used as a crude estimate of region-specific relative risk.

An alternative approach is to utilize a disease or group of diseases which is thought to represent the 'at risk' population in the area but is usually unaffected by the type of pollution being considered. This approach is designed for point event data where a 'background' point event map of a 'control' disease is available. This method could also be used with count data, where counts of 'case' and 'control' diseases are available.

The goal is to find a 'control' disease which affects the same population with respect to possible confounding variables (e.g. age, occupation, smoking, etc.) yet is unrelated to the exposure of interest. While the existence of such a 'control disease' is subject to epidemiological debate, if such data are available, the statistical foundation of the methods is sound.

In many non-geographical studies in epidemiology, it is common to assign *individual* controls to cases, i.e. each case has an individual control who is matched to the case on a selection of variables such as age, gender or exposure history. Such matched case-control studies can be implemented within a geographical setting, and details of the statistical issues relating to these studies and putative source examples are available.

7.3 PROBLEMS OF INFERENCE

The primary inferential problems arising in putative source studies are (a) post hoc analyses, and (b) multiple comparisons.

The well-known problem of post hoc analysis arises when prior knowledge of reported disease incidence near a putative source leads an investigator to carry out statistical tests or fit models to data to 'confirm' the evidence. Essentially, this problem concerns bias in data collection and prior knowledge of an apparent effect. Both modelling and study region definition can be biased by this problem. However, if a study *region* is noted a priori to be of interest because it includes a pollution source, one does not suffer from post hoc analysis problems if the internal spatial structure of disease incidence did not influence the choice of region.

Although much recent work examines the statistical methodology appropriate for analysis for single disease types, there is little consideration of how to accommodate multiple 'health markers' in the investigation of putative sources.

The multiple comparison problem has been addressed in several ways. Bonferonni's inequality may be used to adjust critical regions for multiple comparisons but the conservative nature of such an adjustment is well-known. Multiple comparison problems have also been addressed by the use of cumulative p-value plotting to assess the number of diseases yielding evidence of association with a particular source.

7.3.1 Exploratory techniques

The use of exploratory techniques is widespread in conventional statistical analysis. However, in putative source analysis one must exercise care about how subsequent model design is influenced by exploratory or diagnostic findings. For example, if exploratory analysis isolates a cluster of events located near a pollution source, then this knowledge could lead to a post hoc analysis problem: i.e. inference based on a model specifically including such a cluster is suspect. As long as an analyst predefines the sources of interest and does not include a source simply because of its proximity to a cluster detected during the exploratory phase, many post hoc inference problems may be avoided.

In the case of count data, a variety of exploratory methods exist. One can use representation of counts as surfaces and incorporate expected count standardization (e.g. through a standardized mortality/morbidity ratio (SMR)).

While mapping regional SMRs can help isolate excess incidence, estimates of SMRs from counts in small areas are notoriously variable, especially for areas with few persons at risk. Various methods have been proposed to stabilize these small area estimates (see Section 1.2). Two different approaches are nonparametric smoothing and empirical or full Bayes estimation.

Various implementations of geostatistical prediction (kriging) to obtain a risk surface, have been proposed, although some key assumptions implicit in the methodology may not hold for disease data. Two disadvantages of standard kriging estimators is that they can produce *negative* interpolant values, which are invalid for relative risk surfaces, and that they assume a constant variance in the spatial field. Many alternative forms of smoothing could be used. In Section 6.2 a simple example of the use of Bayesian kriging is given.

Many researchers have proposed empirical Bayes estimates of regional rates. The methods are similar to those used for small area analysis. The empirical Bayes methods stabilize estimates of SMRs in small areas by adding parameters with spatially-correlated prior distributions, or adding uncorrelated random effects to models of disease counts. Application of empirical Bayes methods including approximations to likelihoods have been made in the context of putative source analyses.

Markov chain Monte Carlo (MCMC) algorithms such as the Gibbs sampler allow a fully Bayesian approach. While Bayesian implementations can involve complicated parametric models of disease rates, one could use simple models incorporating only regional heterogeneity and spatial autocorrelation for exploratory purposes.

Results of the various exploratory techniques provide a starting point for model fitting and assessment. As event data around pollution sources are typically available as either point locations or as regional counts, we address modelling issues for the two types of data.

7.4 MODELLING THE HAZARD EXPOSURE RISK

Before considering the detailed modelling of different types of data, it is appropriate to consider the types of evidence, and hence model ingredients, important in the specification of models of risk around putative sources of health hazard. These model components can be included under any data type.

Usually a fundamental issue in the conceptualization of these issues is the assumption that risk at a location or within a tract is related to risk variables measured at the location or interpolated to the location or to represent the tract. This assumption has continued to be made in studies of particular putative sources, and this leads to the formulation of putative source problems as ecological regression studies. That is, the hazard measurements are regarded as explanatory variables, and the analysis proceeds by the assessment of the relation between these variables and the disease incidence. The particular feature of this ecological approach is that only a restricted set of explanatory variables is usually examined, i.e. those variables having a well-defined association with health risk. For example, in a prospective study of respiratory disease morbidity around a waste-product incinerator it may be useful to monitor air pollution at a network of sites around the incinerator. The relation between disease incidence and air pollution could then be examined, e.g. by interpolation of air pollution to case locations or averaging of pollution over tracts. Alternatively, if a retrospective study is to be carried out then some surrogate pollution measures may be required (as direct measurement may not have been made). Surrogate measures commonly used in this connection are: distance from source, direction around source, and functions of these measures. It is also appropriate to employ ecological variables to help to estimate the background 'at risk' population. However, these variables are not usually regarded as surrogate for pollution measurements.

Exposure modelling here concerns the specification of variables and functions of variables which provide evidence for a link with pollution source or sources. Different potential sources of pollution or health risk can give rise to different forms of exposure evidence. For example, waste dump sites or nuclear power stations, may by the nature of the potential pollution risk display only a distance-related effect to cases of disease. That is, only distance of cases from the source (or some function of distance) would be appropriate. This may also be true for electromagnetic fields which may be thought to act without any directional preference. In the case of sources which emit effluent into air or water bodies then dispersion effects related to the movement of the host body take effect. In the case of air pollution, this means that wind direction and speed must play a role in the modelling of exposure.

In prospective studies, direct measurements of pollution can be made, and so there is less need to consider surrogates and their modelling. However, even in prospective studies, the lack of complete observation of the pollution process and the uncertainty of the aetiology of the disease in the particular example under study could lead to the consideration of exposure modelling to augment the information already available. In retrospective studies, surrogate measures are often the only available evidence and it is then essential to specify the exposure evidence which is to be considered with a model.

Here, we consider three basic forms of exposure evidence: distance-based, direction-based, and distance–direction interaction. While distance-only effects may be appropriate for waste dump sites, electromagnetic (EM) fields or nuclear installations, the inclusion of directional effects and also distance–direction interaction is important for any sites which could have an air pollution risk associated with them.

The distance relations described in Figure 7.1 can be regarded as the models for the possible distance–risk relation when a spatially-homogeneous background is present. The patterns are all possible types to be expected around, for example, an air pollution source. Monotonic distance relations are by no means the only patterns possible, and indeed, the results of empirical studies and theoretical studies of dispersal around sources support the possibility of peak-then-decline behaviour with increasing distance. The assumption often made, that monotonic decline should be assessed (alone), is therefore potentially quite misleading.

Directional effects are also likely when a wind regime applies, e.g. with air pollution related to incinerator outfall. Time-averaged wind effects could lead to peaks of concentration in certain directions (possibly downwind of the dominant direction). Peaks downwind of the subdominant direction may also be

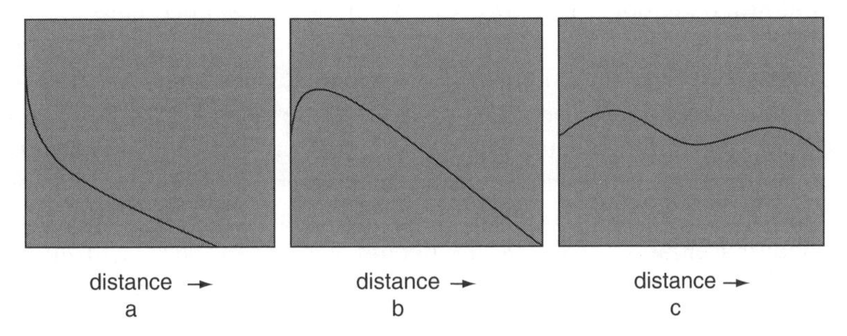

Figure 7.1 Possible exposure patterns over distance from a source: (a) monotone, (b) peaked, (c) clustered.

possible. These types of patterns are typically predicted from dispersal models from outfall sources.

One major issue relating to the choice of a small set of explanatory pollution surrogates, is that after fitting such variables, much unexplained residual structure may remain in the data. This residual structure is likely to be related to the fact that only a small number of effects are being fitted and no attempt to fully describe the spatial pattern is being made. Hence if no further attempt is made to provide a description based on known explanatory variables (for example, trend surface components), then there is likely to exist considerable residual effects. These effects could be modelled as unobserved heterogeneity via random effect modelling, and some consideration should be given to this approach in such studies. However, not all residual *nuisance* structure will be removed by such modelling if long-range effects (trends) are present in the observations. Hence, it may be necessary to model a variety of spatial range effects (both long- and short-range) as well as pollution surrogates, if the underlying nuisance structure is to be properly isolated. Of course, if considerable nuisance structure remains, then the two main results of this would be to (a) *lower the power of hypothesis tests* employed to assess the role of pollution surrogates and (b) *increase the variance of parameter estimates* associated with these surrogates. In addition, any pointwise residual analysis carried out will be marred by the presence of nuisance effects confounded with any pure error present.

7.4.1 The specification of the exposure model: $f_1(r, \theta)$

Define the location of the source as $\mathbf{x}_0$. Usually the spatial relation between the source and disease events is based on the polar coordinates of events from the source: $\{r, \theta\}$ where $r = \|\mathbf{x} - \mathbf{x}_0\|$, $\mathbf{x}$ is the centroid of the small area and θ is the angle measured to the source. The function $f_1(r, \theta)$ describes our beliefs about the linkage between the risk in small areas and the source. It is important to consider how $f_1(r, \theta)$ can be used in models describing pollution effects on surrounding populations. In many studies, only the distance measure (r) has been used as evidence for association between a source and surrounding populations. However, it is dangerous to pursue distance-only analyses when considerable directional effects are present. The reason for this is based on elementary exposure modelling ideas: directional preference or anisotropy can lead to marked differences in exposure in different directions and hence to different distance–exposure profiles. Hence the collapsing of exposure over the directional marginal of the distribution could lead to considerable misinterpretation, and in the extreme to *Simpson's* paradox. In the extreme case, a strong distance relationship with a source may be masked by the collapsing over directions, and this can lead to erroneous conclusions. Further, if the analysis of a large number of putative source sites is carried out by pooling between sites ignoring local directional effects at each site, then these studies should also be regarded with caution.

The importance of the examination of a *range* of possible indicators of association between sources and health risk in their vicinity is clear. The first criterion for association is usually assumed to be evidence of a decline in disease incidence with increased distance from the source. Without this distance-decline effect, there is likely to be only weak support for an association. However, this does not imply that this effect should be examined in isolation. As noted above, other effects can provide evidence for association, or could be nuisance effects which should be taken into consideration so that correct inferences be made. In the former category are directional and directional–distance correlation effects, which can be marked with particular wind regimes. In the latter category are peaked incidence effects, which relate to *increases* of incidence with distance from the source. While a peak at some distance from a source can occur, it is also possible for this to be combined with an overall underlying decline in incidence, and hence is of importance in any modelling approach. This peaked effect is a nuisance effect, in terms of association, but it is clearly important to include such effects. If they were not included then inference may be erroneously made that no distance-decline is present, when in fact a combination of distance-decline and peaked incidence is found. Further nuisance effects which may be of concern are, for example, random effects related to individual *frailty*, where individual variation of susceptibility is directly modelled or where general heterogeneity is admitted.

Often $f_1(r, \theta)$ will consist of functions of a selection of the variables:

$$\{r,\ \log(r),\ \cos(\theta),\ \sin(\theta),\ r\cos(\theta),\ r\sin(\theta),\ \log(r)\cos(\theta),\ \log(r)\sin(\theta)\}.$$

The first four variables represent distance-decline, peakedness, and directional effects, while the latter variables are directional–distance correlation effects. The directional components can be fitted separately and transformations of parameters can be made to yield corresponding directional concentration and mean angle. Figure 7.1 displays different distance-related exposure models which could be used to specify $f_1(r, \theta)$. Note that in Figure 7.1, nuisance effects of peakedness and heterogeneity appear in (b) and (c).

Further examination of dispersal models for air pollution, suggests that the spatial distribution of outfall around a source is likely to follow a convolution of Gaussian distributions where in any particular direction there could be a separate mean level and lateral variance of concentration (dependent on r). As a parsimonious representation of these effects it is possible to use a subset of the exposure variables listed above to describe this behaviour.

Some simple models which can be proposed lead to the specification of $f_1(r, \theta)$ as follows:

(1) distance-decline: $f_1(r, \theta) = 1 + \exp\{-\alpha_1 r\}$;
(2) *peaked* distance-decline: $f_1(r, \theta) = 1 + \exp\{\alpha_1 \log r - \alpha_2 r\} \quad (\alpha_1, \alpha_2 > 0)$;
(3) *peaked* distance-decline with angular concentration:
$f_1(r, \theta) = 1 + \exp\{\alpha_1 \log r - \alpha_2 r + \alpha_3 \cos(\theta) + \alpha_4 \sin(\theta)\}$;
(4) same as (3) except linear angular correlation is added:
$f_1(r, \theta) = 1 + \exp\{\alpha_1 \log r - \alpha_2 r + \alpha_3 \cos(\theta) + \alpha_4 \sin(\theta) + \alpha_5 r \cos(\theta) + \alpha_6 r \sin(\theta)\}$;
(5) *peaked* decline which varies with angle:
$f_1(r, \theta) = 1 + \exp\{\delta \log r - \alpha_2 r\}$, where $\delta = \alpha_1 + \delta_1 * \cos(\theta - \mu)$, and μ is the mean angle.

The specifications above appear to be flexible enough to model a variety of possible outfall patterns. Model (1) has been often applied, while variants with an inner area of constant risk have also been proposed. The use of these simple decline models alone does not appear to be supported by any realistic exposure model for air pollution. In particular, an inner concentric zone of constant risk appears to have little epidemiological foundation a priori.

While the models listed above are not the only possible specifications which can describe potential radial and angular variation in risk, they do provide parsimonious descriptions of the qualitative features of exposure zones around pollution sources. Models where it is assumed that a zone of constant risk occurs around the source for a fixed distance, and then the effect declines, do not seem to have any a priori epidemiological justification and are artifice. Traditional epidemiological studies often used to compare risk between two different zones and test if one was higher than another. However, this was a crude testing-based method which has long been superseded by more sophisticated and sensible models.

While nonparametric models for the risk variation around a source are attractive in that they do not force the use of a parameterized model, they have the disadvantage that they have no parameters which can be identified with evidence for a link with the source. Parametric models usually employ regression coefficients for distance-decline or directional components, and these do provide a means of assessing the strength of the link.

7.5 MODELS FOR COUNT DATA

As mentioned above, outcome data may be available only as counts for small census regions rather than precise event locations, for a variety of reasons. As a result, a considerable literature has developed concerning the analysis of such data.

Let y_i: $i = 1, \ldots, m$ denote the count of disease (or other outcome) events within m arbitrary regions or tracts. Define n_i as the total population size of the

ith region. We assume the study window includes all m region centres. Other sampling rules may lead to size biases in selection of regions. For example, the inclusion rule: 'all regions intersecting the window' (plus sampling) leads to a bias towards larger regions.

The usual model adopted for the region counts assumes $\{y_i\}$ to be independent Poisson random variables with parameters $\{\lambda_i\}$. Many studies of count data assume that λ_i is constant within region a_i so spatial variation between regions follows a step function. When λ_i is parameterized as a log-linear function, one often treats explanatory variables (in particular exposure or radial distance from a pollution source) as constants for the subregions or as occurring at region centres only. While such log-linear models can be useful in describing the global characteristics of a pattern, the differences between a_is and any continuous variation in the intensity between and within regions is largely ignored. Second, the underlying process of events may not be a heterogeneous Poisson process, in which case the independence assumption may not hold or the Poisson distributional assumption may be violated. Analysis based on regional counts is ecological in nature and inference can suffer from the well-known 'ecological fallacy' of attributing effects observed in aggregate to individuals.

While the above factors should be taken into consideration, the independent Poisson model is a useful starting point from which to examine effects of pollution sources. One often uses a log-linear model with a modulating function e_i, say, which acts as the link of the population of subregion i to the expected deaths in subregion $i, i = 1, \ldots, m$. Usually the expected count is modelled as

$$E(y_i) = \lambda_i = e_i\theta_i^* = e_i g(\exp(F\alpha)).$$

Here, the $e_i, i = 1, \ldots, m$ act as a background rate for the m subregions and we now use θ_i^* for the relative risk to distinguish it from the angular measure θ_i. The function $g(\cdot)$ represents a link to spatial and other covariates in the $m \times q$ design matrix F. The parameter vector α has dimension $q \times 1$. Define the polar coordinates of the subregion centre as (r_i, θ_i), relative to the pollution source. Often, the only variable to be included in F is r, the radial distance from the source. When this is used alone, an additive link such as $g(\cdot) = 1 + \exp(F\alpha)$, is appropriate since (for radial distance decline) the background rate (e_i) is unaltered at great distances. However, directional variables (e.g. $\cos\theta$, $\sin\theta, r\cos\theta, \log(r)\cos\theta$, etc.) representing preferred direction and angular–linear correlation can also be useful in detecting directional preference resulting from preferred directions of pollution outfall.

One may extend this model to include unobserved heterogeneity between regions by introducing a prior distribution for the log relative risks $(\log\theta_i^*, i = 1, \ldots, m)$. This could be defined to include spatially uncorrelated or correlated heterogeneity. Bayesian models available on WinBUGS can include such terms.

7.5.1 Estimation

One may estimate the parameters of a log-linear model via maximum likelihood through standard generalized linear modelling packages, such as S-Plus or R. Usually, one treats the log of the known background hazard for each subregion, $e_i, i = 1, \ldots, m$, as an 'offset'. A multiplicative (log) link can be directly modelled in this way, while an additive link can be programmed via user-defined macros.

Log-linear models are appropriate if due care is taken to examine whether model assumptions are met. For example, it has been suggested that the violation of asymptotic sampling distributions be avoided by the use of Monte Carlo tests for change of deviance. If a model fits well then the standardized model residuals should be approximately independently and identically distributed (i.i.d.). One may use autocorrelation tests, again via Monte Carlo, and make any required model adjustments. If such residuals are not available directly, then it is always possible to compare crude model residuals to a simulation envelope of m sets of residuals generated from the fitted model.

7.6 BAYESIAN MODELS

As in the case of disease map reconstruction, the likelihood models discussed above can be extended to include prior distributions for parameters and also random effects which can account for heterogeneity in the data. Both correlated and uncorrelated heterogeneity could exist in focused clustering studies. Simple log-linear models ignore these factors.

A number of studies have proposed Bayesian methods in this area. Lawson (1994) first suggested employing a spatial correlation prior distribution in an empirical Bayes analysis of a focused clustering problem, while Lawson *et al.* (1996) described the use of MCMC methods in a full Bayes analysis of such problems. Recently, Wakefield and Morris (2001) have presented another Bayesian analysis. Usually the form of model which is employed in this analysis, in its fullest form, has components representing covariate terms, random effects and a parameterized function relating the counts in small areas to the source location either via distance or directional components.

An example of the full model, with polar coordinates (r_i, θ_i), could be:

$$E(y_i) = e_i.f_1(r_i, \theta_i). \exp\{F_c\alpha_c + u_i + v_i\},$$
$$f_1(r_i, \theta_i) = [1 + \exp(F_r\alpha_r)]. \exp\{F_\theta\alpha_\theta\},$$

where F_r is a set of functions of r_i, F_θ is a set of functions of θ_i, and F_c is a set of covariates.

The function $f_1(.,.)$ is intended to be a function of distance and directional components but these could include directional–distance interactions. Clearly directional components are fundamental if an air pollution source is suspected (see, for example, Esman and Marsh, 1996). In other cases, directional components may yield evidence for anisotropic patterns in the risk around a source and therefore could be important in focusing further investigations. Both empirical and theoretical studies of air pollution emission, display directional anisotropies (Panopsky and Dutton, 1984). The ignoring of directional components, which has been a feature of some studies, even recently, for example, by Wakefield and Morris (2001), is quite peculiar.

7.7 FOCUSED CLUSTERING IN WinBUGS

In this section we will first highlight the fitting of a variety of models, of increasing complexity, to a simple (*toy*) dataset: respiratory cancer for a fixed time period (1978–1983) in 26 census tracts in Falkirk, central Scotland. A putative source of health hazard is located within the map area. This example is for illustration of the methods and a second dataset is used later to provide a fuller example. That dataset is the Ohio respiratory cancer dataset used in Section 6.1.6 with the Fernald facility as the putative source location. The Fernald facility lies in Hamilton county in the south-west corner of Ohio. The facility recycles depleted uranium for the US Department of Energy and Department of Defense. Concern over respiratory cancer incidence in the surrounding area focuses on potential links between such cancer and uranium inhalation exposure.

7.7.1 Respiratory cancer in Falkirk

Figure 7.2 displays the census geographies for this example, including the putative source location. The putative source is a foundry which may have been an air pollution hazard before the time period of the study. Due to the latency effect it is conceivable that operation of the foundry in the early 1970s could have an impact on the respiratory cancer experience of those living in areas adjacent to the source.

The counts are assumed to be Poisson distributed, i.e. $y_i \sim Poisson(e_i\theta_i^*)$ and $\log(\theta_i^*)$ is modelled with different components. Within each region the following variables are available: deprivation score (d_{ei}) (Carstairs, 1981), expected rate (e_i), distance from the source (r_i), angle in radians around the source (θ_i). Functions of (r_i, θ_i) are also available: $\log(r_i)$, $\cos(\theta_i)$, $\sin(\theta_i)$. Random effects are denoted by $\{v_i\}$ for uncorrelated heterogeneity (UH) and $\{u_i\}$ for correlated heterogeneity (CH). The following models are only a selection of possible models and have been chosen mainly for demonstration purposes.

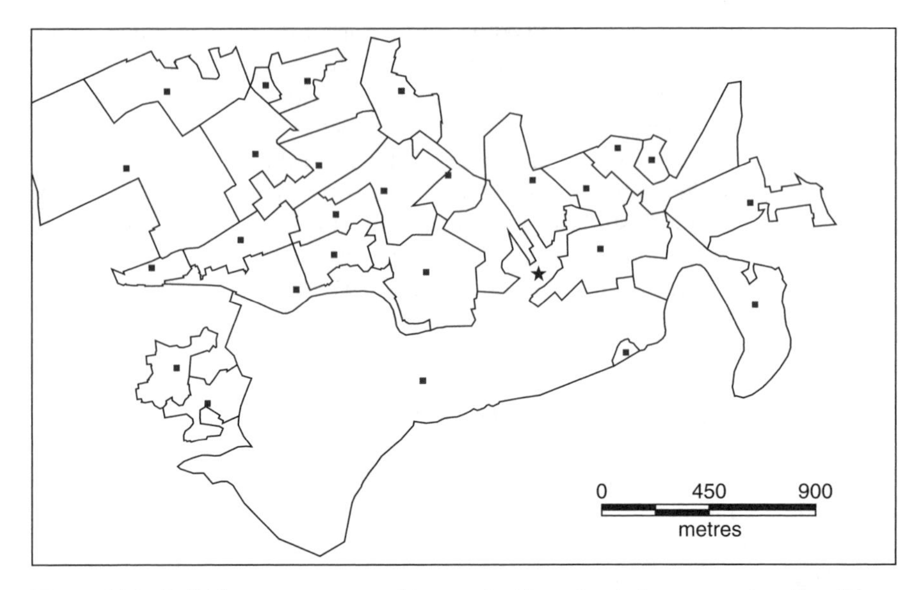

Figure 7.2 Falkirk: census geographies, centroids and putative source location (✱).

Initially, in the following, we display the model code, summary results and densities for each parameter and the final posterior expected relative risk map. In later models we only display the details most relevant to the example. If only small changes to the model formulae are made then we do not include the code, for the sake of brevity. The model code for all models in this section appears in Appendix 1.

7.7.1.1 *Model 1.* $\log(\lambda_i) = \log(e_i) + \alpha_0$.

Here α_0 is the intercept (overall rate) and has a flat prior distribution. The code for this model is listed in Figure 7.3.

```
model{for(i in 1 : 26){
        num[i]      ~ dpois(mu[i])
        log(mu[i])<- log(e[i]) + alpha0
        RR[i]  <- mu[i] / e[i]
    }
    OR<-exp(alpha0)
    alpha0 ~ dflat()

  }
  list(alpha0=0.01)
```

Figure 7.3 WinBUGS code for model 1.

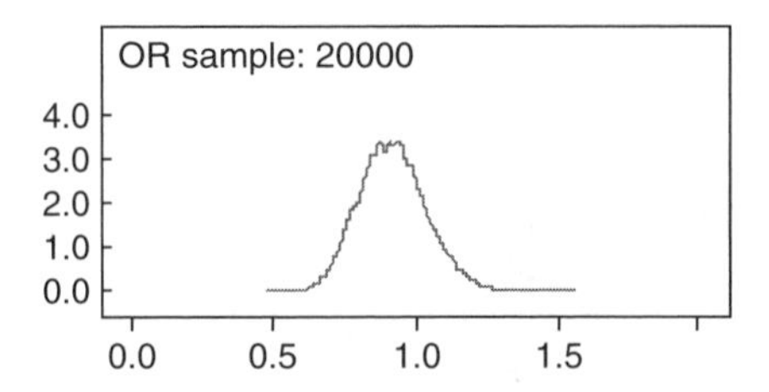

Figure 7.4 Posterior marginal density of the converged sample of the overall rate (OR).

Table 7.1 Sample summary statistics for OR.

node	mean	sd	MC error	2.5%	97.5%
OR	0.9137	0.1175	7.902E−4	0.9091	1.159

The estimate of the relative risk (OR in the code) in each area is 0.9137 which is, within a Monte Carlo approximation, the overall rate for the study region $(\sum y_i / \sum e_i)$. Figure 7.4 displays the posterior marginal sample distribution for OR, while Table 7.1 displays the summary statistics.

7.7.1.2 *Model 2*. $\log(\lambda_i) = \log(e_i) + \alpha_0 + \alpha_1.d_{ei}$

This model includes an intercept and deprivation (in code: dep[ ]) as a covariate (see Figure 7.5). Parameters α_0 and α_1 have prior distributions. Figure 7.6 displays the resulting relative risk posterior expected surface.

7.7.1.3 *Model 3*. $\log(\lambda_i) = \log(e_i) + \alpha_0 + \alpha_1.d_{ei} + \log(1 + \exp(-\alpha_2.r_i))$

This model is the same as model 2 but with an additive-link distance effect included. The distance variable (r_i) is dist[ ] in the code (see Figure 7.7). Note that

```
model
  {
    for (i in 1 : 26){
      num[i]      ~ dpois(mu[i])
      log(mu[i])<- log(e[i]) + alpha0+ alpha1* dep[i]
      RR[i] <- mu[i] / e[i]
    }
    alpha1 ~ dnorm(0.0,1.0E-5)
    alpha0 ~ dflat()
  }
  list(alpha0=0.01alpha1=0.01)
```

Figure 7.5 WinBUGS code for model 2.

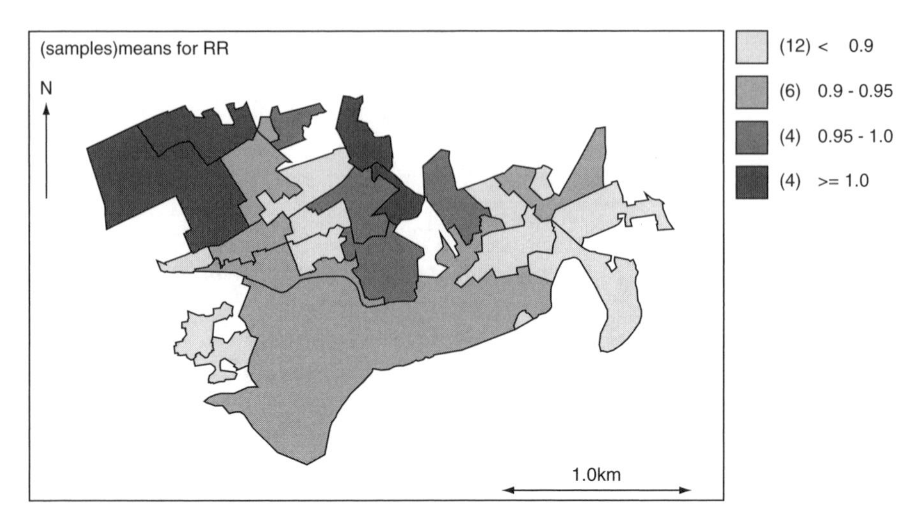

Figure 7.6 Falkirk: posterior expected relative risk map for model 2.

```
model{
        for (i in 1 : 26) {
           num[i]        ~ dpois(mu[i])
    f[i]<-1+exp(-alpha2*dist[i])
           log(mu[i])<- log(e[i])+alpha0+ alpha1* dep[i]+log(f[i]
           RR[i] <- mu[i] / e[i]
        }

    alpha2~ dnorm(0.0,1.0)
       alpha1 ~ dnorm(0.0,1.0E-5)
       alpha0 ~ dflat()
    }
    list(alpha0=0.01alpha1=0.01,alpha2=0.01)
```

Figure 7.7 WinBUGS code for model 3.

if the estimate of α_2 is positive then there is a decline with distance, which might be interpreted as significant if the α_2 is well estimated. Note that for the additive $f_1(\cdot)$ form used here, the prior for α_2 must be constrained as numerical instability can occur if a diffuse prior distribution is used.

The model fit suggests that the distance parameter is positive in this case. Both the 2.5% and 97.5% quantiles of the final sample are positive although the variability is quite high. This suggest weak evidence of a distance decline. The relative risk map still shows some structure with areas in the south showing risks above 1.0. However, the variance of these risks is relatively small. Figure7.8 displays the resulting posterior expected relative risk surface, while Table 7.2 displays the summary statistics.

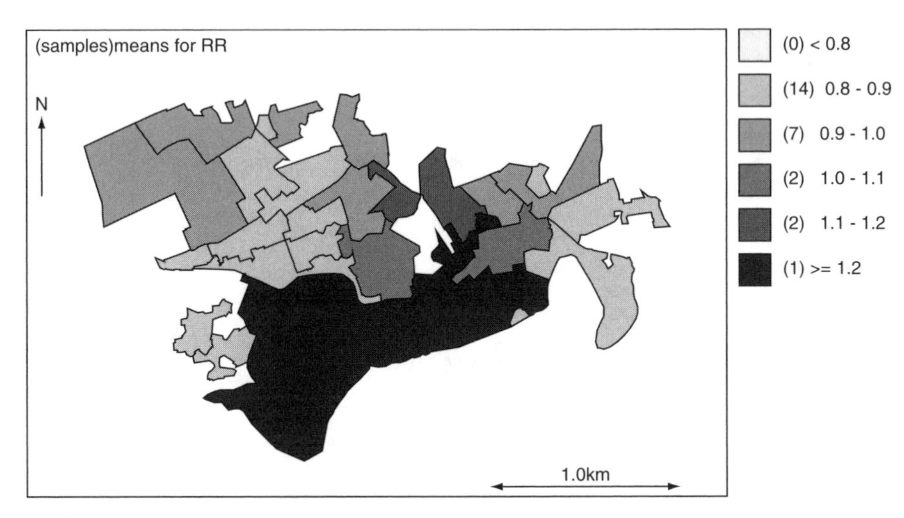

Figure 7.8 Falkirk: posterior expected relative risk map for model 3.

Table 7.2 Summary statistics for parameters in model 3.

node	mean	sd	MC Error	2.5%	97.5%
alpha0	−0.2078	0.1874	0.006818	− 0.6793	0.0921
alpha1	−0.0245	0.0667	5.106E−4	− 0.1536	0.1065
alpha2	0.632	0.4953	0.01514	0.0303	1.937

7.7.1.4 *Model 3b.* $\log(\lambda_i) = \log(e_i) + \alpha_0 + \alpha_1.d_{ei} + \log(1 + \alpha_8 \exp(-(r_i/\alpha_2^*)^2))$ where $\alpha_2 = 1/\alpha_2^*$

This is a variant of model 3 but with the overparameterized Gaussian-type distance-risk model of, among others, Wakefield and Morris (2001). The code is given in Figure 7.9, while Table 7.3 displays the summary statistics.

The qualitative features of this model are similar to the simple distance model (model 3). However, the distance parameter is now multimodal and is less well estimated (see Figure 7.10). Note that we have not included the extra link parameter (α_8) found in Wakefield and Morris (2001), as this is not parsimonious. In addition, the inclusion of the extra parameter causes instabilities in the numerical computation. Figure 7.11 displays the posterior expected relative risk surface.

7.7.1.5 *Model 4.* $\log(\lambda_i) = \log(e_i) + \alpha_0 + \alpha_1.d_{ei} + \log(1 + \exp(\alpha_3 \log(r_i) - \alpha_2.r_i))$

This model is the same as model 3 but with a peaked intensity effect. This is a gamma-like function which allows for a supremum of risk at some distance

```
model{
    for (i in 1 : 26) {
      num[i]       ~ dpois(mu[i])
  ab[i]<-alpha2*dist[i]
  f[i]<-1+exp(-ab[i]*ab[i])
      log(mu[i])<- log(e[i])+alpha0+ alpha1* dep[i]+log(f[i]
      RR[i] <- mu[i]/ e[i]
    }

  alpha2~ dnorm(0.0,1.0)
    alpha1 ~ dnorm(0.0,1.0E-5)
    alpha0 ~ dflat()
  }
  list(alpha0=0.01alpha1=0.01,alpha2=0.01)
```

Figure 7.9 WinBUGS code for model 3b.

Table 7.3 Summary statistics for parameters in model 3b.

node	mean	sd	MC error	2.5%	97.5%
alpha0	−0.2026	0.1945	0.00806	− 0.7285	0.0983
alpha1	−0.0229	0.0669	5.35E−4	−0.1532	0.1088
alpha2	−0.0035	0.7411	0.01372	−1.5880	1.5630

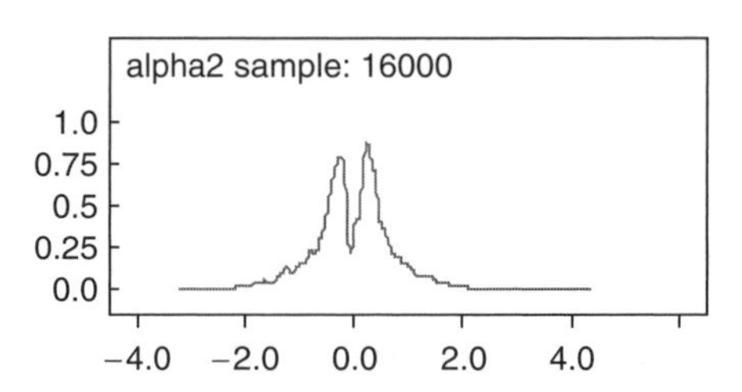

Figure 7.10 Posterior marginal sample distribution of α_2.

from the source as long as both α_2 and α_3 are both positive. If α_3 is negative then there is a simple distance-decline. The code for this model appears in Figure 7.12, while Table 7.4 displays the summary statistics.

Here the extra peaked effect term has a large variance and is negative, which suggests that only a monotonic effect is present.

7.7.1.6 *Model 5.* $\log(\lambda_i) = \log(e_i) + \alpha_0 + \alpha_1.d_{ei} + \alpha_4 \cos(\theta_i) + \alpha_5 \sin(\theta_i)$

This model is the same as model 2 but with a simple directional effect and no distance effect. The angle (θ_i) is represented by ang[ ] in the code (Figure 7.13). Table 7.5 displays the summary statistics.

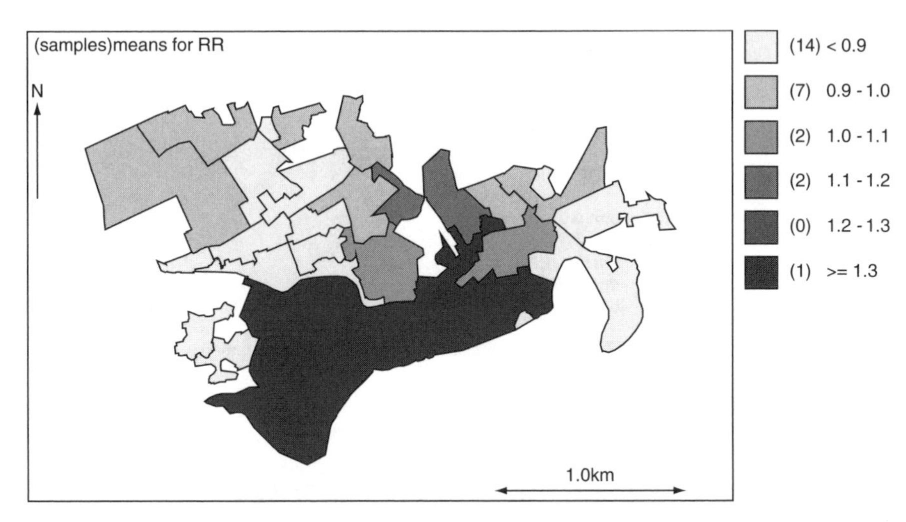

Figure 7.11 Falkirk: posterior expected relative risk map for model 3b.

```
model{
      for (i in 1 : 26) {
        num[i]      ~ dpois(mu[i])
   f[i]<-1+exp(alpha3*log(dist[i])-alpha2*dist[i])
        log(mu[i])<- log(e[i])+alpha0+ alpha1* dep[i]+log(f[i]
        RR[i] <- mu[i]/ e[i]
      }
  alpha3~ dnorm(0.0,0.5)
    alpha2~ dnorm(0.0,1.0)
      alpha1 ~ dnorm(0.0,1.0E-5)
      alpha0 ~ dflat()
    }
    list(alpha0=0.01alpha1=0.01,alpha2=0.01,alpha3=0.01
```

Figure 7.12 WinBUGS code for model 4.

Table 7.4 Summary statistics for parameters in model 4.

node	mean	sd	MC error	2.5%	97.5%
alpha0	−0.1993	0.1851	0.00487	−0.6253	0.0968
alpha1	−0.0239	0.0671	6.219E−4	−0.1523	0.1096
alpha2	0.6480	0.5550	0.01439	−0.2098	2.0220
alpha3	−0.1766	1.3610	0.03699	−2.9940	2.3920

This model considers the directional components of the risk variation. However, it appears that there is little evidence for directional effects (alone) as the variances are high for both components.

```
model
  {
    for (i in 1 : 26) {
      num[i] ~ dpois(mu[i])
  f[i]<-exp(alpha4*cos(ang[i])+alpha5*sin(ang[i]))
      log(mu[i])<- log(e[i])+alpha0+ alpha1* dep[i]+log(f[i])
      RR[i] <- mu[i] / e[i]
    }
  alpha5~ dnorm(0.0,1.0)
 alpha4~ dnorm(0.0,1.0)
  alpha2~ dnorm(0.0,1.0)
    alpha1 ~ dnorm(0.0,1.0E-5)
    alpha0 ~ dflat()
  }
  list(alpha0=0.01,alpha1=0.01,alpha2=0.01,alpha4=0.01,alpha5=0.
```

Figure 7.13 WinBUGS code for model 5.

Table 7.5 Summary statistics for parameters in model 5.

node	mean	sd	MC error	2.5%	97.5%
alpha4	0.1397	0.1857	0.004767	−0.2419	0.4969
alpha5	−0.3099	0.3000	0.006975	−0.9058	0.3038

7.7.1.7 Model 6. $\log(\lambda_i) = \log(e_i) + \alpha_0 + \alpha_1.d_{ei} + \log(1 + \exp(-\alpha_2.r_i)) + \alpha_4 \cos(\theta_i) + \alpha_5 \sin(\theta_i)$

This model is the same as model 2 but with simple distance and directional effects added. The code for this model appears in Figure 7.14, while Table 7.6 displays the summary statistics.

```
model{
      for (i in 1 : 26) {
        num[i] ~ dpois(mu[i])
a[i]<-(1+exp(-alpha2*dist[i]))
f[i]<-a[i]*exp(alpha4*cos(ang[i])+alpha5*sin(ang[i]))
log(mu[i])<- log(e[i])+alpha0+ alpha1* dep[i]+log(f[i]
        RR[i] <- mu[i] / e[i]
      }
    alpha5~ dnorm(0.0,1.0)
  alpha4~ dnorm(0.0,1.0)
    alpha2~ dnorm(0.0,1.0)
      alpha1 ~ dnorm(0.0,1.0E-5)
      alpha0 ~ dflat()
    }
list(alpha0=0.01,alpha1=0.01,
 alpha2=0.01,alpha4=0.01,alpha5=0.01)
```

Figure 7.14 WinBUGS code for model 6.

Table 7.6 Summary statistics for parameters in model 6.

node	mean	sd	MC error	2.5%	97.5%
alpha2	0.6334	0.5303	0.01711	0.0026	2.000
alpha4	0.0913	0.1863	0.00488	−0.2827	0.4446
alpha5	−0.2479	0.2870	0.00643	−0.8007	0.3264

In this model there is a distance effect and directional components. None of the effects have low estimated variability. The relative risk variation under this model is larger than under a simple distance model. The posterior expected relative risk surface is displayed in Figure 7.15.

7.7.1.8. *Model 7.* $\log(\lambda_i) = \log(e_i) + \alpha_0 + \alpha_1.d_{ei} + \log(1 + \exp(-\alpha_2.r_i)) + v_i$ where $\{v_i\} \sim N(0, d^*)$

This model is an example of using a random effect in the analysis. The node dstar is used in the code to represent d^*. In this case a simple UH effect with zero mean normal prior distribution is assumed, and a simple distance effect is fitted. Figures 7.16 to 7.19 show the posterior sample densities. Figure 7.20 displays the posterior expected relative risk surface, while Figure 7.21 displays the posterior expected value surface of the v_i. Table 7.7 displays the summary statistics.

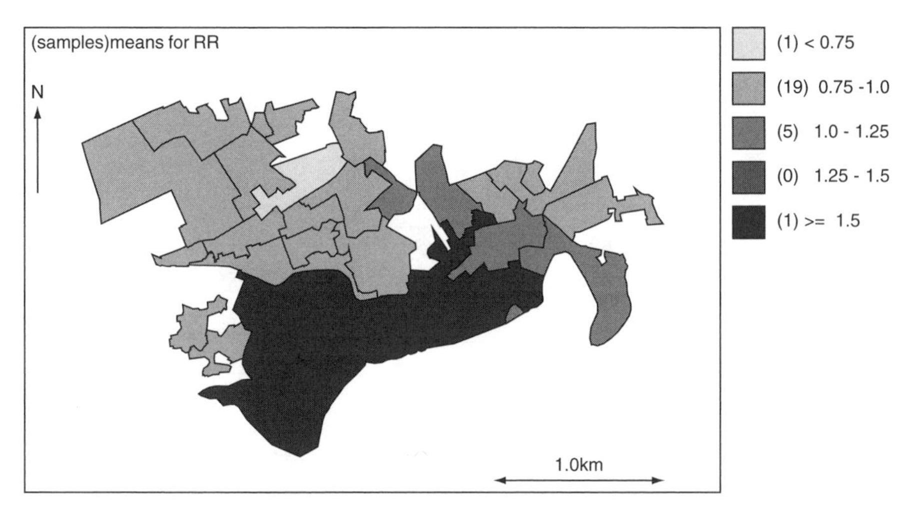

Figure 7.15 Falkirk: posterior expected relative risk map for model 6.

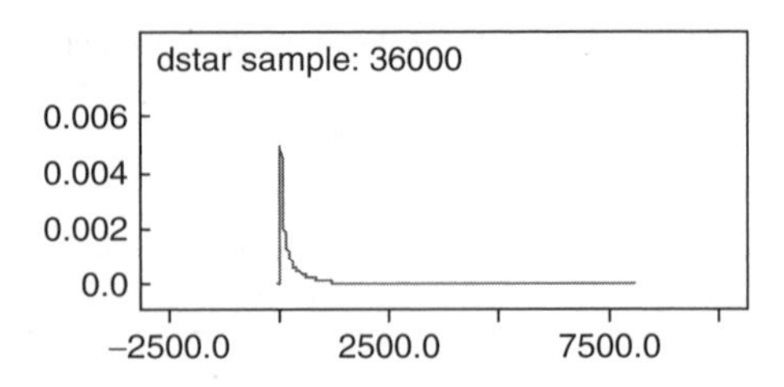

Figure 7.16 Posterior sample density of *dstar*.

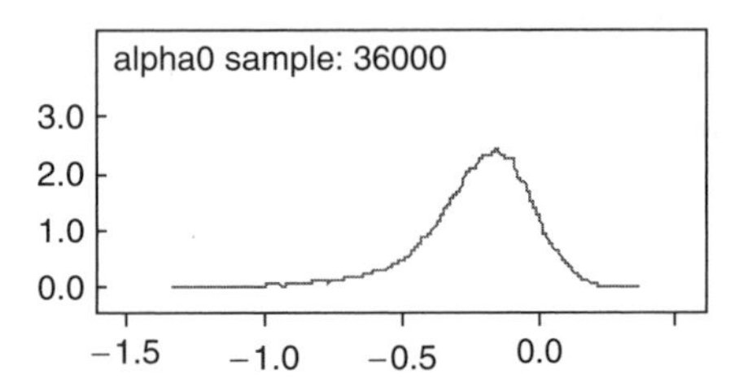

Figure 7.17 Posterior sample density of *alpha0*.

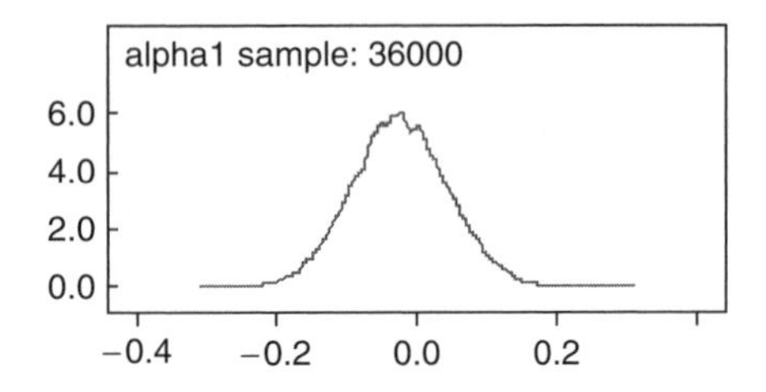

Figure 7.18 Posterior sample density of *alpha1*.

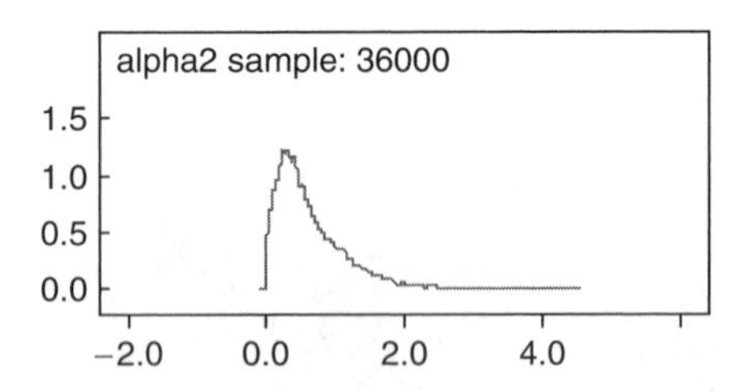

Figure 7.19 Posterior sample density of *alpha2*.

7.7.1.9 Model 8

The final model here is as for model 7 but with correlated heterogeneity included (see Figure 7.22). The posterior expected value of the distance component (α_2) is clearly positive (see Table 7.8). However, the variability of the value is such that it cannot be relied on. The interpretation of this result would be

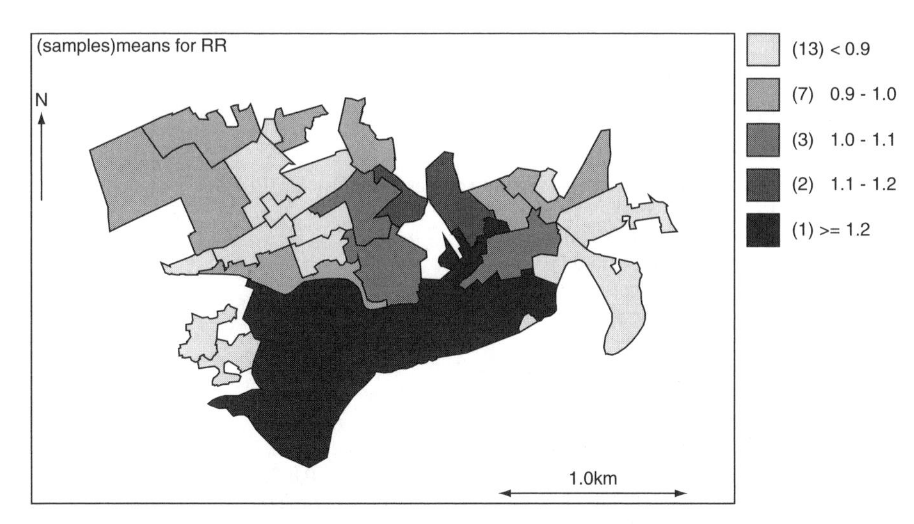

Figure 7.20 Falkirk: posterior expected relative risk map for model 7.

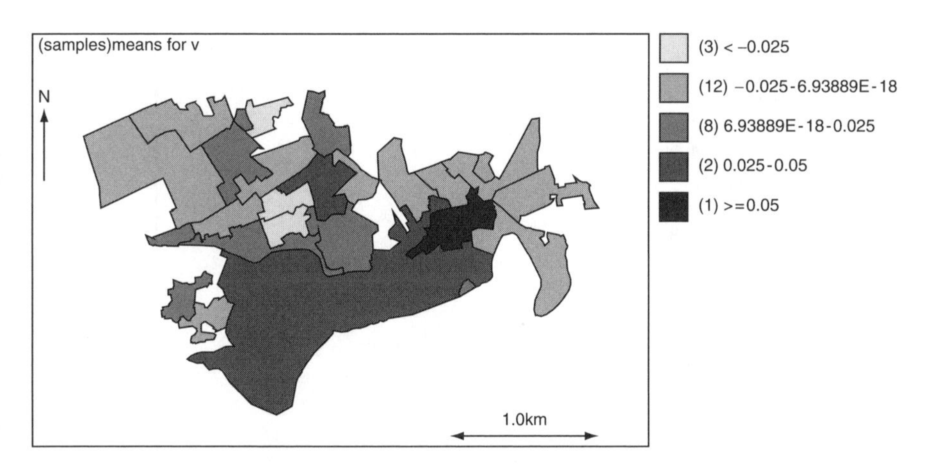

Figure 7.21 Falkirk: posterior expected map of the uncorrelated random effect $\{v_i\}$ for model 7.

Table 7.7 Summary statistics for parameters in model 7.

node	mean	sd	MC error	2.5%	97.5%
dstar	318.3	483.5	14.03	8.082	1670.0
alpha0	−0.2260	0.2022	0.00623	−0.7391	0.0941
alpha1	−0.0251	0.0683	3.90E−4	−0.1580	0.1109
alpha2	0.6253	0.5099	0.01058	0.0206	1.9490

```
   model      {
u[1 : 26] ~ car.normal(adj[],wei[],nn[],tau)
     for(i in 1 : 26) { v[i]~dnorm(0.0,dstar)
     num[i]~dpois(mu[i])
  f[i]<-(1+exp(-alpha2*dist[i]))
    log(mu[i])<- log(e[i])+alpha0 + alpha1 * dep[i]+log(f[i])+v[i]+u[i]
      RR[i]<- mu[i]/e[i]  }
   tau ~ dgamma(0.001,0.001)
    dstar ~dgamma(0.001,0.001)
     alpha2 ~dnorm(0.0,1.0)
   alpha1 ~dnorm(0.0, 1.0E-5)
    alpha0 ~dflat()
for(i in 1 : 104) {wei[i]<-1}
   }
#inits
  list(alpha0=0.01, alpha1=0.01, alpha2=0.01,dstar=0.001,tau=0.01,
     v=c(0,0,0,0,0,0,0,0,0,0,0,0,0,0,0,0,0,0,0,0,0,0,0,0,0,0),
  u=c(0,0,0,0,0,0,0,0,0,0,0,0,0,0,0,0,0,0,0,0,0,0,0,0,0,0))
#adjacency data
list(nn = c(1, 8, 3, 4, 5, 4, 3, 5, 1, 2,
1, 3, 4, 3, 3, 2, 6, 7, 6, 6,
4, 4, 5, 6, 5, 3),
adj = c(
2,
20, 19, 10, 6, 5, 4, 3, 1,
22, 4, 2,
8, 5, 3, 2,
8, 7, 6, 4, 2,
8, 7, 5, 2,
8, 6, 5,
9, 7, 6, 5, 4,
8,
11, 2,
10,
19, 18, 13,
18, 17, 14, 12,
17, 15, 13,
17, 16, 14,
17, 15,
25, 18, 16, 15, 14, 13,
25, 24, 21, 19, 17, 13, 12,
24, 21, 20, 18, 12, 2,
24, 23, 22, 21, 19, 2,
24, 20, 19, 18,
26, 23, 20, 3,
26, 25, 24, 22, 20,
25, 23, 21, 20, 19, 18,
26, 24, 23, 18, 17,
25, 23, 22
))
```

Figure 7.22 WinBUGS code for model 8.

Table 7.8 Summary statistics for parameters in model 8.

node	mean	sd	MC error	2.5%	97.5%
alpha2	0.6439	0.5170	0.01625	0.0376	2.0090

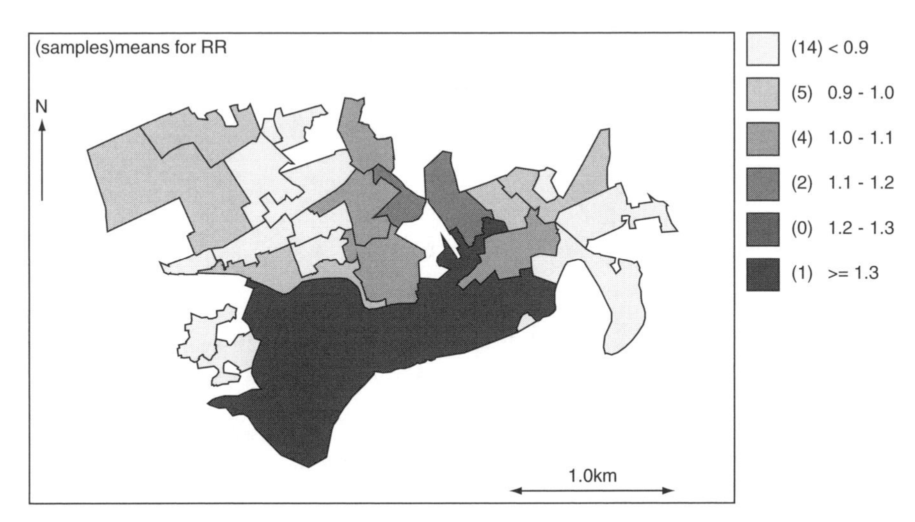

Figure 7.23 Posterior expected relative risk estimates for the model with UH and CH (model 8).

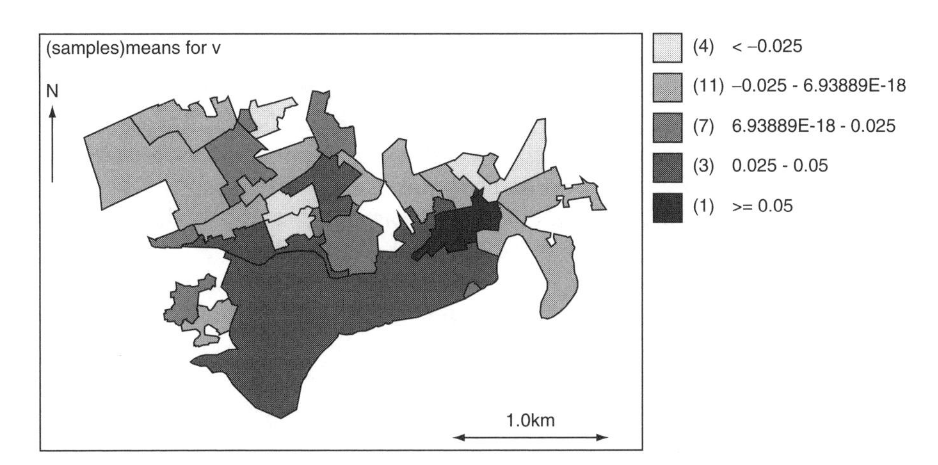

Figure 7.24 Posterior expected relative risk estimates for the UH component $\{v_i\}$.

that there is evidence of a distance-decline effect but it is only weak evidence. Figures 7.23, 7.24 and 7.25 display, respectively, the posterior expected relative risk surface, v_i surface and u_i surface.

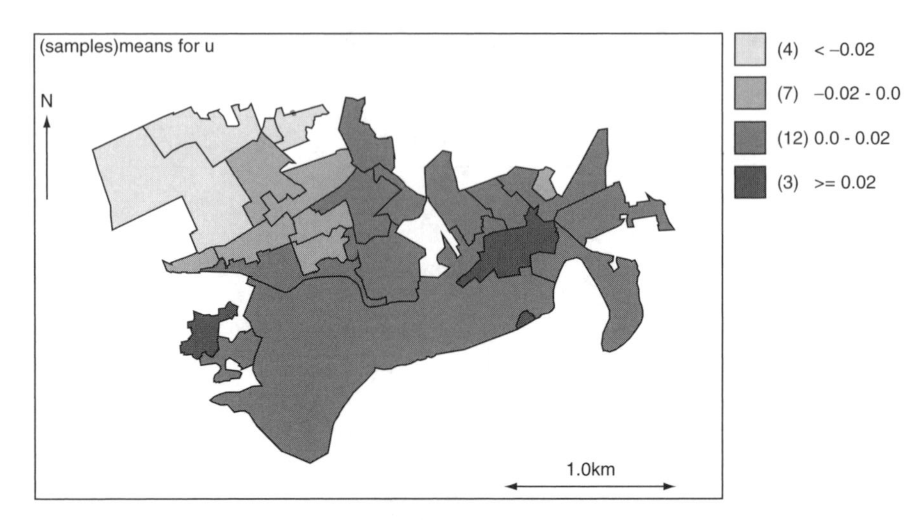

Figure 7.25 Posterior expected relative risk estimates for the CH component $\{u_i\}$.

In the next section we will examine a more substantial data example, and will examine a wide range of models for the data, including both uncorrelated and correlated heterogeneity models and goodness of fit via the DIC criterion.

7.7.2 The Ohio data example

In this example, respiratory cancer incidence for the counties of Ohio, USA, are available for the years 1979–1988. The original analysis of these data was carried out by Waller *et al.* (1997), and subsequently by Knorr-Held and Besag (1998) and Zia *et al.* (1997), and is also discussed by Carlin and Louis (1996). This data set is spatiotemporal in that observations are available for a variety of time periods. The original purpose of the analysis of this data set was the assessment of risk around a putative source of hazard, and it is that purpose which we pursue here. In Chapter 6, on relative risk estimation, this dataset was examined via two different spatiotemporal models. In that situation, the focus was primarily on the description of the spatiotemporal variation in relative risk. Here our focus is on the assessment of evidence for a link with a source, and the associated parsimonious modelling of the background variation in risk. In what follows, for brevity we do not display the code for each model. These are provided in Appendix 1.

Using the notation from Section 6.1.6, we can specify the following basic terms: y_{ik} is the count of disease in the ith region in the kth time period. We assume that the counts are Poisson distributed with expectation $e_{ik}\theta^*_{ik}$, that is

$$y_{ik} \sim Poisson(e_{ik}\theta^*_{ik}).$$

Here the focus is on modelling the relative risks $\{\theta^*_{ik}\}$ in the different areas and time periods. Once again models for $\log(\theta^*_{ik})$ are usually specified. In this case we would want to include covariate terms specifically relating to the hazard source. Here we will want to employ (r_i, θ_i), the polar coordinates of the region centroids from the putative source. A small set of possible models is examined here. There are no other covariates considered in the analysis.

7.7.2.1 *Model 1.* $\log(\lambda_{ik}) = \log(e_{ik}) + \alpha_0 + v_i + \eta_k$ where $v_i \sim N(0, d^*)$ and $\eta_k \sim N(\eta_{k-1}, c^*)$

This model includes a frailty effect in space and a temporal effect which is allowed to follow a random walk. There are no interaction effects.

The model allows for a smooth variation in time, while allowing spatial UH estimated over all time periods.

The results show that there remains an amount of positive extra variation in Hamilton county (south-west corner), although other counties also show such extra variation (see Figure 7.26). For this model the components of the DIC are

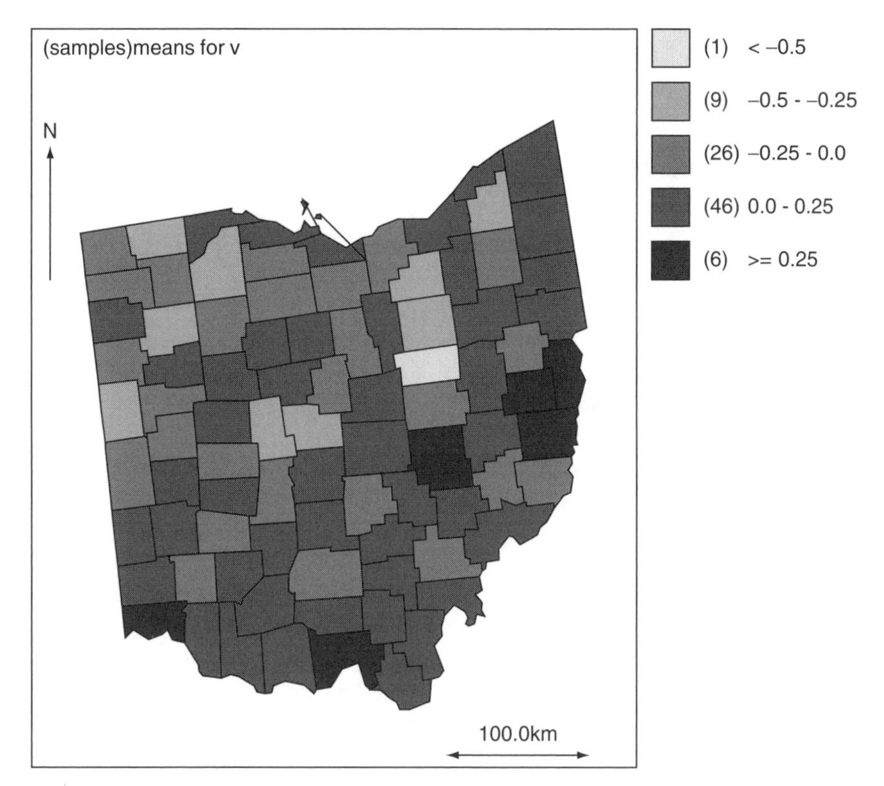

Figure 7.26 Ohio: Plot of the posterior expected uncorrelated random effect (v) for model 1.

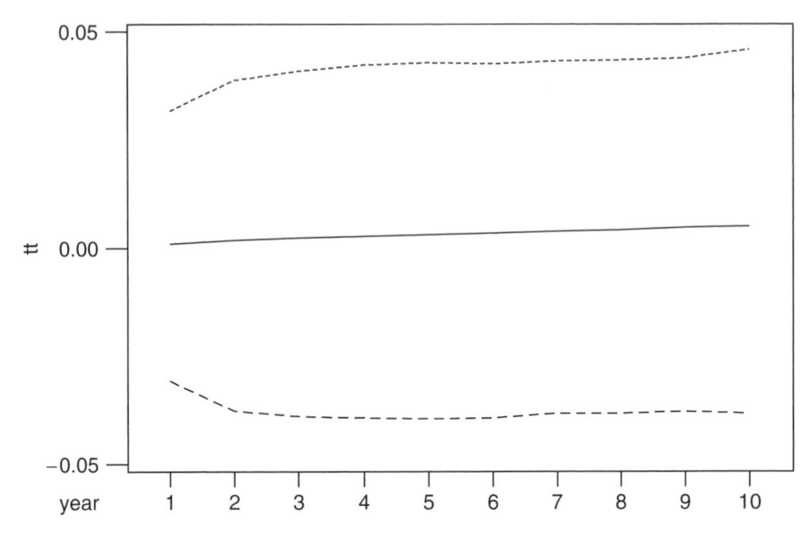

Figure 7.27 Ohio: plot of the temporal random effect (*tt*) for model 1: 95% confidence limits shown.

$\bar{D} = 5684.33, \hat{D} = 5600.61, pD = 83.72, \text{DIC} = 5768.05$. There is an overall increasing temporal trend over the state (see Figure 7.27).

7.7.2.2 Model 2

This model is the same as model 1 but includes a distance effect:

$$log(\lambda_{ik}) = log(e_{ik}) + \alpha_0 + v_i + \eta_k + \log(1 + e^{-\alpha_1 r_i}).$$

This model has the following DIC statistics: $\bar{D} = 5683.730, \hat{D} = 5600.800$, $pD = 82.934$ and $\text{DIC} = 5766.670$. This is a lower DIC than the previous model but only by 1.38. This suggests that adding the distance effect improves the model marginally.

Notice how the trend is still apparent but has a different range once the distance effect is added (Figure 7.29). In this example the distance parameter α_1 is not well estimated and there does not appear to be a distance effect across counties (posterior mean: 0.006901, sd = 0.04936, 2.5%: −0.09361, 97.5%: 0.096 03). The fact that the parameter credible interval crosses zero supports this contention. The spatial distribution of the uncorrelated heterogeneity (UH) component is displayed in Figure 7.28.

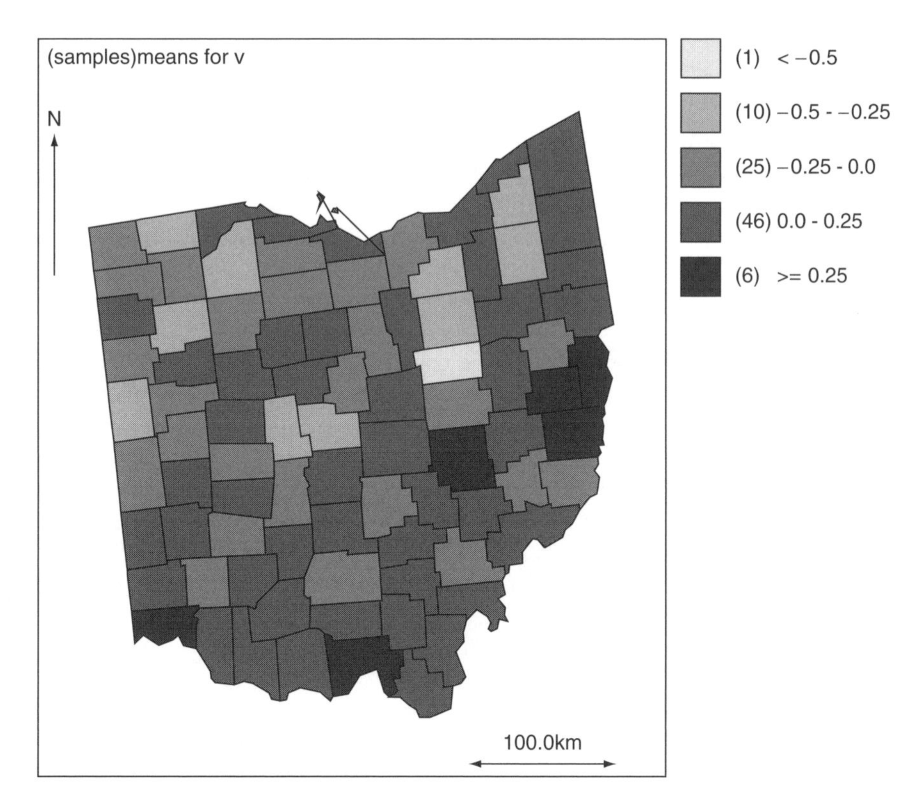

Figure 7.28 Ohio: plot of the posterior expected uncorrelated random effect (v), for model 2.

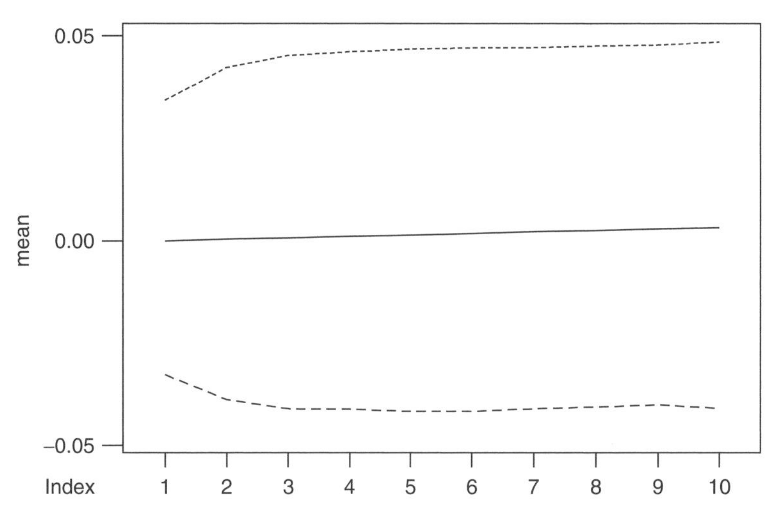

Figure 7.29 Ohio: plot of the posterior expected temporal random effect (tt) for model 2; 95% confidence limits shown.

7.7.2.3. Model 3. $log(\lambda_{ik}) = log(e_{ik}) + \alpha_0 + v_i + u_i + \eta_k + \log(1 + e^{-\alpha_1 r_i})$ where $\{u_i\}$ is the CH term

This is an extension of model 2 to include spatial correlation (CH). There are no interaction terms. The results for this model are as follows. The DIC has components: $\bar{D} = 5683.5, \widehat{D} = 5601.2, pD = 82.28$ and $DIC = 5765.7$. This is lower than the uncorrelated random effect model but the difference is very small (≈ 0.9). Figure 7.30 displays the posterior expected time trend for this model. Figures 7.32 and 7.33 display the posterior expected v_i and area-specific relative risk surfaces.

One feature of this model is that the correlated random effect has behaved like a spatial trend estimator as it has estimated blocks of risk with increasing southward trend. This may provide a partial explanation of the fact that the distance parameter is no longer well estimated: the CH effect has soaked up the spatial trend within the study area (Figure 7.31). This result suggests that the inclusion of a CH term could be confounded with any spatial trend. Hence, if separate spatial trend is to be estimated then there may be difficulty in separating these effects. On the other hand, this supports the results found in large simulations (Lawson *et al.*, 2000) which suggest that CH models are robust against a range of underlying true risks, including basic trend models.

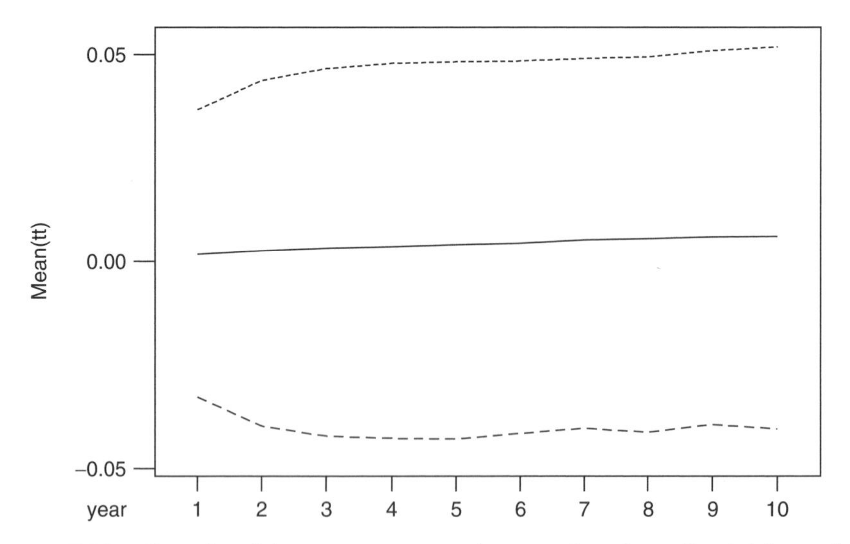

Figure 7.30 Ohio: plot of the posterior expected temporal random effect (tt) for model 3; 95% confidence limits shown.

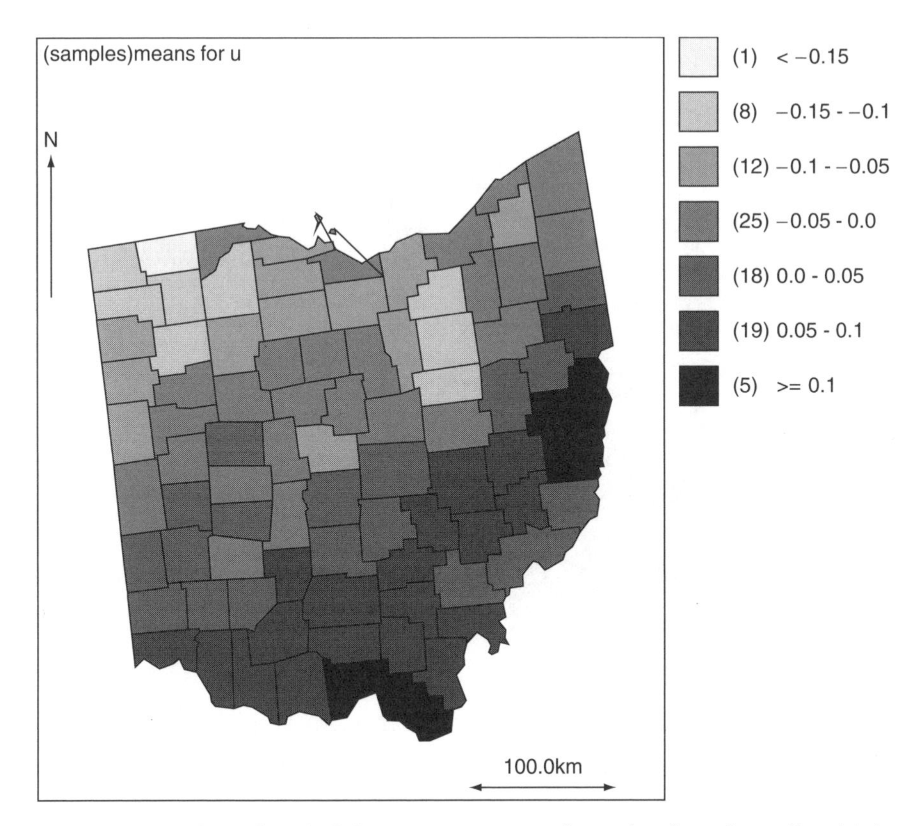

Figure 7.31 Ohio: plot of of the posterior expected correlated random effect (u) for model 3.

7.7.2.4. *Model 4.* $log(\lambda_{ik}) = log(e_{ik}) + \alpha_0 + v_i + u_i + \eta_k + \zeta_{ik} +$ $\log(1 + e^{-\alpha_1 r_i})$ where ζ_{ik} is space−time interaction term and $\zeta_{ik} \sim N(0, b^*)$

In this model $\bar{D} = 5599.9, \widehat{D} = 5447.2, pD = 152.8$ and the $DIC = 5752.8$. This suggests that the inclusion of the S-T interaction term produces a significant increase in goodness of fit. However, the posterior expected value for $\alpha_1 = -0.0088$ with 2.5% quantile of −0.0749 and 97.5% quantile of 0.08565 suggests that there is little evidence for a distance-decline effect in the data. Of course this could also be due to the fact that a spatial correlation term has been included in the analysis which is not zero-centred (see Figure 7.35). Figures 7.36 and 7.34 display the uncorrelated heterogeneity and temporal relative risk components averaged over the posterior sample.

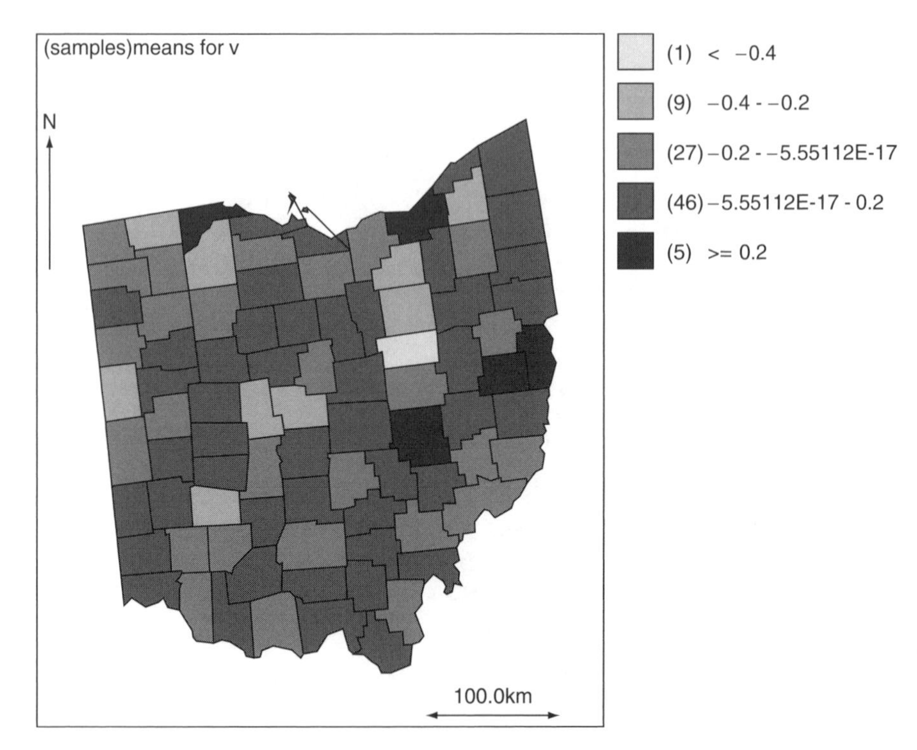

Figure 7.32 Ohio: plot of the posterior expected uncorrelated random effect (v) for model 3.

As a variant of the above model, we have examined the same model but without the correlated heterogeneity term. This yields a DIC of 5754.1, which is marginally higher than the previous model. This suggests that, given parsimony considerations, the current model would be favoured. The distance parameter α_1 still has a negative expected value (− 0.00326) and wide range. However, the distribution has a marked peak around 0.02. It would appear that this model allows more of this spatial variation to be absorbed by the distance parameter.

Figure 7.37 displays the spatial distribution of the UH component which now has slightly increased levels in the southern areas (including Hamilton county). This is also apparent on the map of the spatial relative risk (Figure 7.38). Knorr-Held (2000) has also proposed an extension of this model where the interaction term has itself a temporal dependence and is defined by $\zeta_{ik} \sim N(\zeta_{ik-1}, b^*)$. This form could also be fitted here but has not been pursued for the sake of brevity.

Overall there appears to be evidence for a weak distance effect, but the effect does not persist once random effects are included. This more detailed analysis

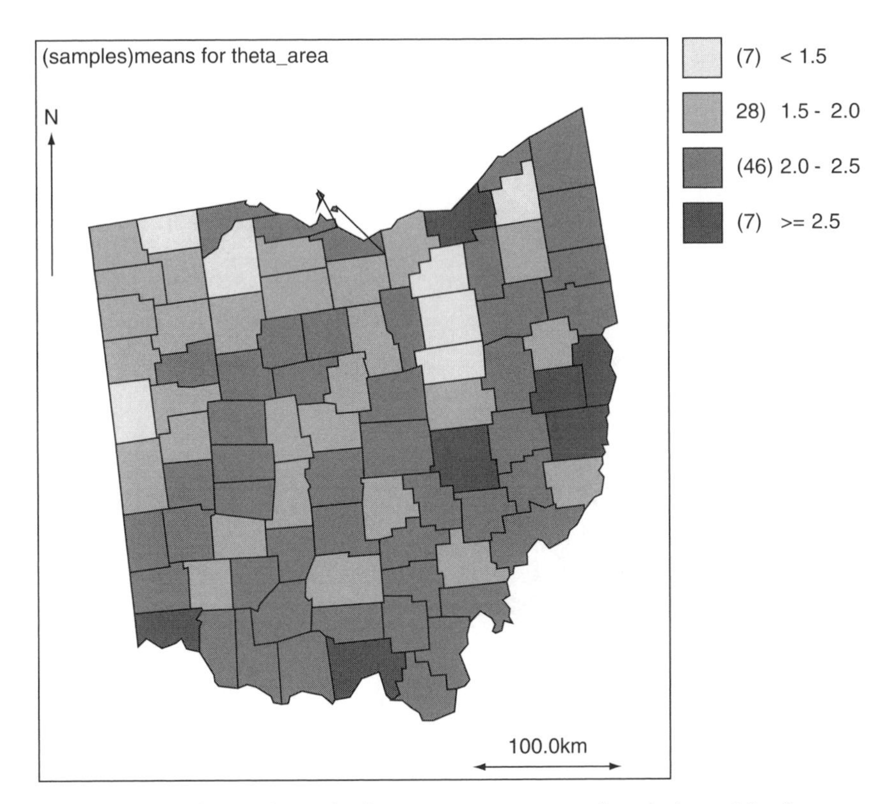

Figure 7.33 Ohio: plot of the posterior expected relative risk for areas ($\theta_i = \exp(v_i + u_i + \log(f_i))$) for model 3.

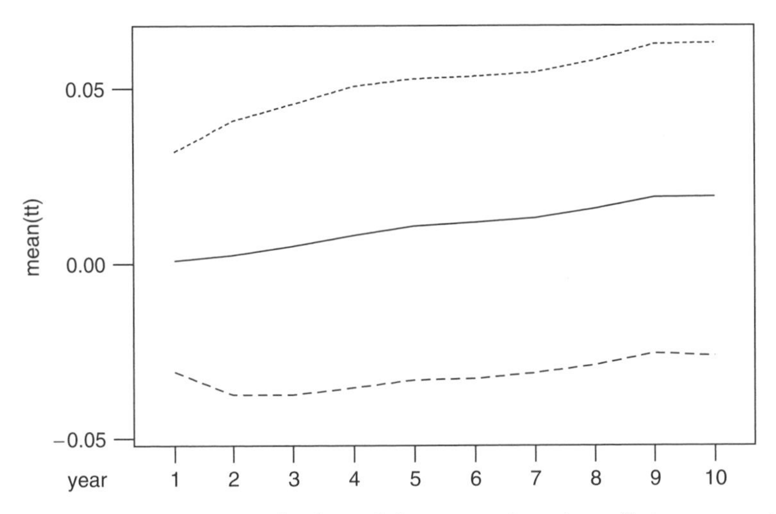

Figure 7.34 Posterior expected values of the temporal random effect component $\{tt\}$ over 10 time periods for model 4.

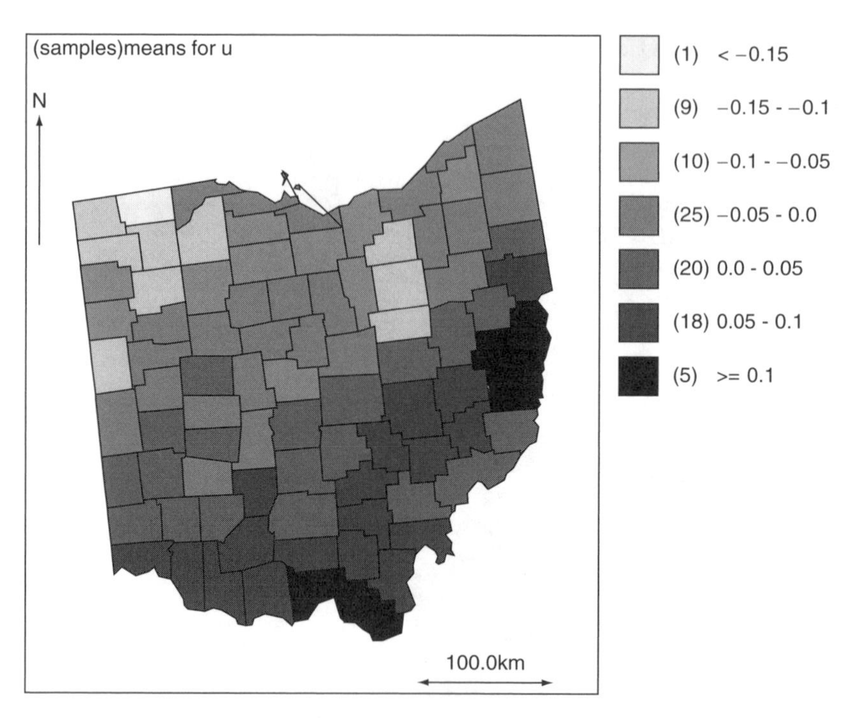

Figure 7.35 Ohio: posterior expected value for the correlated heterogeneity component for model 4.

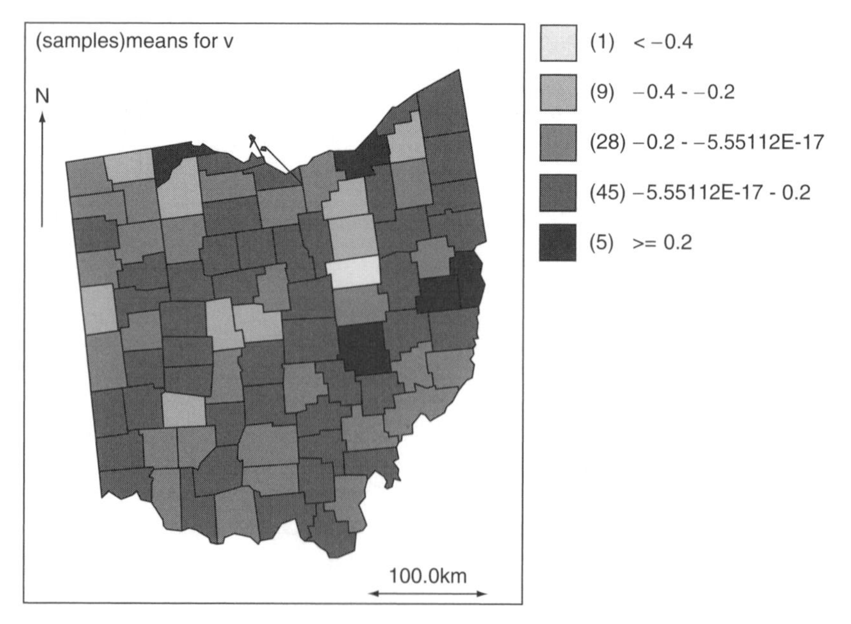

Figure 7.36 Posterior expected v_i surface for model 4.

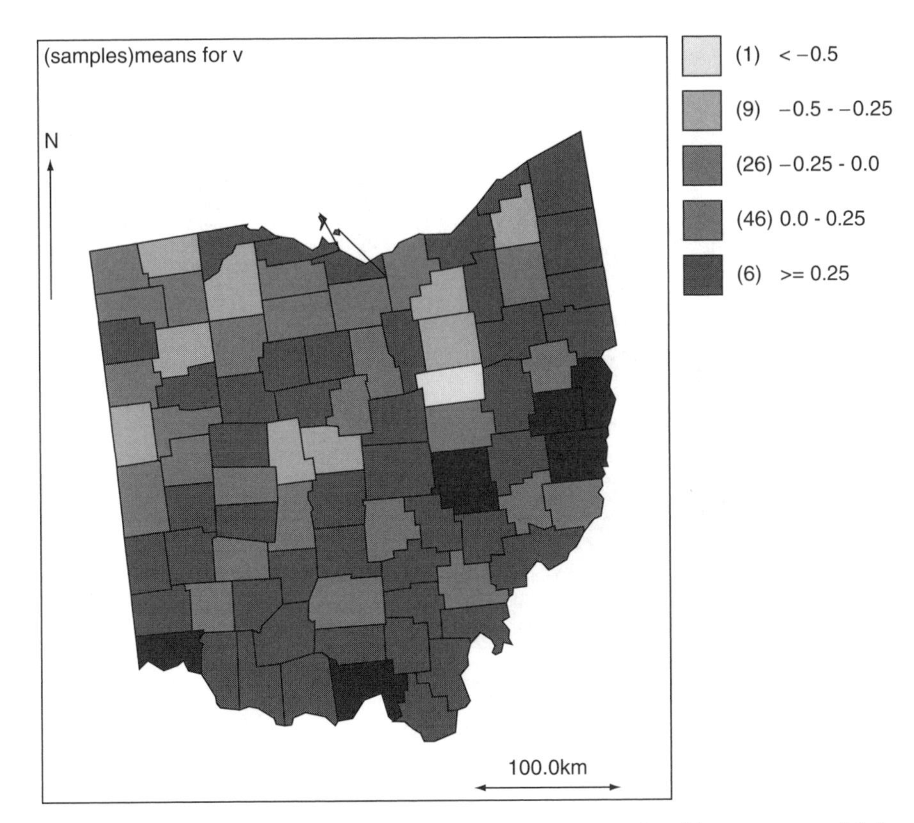

Figure 7.37 Ohio: posterior expected value for the uncorrelated heterogeneity $\{v\}$ for model 4b.

does not strongly support the results of previous analyses in the literature which purport that an excess is found in the Hamilton area. Even if this excess exists, there is no strong distance effect evident. It should be borne in mind that we have only examined a small selection of models for the space–time variation and others may be found which yield better fits to this data set. It is possible to also fit directional components or more complex models with spatial effects nested within time, for example. Our aim has been to illustrate some of the possible forms of analysis which are available in this case.

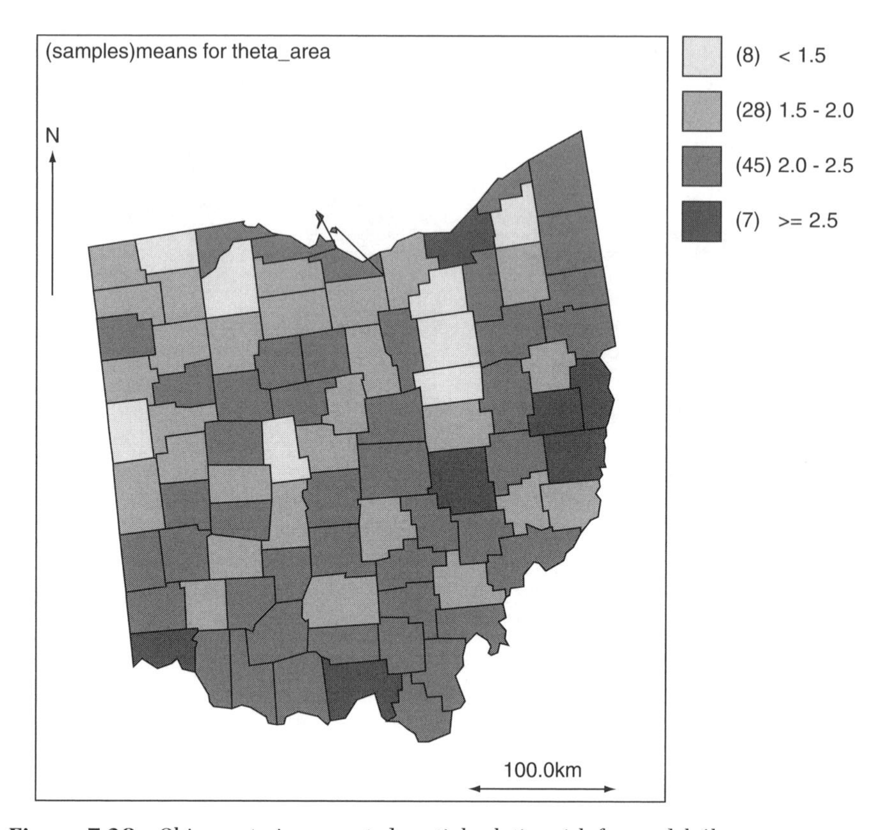

Figure 7.38 Ohio: posterior expected spatial relative risk for model 4b.

7.8 FOCUSED CLUSTERING IN MLwiN

Focused clustering is the study of the effect of point pollution sources on the distribution of a disease. Here we are aiming to explain 'clusters' of higher than expected area counts of a disease by relating the cluster to a point source that may lie within or close to the cluster. Typically we will model the effect of the point source by fitting a model that relates the counts to the distance and direction the areas are from the point source. The functional relationships that are fitted are often complex and nonlinear and are consequently beyond the scope of MLwiN's current modelling facilities and so in this section we will demonstrate how far MLwiN can be applied to focused clustering.

7.8.1 An analysis of the Falkirk dataset using MLwiN

We will begin by considering the Falkirk dataset which consists of 26 census tracts in the locality of Falkirk, central Scotland. The response of interest is the number of cases of respiratory cancer in the time period 1978–1983, and it is of interest to relate these counts to the proximity of the tracts to a foundry that ran in the early 1970s and may have been an air pollution hazard. If we load the worksheet '*falkirk.ws*' into MLwiN we will see the following:

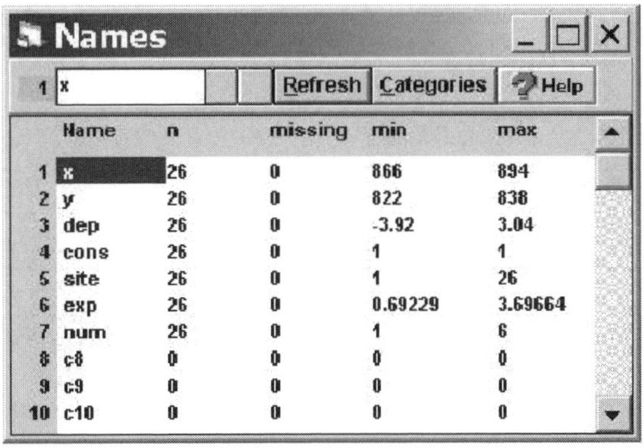

Names

1 x Refresh Categories Help

	Name	n	missing	min	max
1	x	26	0	866	894
2	y	26	0	822	838
3	dep	26	0	-3.92	3.04
4	cons	26	0	1	1
5	site	26	0	1	26
6	exp	26	0	0.69229	3.69664
7	num	26	0	1	6
8	c8	0	0	0	0
9	c9	0	0	0	0
10	c10	0	0	0	0

7.8.2 Preliminary analysis

The data consists of, for each census tract, '*site*', the observed count of respiratory cancer, '*num*', and expected count, '*exp*'. As a predictor variable we have the rate of deprivation based on the Carstairs index, '*dep*' and for each tract we have the x (East) and y (North) coordinates, '*x*' and '*y*' of the centroid of the tract. We know that the foundry site is towards the south-east of the region at coordinates $x = 885, y = 829$. We can therefore transform our absolute coordinates to coordinates that are relative to the foundry and also calculate the standardized mortality ratio (SMR). These new variables can be constructed via the **Calculate** or **Command Interface** windows by typing the following commands:

```
calc c8 = 'x' - 885
name c8 'xcent'
calc c9 = 'y' - 829
name c9 'ycent'
calc c10 = 'num' / 'exp'
name c10 = 'smr'
```

We can now have a preliminary look at the data by plotting '*smr*' against both '*xcent*' and '*ycent*'. Such a point plot can be constructed using the **Customised Graph** window and the resulting plots will look similar to the following:

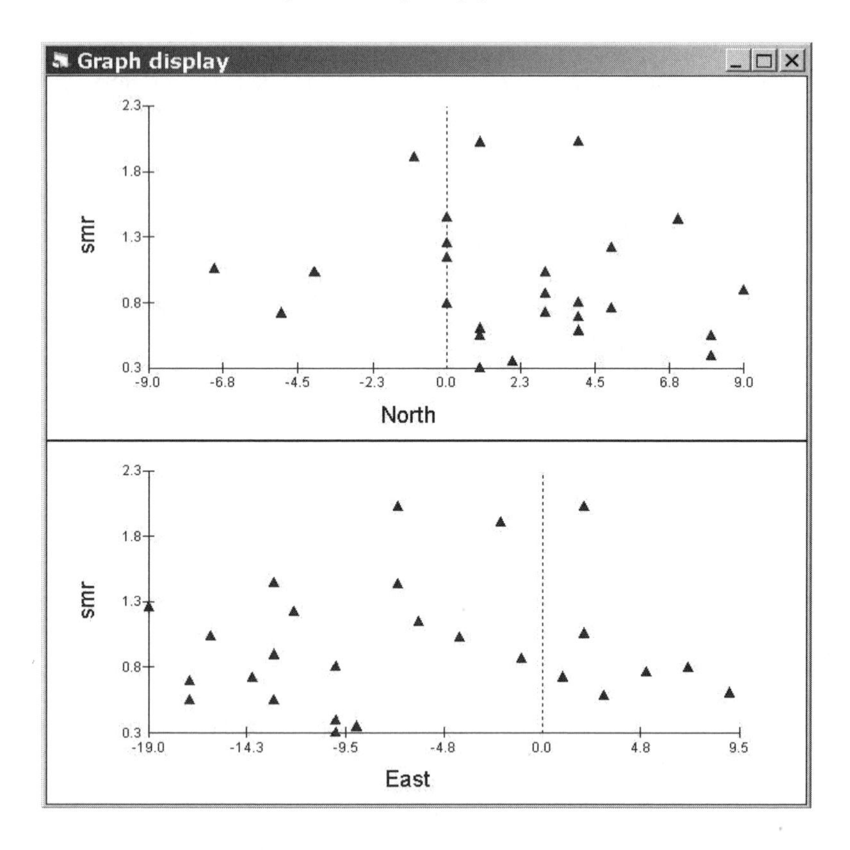

Here we see that the three tracts with largest SMR are fairly close to the foundry in both a north–south and east–west direction and of the outlying regions we see that many have low SMRs. It may of course make more sense to transform the north–south and east–west directions into a distance and direction measure. This is sensible as we believe the foundry to be a source of airborne pollution and so the rate of dispersal will be influenced by the prevailing winds. We therefore need to create two variables to represent distance and direction. The distance is calculated via the Pythagoras formula and the direction by taking the function $\tan^{-1}(y/x)$. Of course the pairings of x, y and $-x, -y$ will give the same answer even though they are in opposite directions so we need to adjust for this by adding π to values where x is negative and y is positive and subtracting π from values where both x and y are negative. This will result in the direction taking values of 0 for points directly east of the foundry, $\pi/2$ if directly north, $-\pi/2$ if directly south and π or $-\pi$ if directly west. The commands are as follows:

```
calc c11 = sqrt ('xcent' * 'xcent' +'ycent' * 'ycent' )
name c11  'distance'
calc c12 = atan ('ycent' / 'xcent' )+
((' xcent '<= 0) * ((3.1415)* ('ycent' >=0) - (3.1415* ('ycent' <0))))
name c12 'direction'
```

Now if we plot the SMRs against both distance and direction we will get the following:

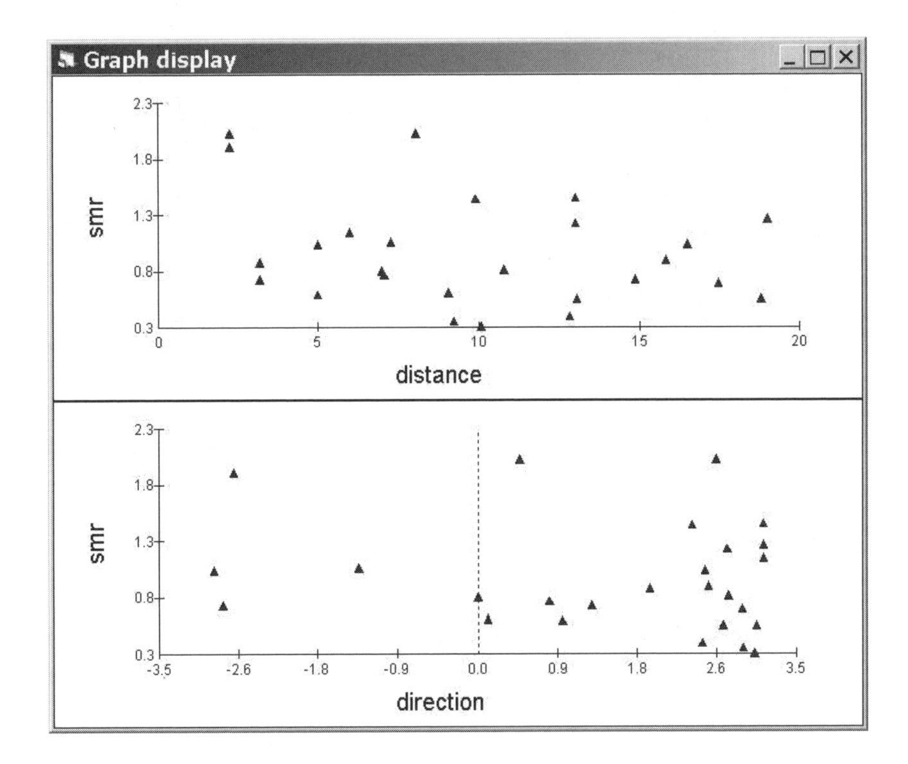

It here appears that there is a negative relationship between distance and SMR, i.e. the closer to the foundry a tract is the higher the SMR, but no really obvious pattern with regard to the direction relationship.

7.8.3 Statistical modelling

Before we begin modelling we will need to create an offsets column (via the **Command Interface** window) that contains the logs of the expected counts. We will use the first free column, *c13*, and name it '*logexp*'. After constructing this column the **Names** window should look as follows:

	Name	n	missing	min	max
1	x	26	0	866	894
2	y	26	0	822	838
3	dep	26	0	-3.92	3.04
4	cons	26	0	1	1
5	site	26	0	1	26
6	exp	26	0	0.69229	3.69664
7	num	26	0	1	6
8	xcent	26	0	-19	9
9	ycent	26	0	-7	9
10	smr	26	0	0.3069641	2.033512
11	distance	26	0	2.236068	19
12	direction	26	0	-2.896521	3.1415
13	logexp	26	0	-0.3677503	1.307424
14	C14	0	0	0	0
15	C15	0	0	0	0

Perhaps one of the simplest models to fit now is a Poisson regression with just an intercept term. This model can be set up in the **Equations** window by specifying '*num*' as the response variable, *Poisson* as the distribution type, 1 level for the data structure with '*site*' as the level 1 identifier, '*logexp*' as the offset column and finally specifying '*cons*' as a fixed effect. It is interesting to look at whether we have any additional variation other than that expected by the Poisson assumption. We will therefore allow extra-Poisson variation. The model when set up and run using first-order MQL will look as follows:

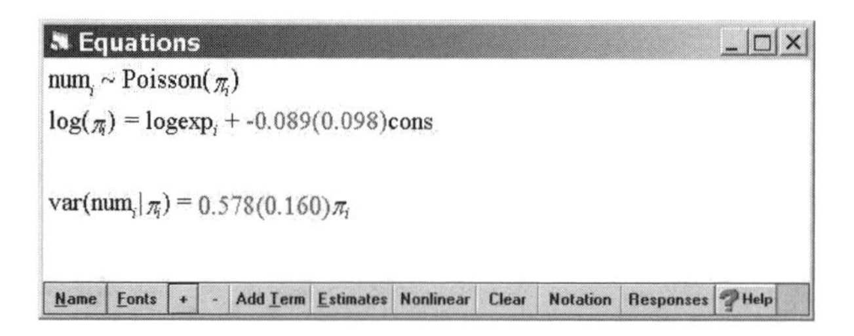

As we can see here there is in fact less variation in the dataset than would be expected from a Poisson distribution. This is a problem as the assumption of a Poisson distribution is not correct. We may, however, still think that there is a spatial effect due to the distance and direction that the sites are from the foundry. There may also be variation in the counts due to the deprivation in the areas. We could add these predictors individually or, as below, en masse.

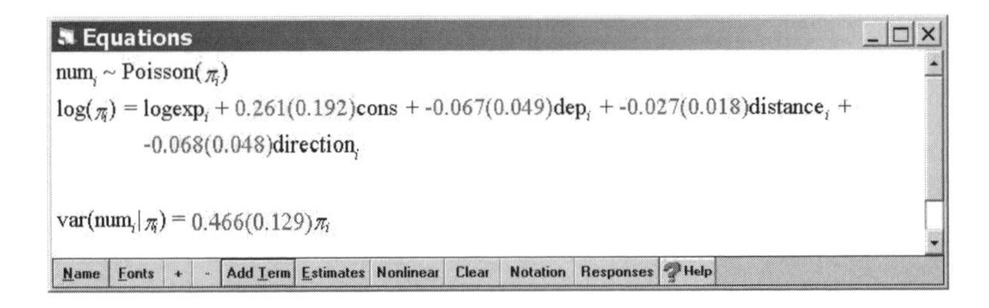

Here we see that none of the effects are statistically significant but that the closer to the foundry the site is the larger the relative risk. Similarly a negative direction which equates to a southern and western direction corresponds to a larger relative risk and finally a smaller deprivation score results in a larger relative risk. We can also see that the scale factor for the extra-Poisson variation has also reduced. It may be more sensible, as we do not have truly Poisson data, to consider the SMRs instead and use a Gaussian response. If we do this here we get the following:

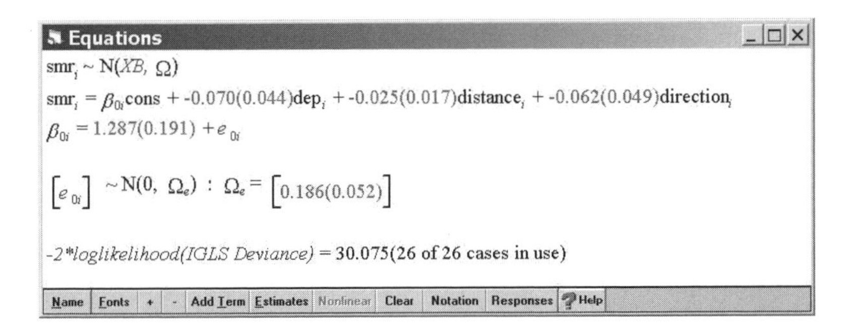

Here we get non-significant effects for distance, direction and deprivation with the same signs as the equivalent coefficients in the Poisson model. The residual variance is reduced from 0.225 to 0.186 by the addition of the three predictors. Therefore, although the graphs show some evidence of spatial effects, we do not, when we fit statistical models, get statistically significant effects. This is possibly because the dataset is quite small and so it is difficult in such a small dataset to pick up spatial effects.

8

Ecological Analysis

8.1 INTRODUCTION

Ecological analysis is closely related to disease map reconstruction. The aim of ecological studies is to describe the relationship between geographical variation in disease incidence or mortality and explanatory covariates (e.g. environmental agents or lifestyle characteristics) measured on groups rather than individuals. Typically the groups are defined by geographical areas such as countries, regions or smaller administrative units. Ecological studies are used in conjunction with simple descriptive studies of geographical variation in an attempt to determine how much of the variation in disease rates is associated with variation in exposure. Ecological analyses include, for example, studies relating cancer mortality in different areas to a variety of environmental and socio-economic factors. While an ecological study can be carried out based on georeferenced data the focus is not usually on the spatial distribution but on the linkage between the dependent variable and the measured covariates.

An important problem affecting ecological analyses is *ecological bias*. This bias arises when the association observed at the group or ecological level is applied to the association between the corresponding variables at the individual level. The magnitude and direction of the association between exposure status and disease risk at individual level could be different from the ecological association (just because the disease rate is higher in regions with a larger exposure does not mean that the exposed individuals are at greater risk than the unexposed individuals). In some cases individual studies can display the reverse relationship between dependent variable and covariate. The underlying problem of ecological bias is that each group is not entirely homogeneous with respect to the exposure. Despite this, ecological studies have some advantages over other studies because they include the ability to study a large population at a low cost and to address questions of environmental health that might be difficult to address at an individual level; they are particularly useful when individual measurements of exposure are either difficult or impossible to obtain (for

Disease Mapping with WinBUGS and MLwiN A. Lawson, W. Browne and C. Vidal Rodeiro
 ISBN: 0-470-85604-1 (HB)

example air pollution) or are measured imprecisely (for example dietary intakes, sunlight exposure).

8.2 STATISTICAL MODELS

The simplest approach to ecological analysis is to use a regression model for disease rates which only allows for Poisson variation.

Let $\{y_1, \dots, y_m\}$ be the observed number of events and $\{e_1, \dots, e_m\}$ the expected number of events for a certain disease in m areas of the region of interest. Assume that y_i is distributed according to a Poisson distribution with mean $\mu_i = e_i\theta_i$, where θ_i is the relative risk in the ith area. As mentioned in previous chapters, the saturated maximum likelihood estimates of θ_i are given by the SMRs. This model can be extended to include a set of explanatory variables $x_1, x_2, \dots, x_p$ in a log-linear formulation

$$\log \mu_i = \log e_i + \sum_{j=1}^{p} \beta_j x_{ij}.$$

This model can include many ecological factors (covariates $\mathbf{x}_j$) but, in many situations, the variation not explained by the ecological variables might exceed that expected from the Poisson model leading to overdispersion. Spatially unstructured extra-Poisson variability can be easily modelled assuming that the true relative risk distribution takes a log-normal form with additive components. An overdispersion parameter is therefore added to the model. The overdispersion parameter relates to the degree of heterogeneity among the true relative risks, given the variables in the model. Unstructured and structured extra-Poisson sources of variability can be accounted for by adopting a model where the extra-Poisson variation is decomposed into two components. The first component is spatially unstructured extra-Poisson variation (unstructured heterogeneity) and the second component varies smoothly across areas (structured heterogeneity or clustering). Introducing these terms in the model represents a way of controlling for unmeasured covariates. The model could be written as follows:

$$\log(\theta_i) = \rho + \mathbf{x}_i'\boldsymbol{\beta} + u_i + v_i,$$

where ρ is a constant, $\mathbf{x}_i$ represents a vector of explanatory variables in the ith region, $\boldsymbol{\beta}$ is a vector of parameters, and the heterogeneity and the clustering components of variation are represented by u_i and v_i. This model was introduced by Clayton and Kaldor (1987) and it was developed by Besag *et al.* (1991) and Clayton *et al.* (1993). In their paper, Clayton *et al.* (1993) stress the importance of including a spatially-correlated term (u_i) in ecological analysis to allow for unobserved variation.

Examples of ecological regression studies that have implemented models of this form include Clayton and Bernardinelli (1992), who analyse the effects of urbanization on breast cancer mortality in Sardinia, and Richardson *et al.* (1995) who analyse geographical variation of childhood leukaemia in the UK in relation to natural radiation. See also Bernardinelli *et al.* (2000) and Bernardinelli *et al.* (1999) for further discussion. For an extensive range of statistical models proposed in the context of ecological regression see Biggeri *et al.* (1999). They discussed the models mentioned above, approximate hierarchical random effects models and nonparametric models for this kind of analysis.

8.3 WinBUGS ANALYSES OF ECOLOGICAL DATASETS

In this section, we provide some examples of ecological analysis on WinBUGS applied to a range of case studies.

8.3.1 Mortality from cancer in South Carolina

The present example concerns 8080 deaths from cancer (malignant neoplasms) in South Carolina in 1999. Usually, the most populated areas of a region register the highest risk of cancer mortality. This led us to consider the introduction into the analysis of an urbanization-related covariate: population density (DEN). This covariate was computed for each county using census information. While expected rates for malignant neoplasm are based on the local 'at risk' population in small areas, it was considered possible that extra variation could be related to the additional effect of urbanization.

We are going to model the covariate effect in the presence of unstructured and spatially-structured extra-Poisson variation. The WinBUGS code for fitting this model is presented in Figure 8.1.

Model fitting was carried out using two separate chains starting from different initial values. Convergence was checked by visual examination of time series plots of samples for each chain and by computing the Gelman and Rubin diagnostic. The first 5000 samples were discarded as a burn-in; each chain was run for a further 10 000 iterations.

Figure 8.2 shows the saturated maximum likelihood estimates of cancer mortality (SMRs) by county.

Table 8.1 reports the results obtained from the fitting of four Bayesian models, all including the covariate *DEN*. For comparison purposes, a model which ignores extra-Poisson variation (Model 1) is also fitted. Model 2 includes the uncorrelated heterogeneity term (v_i). The estimates of the regression coefficient for the covariate *DEN* are very similar in both models, but Model 2 gives a higher standard error. This is expected because Model 1 ignores random influences modelled by the heterogeneity term. Model 3 introduces the correlated

```
model
{
for (i in 1:m)
{
    # Poisson likelihood for observed counts
    y[i]~dpois(mu[i])
    log(mu[i])<-log(e[i])+alpha+v[i]+u[i]+beta*x[i]
    # Relative Risk
    theta[i]<-exp(alpha+v[i]+u[i]+beta*x[i])
    res_theta[i]<-exp(alpha+v[i]+u[i])
    # Posterior probability of RR[i]>1
    PP[i]<-step(theta[i]-1+eps)
    res_PP[i]<-step(res_theta[i]-1+eps)
    # Prior distribution for the uncorrelated heterogeneity
    v[i]~dnorm(0,tau.v)
    # Relative Risk decomposition
    RR_exp[i]<-exp(beta*x[i])
    RR_het[i]<-exp(v[i])
    RR_clust[i]<-exp(u[i])
}

eps<-1.0E-6

# CAR prior distribution for spatial correlated heterogeneity
u[1:m]~car.normal(adj[],weights[],num[],tau.u)

# Weights
for(k in 1:sumNumNeig)
{
    weights[k]<-1
}

# Improper prior distribution for the mean relative risk in the study region
alpha~dflat()
mean<-exp(alpha)

# Prior on regression coefficients
beta~dnorm(0.0,1.0E-5)

# Hyperprior distributions on inverse variance parameter of random effects
tau.u~dgamma(0.5,0.0005)
tau.v~dgamma(0.5,0.0005)
}
```

Figure 8.1 WinBUGS code for ecological regression.

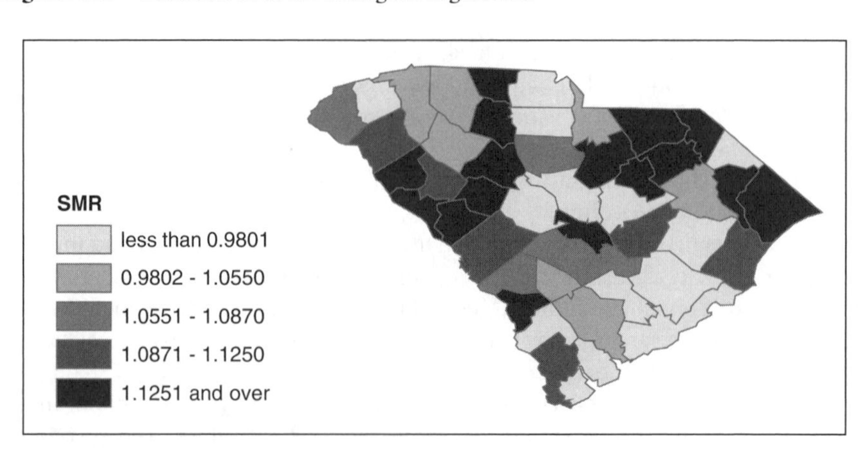

Figure 8.2 Cancer mortality, SMR. South Carolina, 1999.

Table 8.1 Regression coefficient for the covariate for different models.

Model	β	SE(β)	95% credible interval (β)
1. *DEN*	−0.001159	2.435E−4	(−0.001637,−0.000679)
2. $DEN + v_i$	−0.001485	5.198E−4	(−0.002497,−0.000450)
3. $DEN + u_i$	−0.001568	4.715E−4	(−0.002492,−0.000634)
4. $DEN + u_i + v_i$	−0.001543	5.098E−4	(−0.002518,−0.000542)

Table 8.2 Goodness of fit for different models.

	Without covariates		With covariates	
Model	Deviance	DIC	Deviance	DIC
2. v_i	348.976	375.437	347.423	372.027
3. u_i	351.609	377.535	350.587	373.767
4. $u_i + v_i$	349.256	375.767	347.422	371.779

Table 8.3 Estimation of the hyperprior variances for clustering and heterogeneity effects.

	No covariates	With covariates
Clustering	1.0435E−6	1.1617E−6
Heterogeneity	1.1569E−5	1.0459E−5

heterogeneity component (u_i), but neither the estimate of the coefficient of *DEN* or its standard error are greatly affected. Model 4 includes all terms and gives similar results to Models 2 and 3. The point estimate of the regression coefficient indicates a negative (very small) association between the covariate *DEN* and the risk of cancer mortality.

Table 8.2 shows the goodness of fit of the models with and without the covariate *DEN*. The addition of the covariate to the model improves its deviance and DIC. Although DIC is subject to Monte Carlo sampling error, since it is generated under a MCMC sampling scheme, the changes in its value are significant in this case study. The model that yields best results is the one with both uncorrelated and correlated heterogeneity terms.

Table 8.3 summarizes the effects of the clustering and the heterogeneity components. From it, we can see that the inclusion of covariates has only a small impact on the estimation of these random effects.

Since the explanatory variable significantly influences the geographical variation of the relative risk (see Table 8.1), in order to obtain a better estimate of the map, we estimate relative risks allowing for this covariate. A map of the relative risk estimates when the covariate *DEN* is included in the model is presented in Figure 8.3. The map shows a considerable amount of smoothing compared to the SMR (Figure 8.2). Figure 8.4 is an estimate map of variation of the relative risk not explained by *DEN* (residual variation). It was obtained by subtracting the contribution of *DEN* from the fitted relative risks. It shows that the residual relative risk in the north-west is uniformly high.

To aid the interpretation of the model, Figure 8.5 shows the fitted relative risks decomposed into their three components:

(a) that due to the explanatory variable ($\exp(\beta x_i)$),

(b) heterogeneity ($\exp(v_i)$), and

(c) clustering ($\exp(u_i)$).

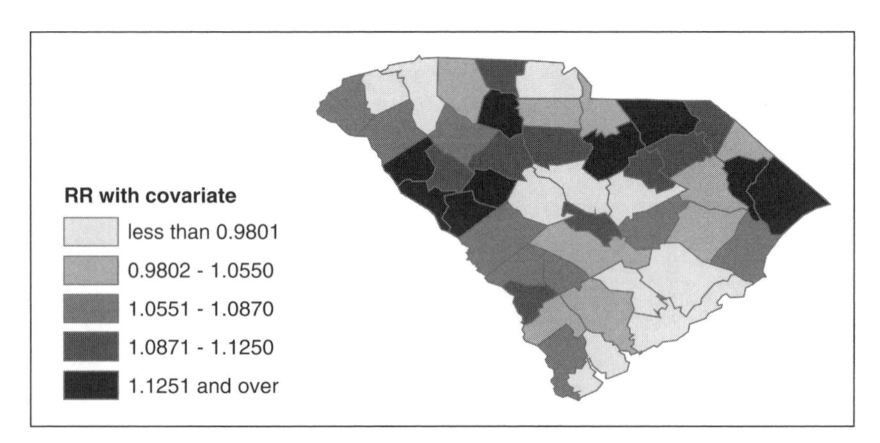

Figure 8.3 Relative risk (model 4 with covariate) for cancer mortality: South Carolina, 1999.

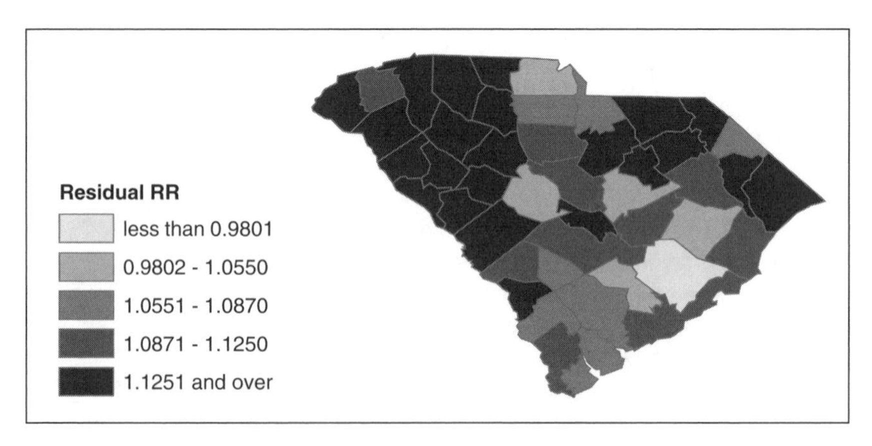

Figure 8.4 Residual relative risk (model 4 with covariate) for cancer mortality: South Carolina, 1999.

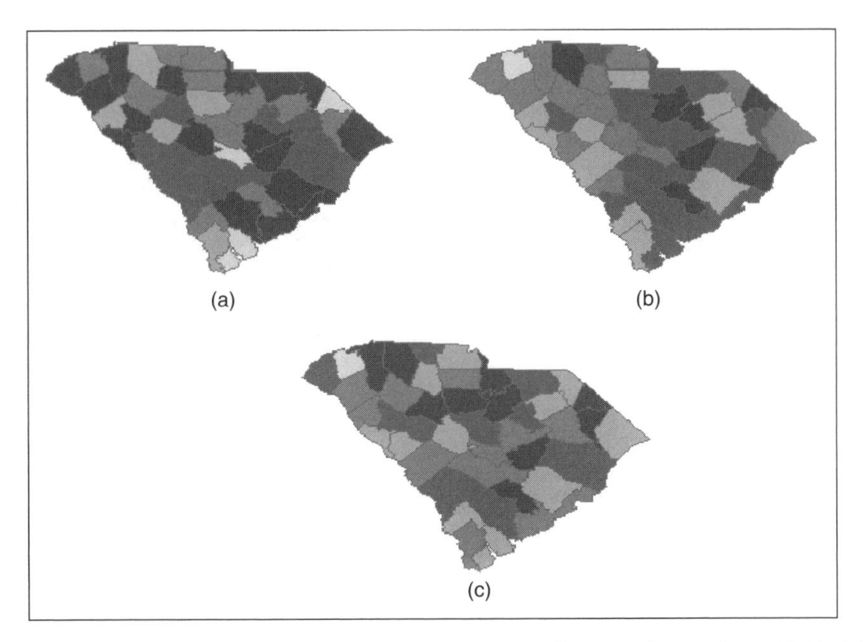

Figure 8.5 Components of variation in cancer mortality: South Carolina, 1999. (a) Explanatory variable, (b) heterogeneity, (c) clustering.

8.3.2 Low birth weight probability in South Carolina: a binomial model example

In this example we are going to analyse the probability of low birth weight babies in South Carolina in the year 2000 together with data about the percentage of population classified by the USA Bureau of Census as in poverty. Figure 8.6 shows a map of the poverty pattern in the state.

The data about births, taken for the South Carolina Department of Health and Environment Control (*http://scangis.dhec.sc.gov/scan/index.html*), contains the number of low weight births and the total number of births, both broken by race, in the 46 counties of South Carolina.

Let y_{ik} be the number of low weight births in county i and race k (white and black) and n_{ik} the corresponding number of total births, where $i = 1, \ldots, 46$ and $k = 1, 2$. We assume

$$y_{ik} \sim \text{ binomial } (p_{ik}, n_{ik}),$$

where the probability of a low birth weight baby in area i and for race k, p_{ik}, is given by

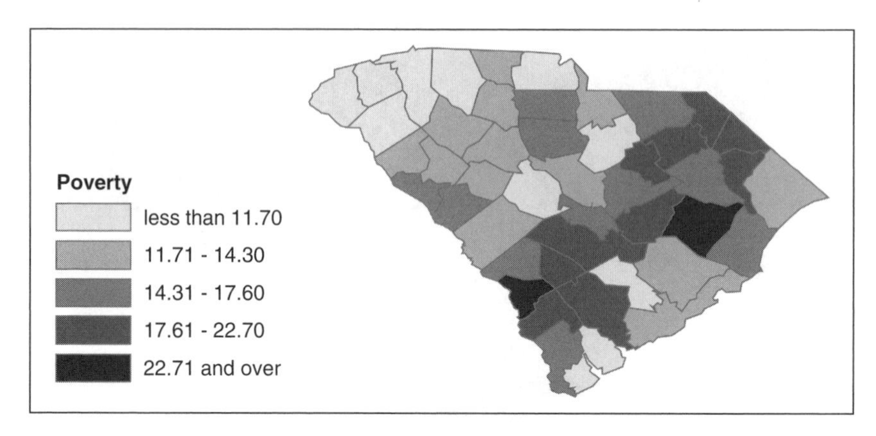

Figure 8.6 Percentage of population classified by the USA Bureau of Census as in poverty: South Carolina, 1999.

$$\text{logit}(p_{ik}) = \alpha + u_i + v_i + \beta_1 \cdot x_i + \beta_2 \cdot r_k,$$

where x_i indicates the poverty in the ith area and r_k indicates the race (0 if white, 1 if black). As usual, u_i and v_i indicate the spatially-correlated and uncorrelated heterogeneity random effects.

The WinBUGS code for fitting this model is presented in Figure 8.7.

Using the above model we run 15 000 MCMC iterations, discarding the first 5000 samples as preconvergence burn-in. Thus, we use a total of 10 000 samples to compute the posterior summaries.

Table 8.4 shows the goodness of fit of various models with race and poverty as covariates. The addition of the covariate poverty improves, although not significantly, the goodness of fit of the models.

When both correlated and uncorrelated heterogeneity terms are added, the model with both covariates yields better results in terms of DIC. Figures 8.8 and 8.9 display the uncorrelated and correlated heterogeneity components in model 4.

The coefficients for the covariates, their standard deviations (sds) and 95% credible intervals for the full model (model 4) are shown in Table 8.5.

Finally, Figures 8.10 and 8.11 map the probability of a low birth weight baby in South Carolina for whites and blacks, respectively. Both maps present similar patterns.

```
model
{
   for (i in 1:m)
   {
      for (k in 1:K)
      {
         y[i,k]~dbin(p[i,k],n[i,k])
         # Probability
         logit(p[i,k])<-alpha+beta1*x[i]+beta2*r[k]+u[i]+v[i]
      }
      # Prior distribution for the uncorrelated heterogeneity
      v[i]~dnorm(0, tau.v)
   }

   # CAR prior distribution for spatial correlated heterogeneity
   u[1:m]~car.normal(adj[],weights[],num[],tau.u)

   # Weights
   for(i in 1:sumNumNeig)
   {
      weights[i]<-1.0
   }

   # Prior on regression coefficients
   beta1~dnorm(0.0,tau.beta1)
   beta2~dnorm(0.0,tau.beta2)

   # Improper prior distribution for the mean relative risk in the study region
   alpha~dflat()

   # Hyperpriors
   tau.v~dgamma(0.1,0.001)
   tau.u~dgamma(0.1,0.001)
   tau.beta1~dgamma(0.1,0.001)
   tau.beta2~dgamma(0.1,0.001)
}
```

Figure 8.7 WinBUGS code for example 8.3.2.

Table 8.4 Value of DIC for the models fitted in this example.

	Covariates	
Model	Race	Race + poverty
1. *no area effects*	1333.370	1335.410
2. u_i	644.749	644.621
3. v_i	642.602	642.381
4. $u_i + v_i$	642.825	642.225

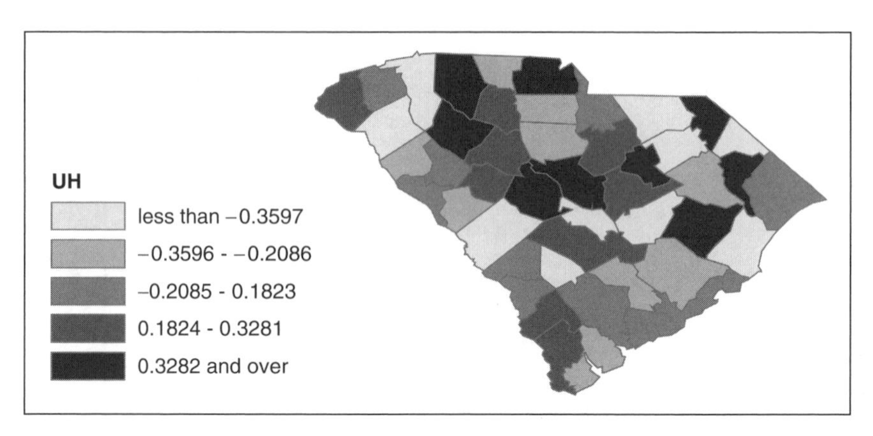

Figure 8.8 Posterior expected value for the uncorrelated heterogeneity component. Model 4.

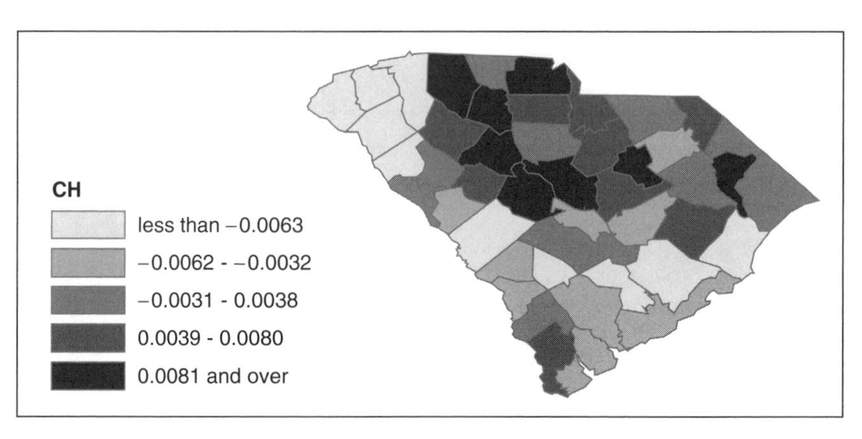

Figure 8.9 Posterior expected value for the correlated heterogeneity component. Model 4.

Table 8.5 Coefficients for covariates in model 4.

Coefficient	sd	95% credible interval
β_1 −0.01272	0.01105	(−0.03695, 0.00775)
β_2 0.02928	0.03347	(−0.03377, 0.09720)

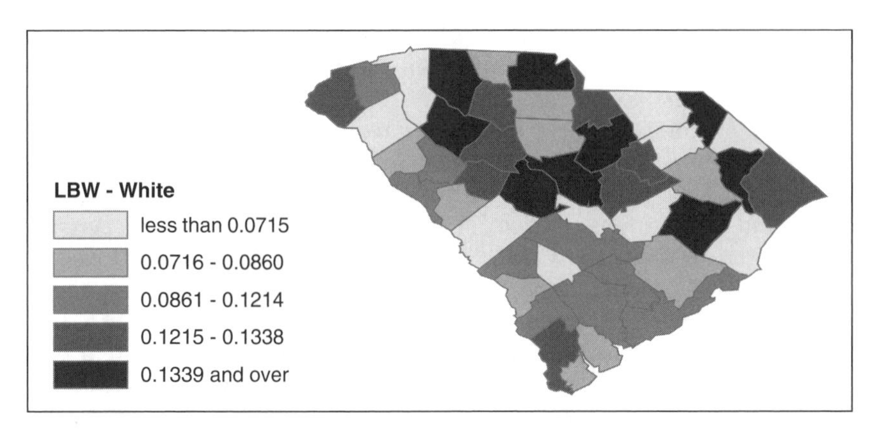

Figure 8.10 Probability of low birth weight babies among white people in South Carolina, 2000.

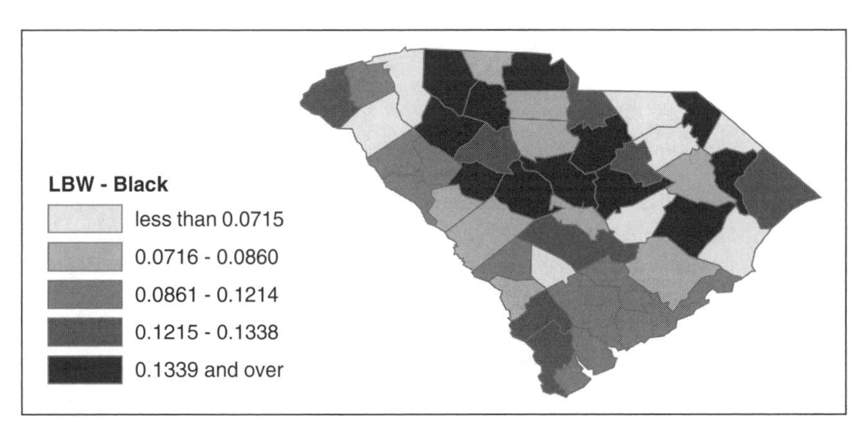

Figure 8.11 Probability of low birth weight babies among black people in South Carolina, 2000.

8.3.3 Lip cancer in the former East Germany: an example of geographically weighted regression

In this example we extend our analysis methods to allow for local variation in the association between the cancer distribution and an associated ecological covariate. In this case we examine lip cancer mortality for the years 1980–1989 in the former East Germany. The percentage of the population employed in agriculture, fisheries and forestry (AFF) is thought to be associated with the incidence of lip cancer. The data consist of counts of disease within the *Landkreise* (219 municipalities), with expected rates calculated from the national rates for 18 age/gender groups.

In regression models where the cases have a geographical location, sometimes regression coefficients do not remain fixed over space. Techniques for exploring this phenomenon are discussed, for example, in Brunsdon *et al.* (1999) and Fotheringham *et al.* (1998) and more recently by Assuncao *et al.* (2002) and Assuncao (2003). The original idea of geographically weighted regression (GWR) was developed as a type of nonparametric regression where the estimate of the local main effect is a function of a local neighbourhood defined by a kernel function. Instead of the estimate being a function of the whole observed dataset, it becomes a function of a spatially-restricted subset of locations. In the original formulation, the local regressions are estimated by least squares and cross-validation. Here we assume a simple Bayesian version of this model where each location has a regression parameter but this parameter has a correlated prior distribution, thereby allowing for similarity at a higher level of the hierarchy.

In this study, the basic model can be written as follows

$$\log(\theta_i) = \alpha_0 + \beta_i x_i,$$

where β_i has a spatial correlation structure, described in section 6.1.5. of this volume. The other models fitted here are extensions of this one when uncorrelated and spatially-correlated heterogeneity terms are included.

The code for fitting the basic model is presented in Figure 8.12. The covariate is defined as AFF. Figures 8.13 to 8.19 display a range of estimated maps relating to a variety of models fitted.

Table 8.6 shows the goodness of fit of the models with and without the covariate with different combinations of heterogeneity terms. The varying coefficient model without heterogeneity terms displayed a DIC of 1363.4. The fixed coefficient model including heterogeneities (u_i, v_i) displayed a DIC of 1181.9.

The addition of the covariate with fixed coefficients improves the goodness of fit of each model. The model that yields best results, for the GWR model, in terms of DIC is the one with both heterogeneity terms. Including the covariate with regression coefficients that are spatially varying increases the value of the DIC. The inclusion of v_i and u_i makes a significant improvement to the GWR model. Hence in this example there is little support for localized regression modelling without random effects. We have also found that model 2 is never better than a fixed coefficient model with random effects (model 1) in a range of examples (e.g. the Scottish lip cancer and London Health Authority examples).

```
model
{
 for(i in 1 : N) {
#h[i]~dnorm(0.0,tau.h)
    O[i]  ~ dpois(mu[i])
    log(mu[i]) <- log(E[i]) + alpha0 + b[i] *AFF[i]
    RR[i] <- exp(alpha0 + b[i] * AFF[i])   # Area-specific relative risk (for maps)
 }

 # CAR prior distribution for random effects:
 b[1:N] ~ car.normal(adj[], weights[], num[], tau)
 for(k in 1:sumNumNeigh) {
    weights[k] <- 1

 }

 # Other priors:
 alpha0 ~ dflat()
 tau ~ dgamma(0.5, 0.0005)    # prior on precision
 sigma <- sqrt(1 / tau)              # standard deviation
}
```

Figure 8.12 WinBUGS code for ecological analysis with spatially varying regression coefficients.

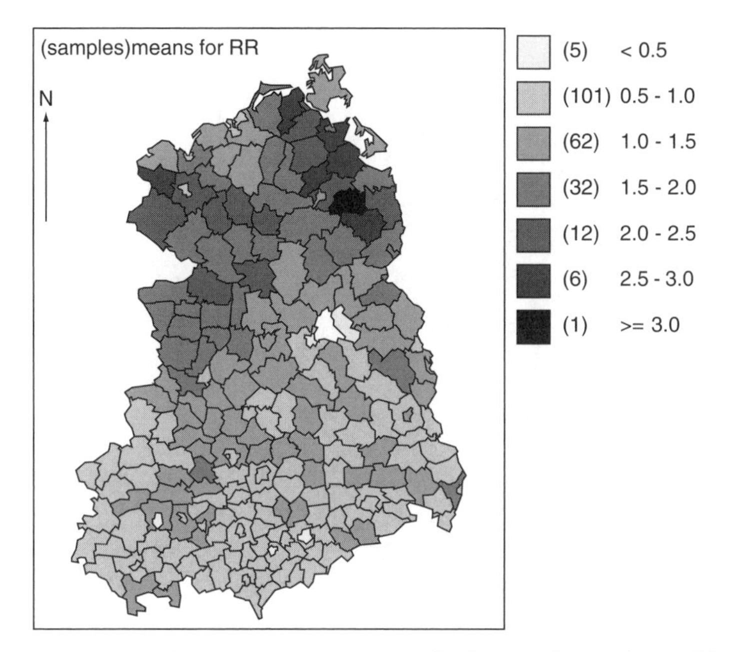

Figure 8.13 Eastern Germany: posterior expected relative risk map for model 1.

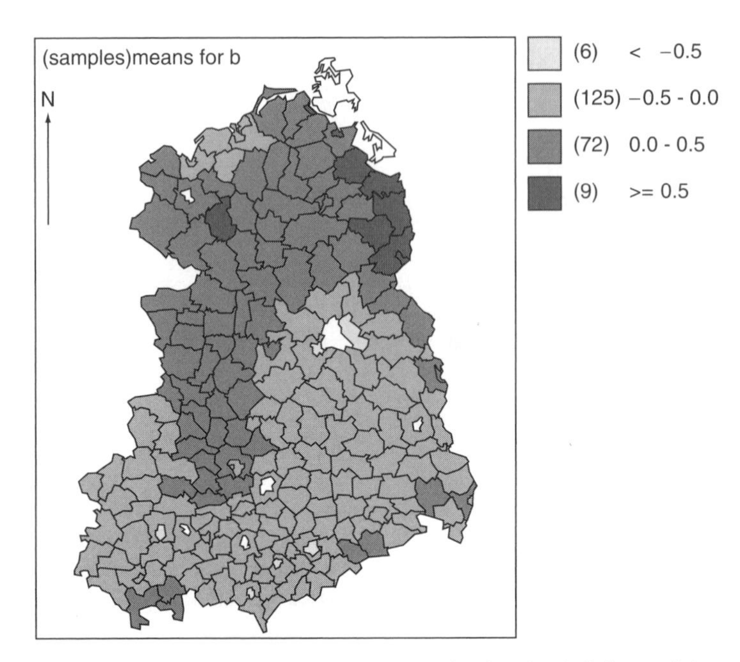

Figure 8.14 Eastern Germany: posterior expected value for $\{u_i\}$ for model 1.

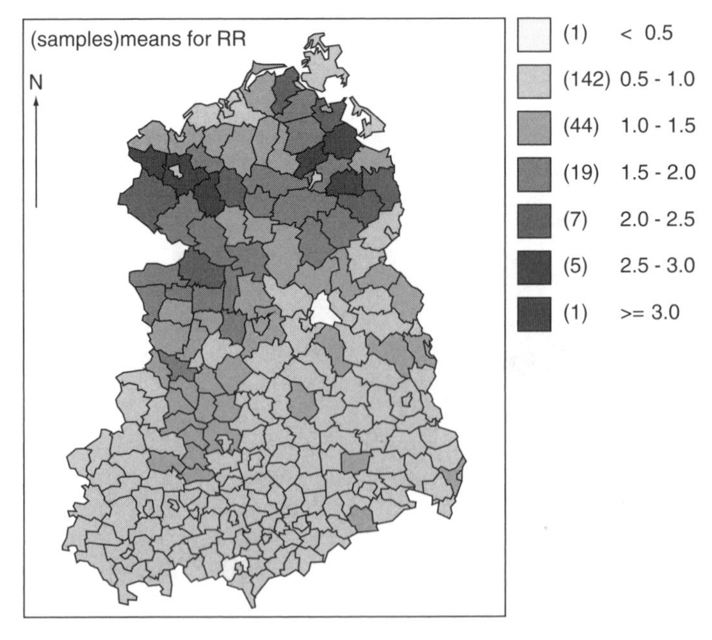

Figure 8.15 Eastern Germany: posterior expected relative risk for GWR model 2.

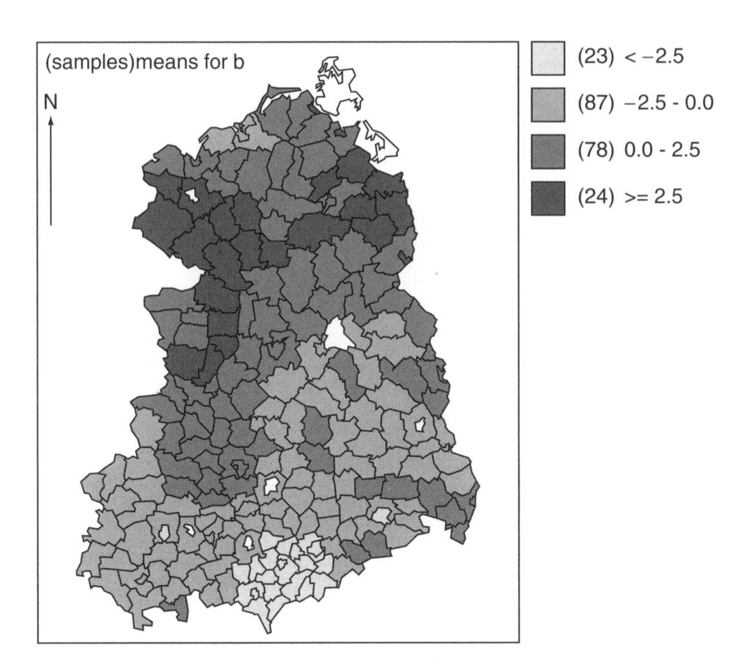

Figure 8.16 Eastern Germany: posterior expected value for $\{\beta_i\}$ for model 2.

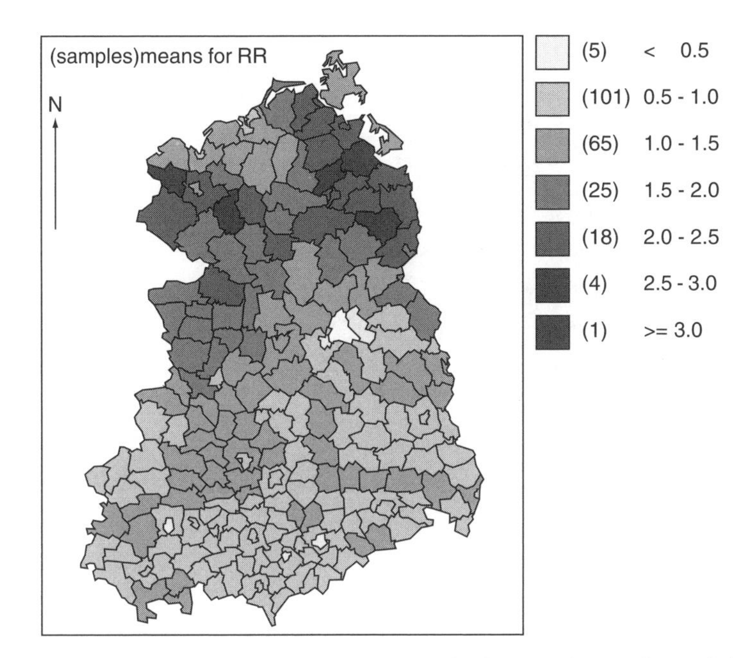

Figure 8.17 Eastern Germany: posterior expected relative risk map for model 4.

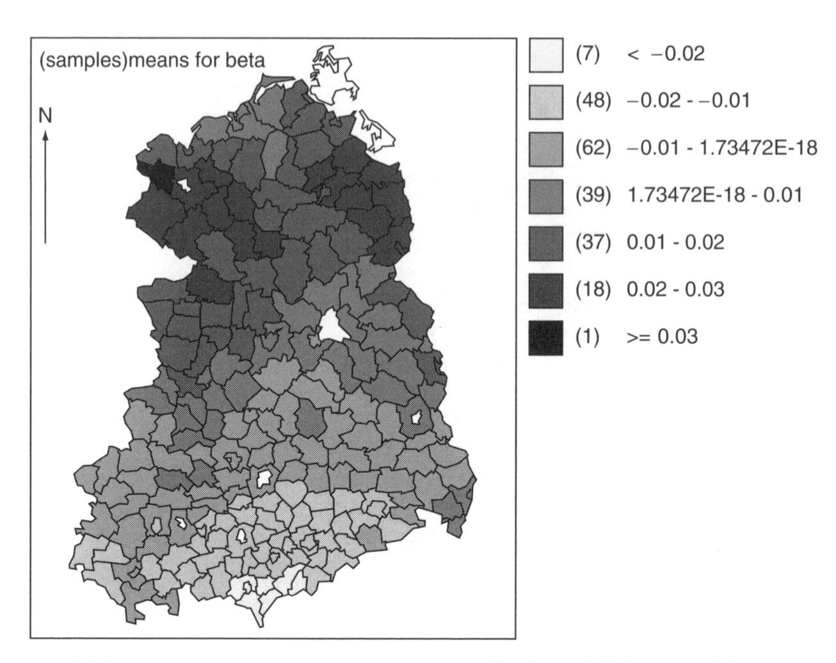

Figure 8.18 Eastern Germany: posterior expected value of $\{\beta_i\}$ for model 4.

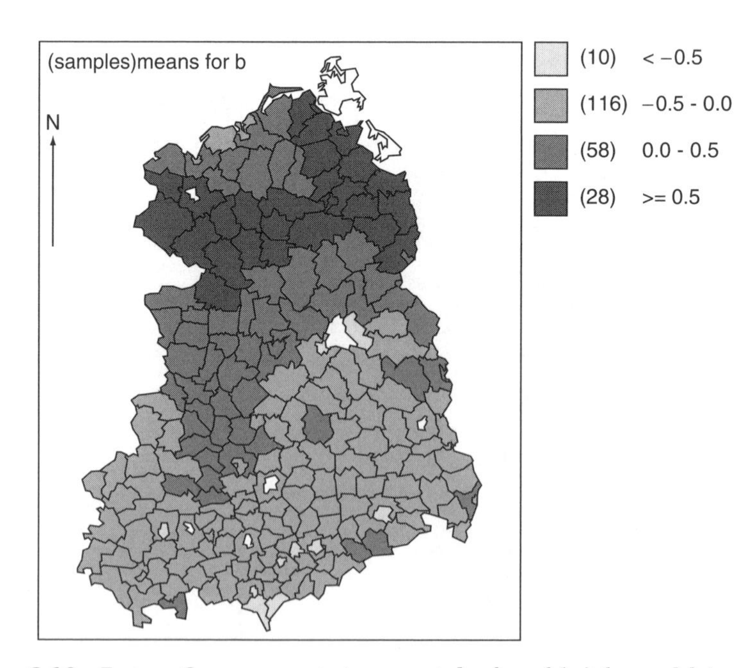

Figure 8.19 Eastern Germany: posterior expected value of $\{u_i\}$ for model 4.

Table 8.6 Value of DIC for the models fitted in example 8.3.3.

Model	DIC
1. $\beta x_i + v_i + u_i$	1181.9
2. $\beta_i x_i$	1363.4
3. $\beta_i x_i + v_i$	1216.0
4. $\beta_i x_i + v_i + u_i$	1190.8

8.3.4 Mortality from congenital abnormality in South Carolina: misaligned data

In this final example we examine the relation of the distribution of counts of congenital abnormality deaths for the year 1990 in the counties of South Carolina. This example is based on one year of data found in Section 1.2. The purpose of this example is to demonstrate the possibility of incorporating covariates in the analysis which are measured with error and further to highlight the possibility of incorporating data which is spatially misaligned. By this we mean that the relevant value of the covariate must be estimated by interpolation to the locations of the dependent variable. Hence some error must be assigned to the interpolated estimate.

For the purposes of this example we have simulated a uranium field based on measured variability of groundwater uranium in the state of South Carolina. The field has been sampled at 30 sites. For the purposes of this simple example we have used a crude white noise simulation of the uranium around a constant surface mean level. While this does not include correlation between sites, it will serve to highlight some of the issues related to the use of a spatially-sampled covariate within an ecological analysis.

Where measures are made at spatial sampling sites, and these measures are to be related to disease risk at other locations or within small areas, the issue of mismatched spatial units must be considered. First, it is important to consider how the covariate could affect the disease incidence. In the case of small area or tract counts, the incidence has been collected over a region and so the individual's risk within these regions could vary across the small area unit or tract. The covariate, on the other hand, is a measure made on a spatial field. Some attempt must be made to match the measurement made to the disease risk in the small area unit. Usually it is appropriate to try to match the covariate using a form of interpolation. Two simple solutions could be considered. First, it could be assumed that the disease count can be 'located' at the small area centroid and the covariate interpolated to that point. An alternative would be to consider that the covariate should be interpolated over the whole of the small area. The second approach is more suited to situations where it is felt that the disease risk and hence the disease count arises from variation over the small area and the total covariate effect over the small area is appropriate to model. These two situations correspond to interpolation based on simple kriging and block kriging in geostatistics (see, for example, Cressie, 1993). In the case of groundwater uranium it is likely that variation over small areas will be small and hence it is adequate to adopt the first approach.

Interpolation of uranium measures to the centroids of small areas can be achieved in a variety of ways. First, simple plug-in interpolation estimators could be used. However, this would not allow the inclusion of the interpolation error in the subsequent analysis. An alternative is to assume a spatial model for the measures which allows interpolation. A variety of possibilities exist for such modelling. Here we explore some simple ideas related to neighbourhood smoothing of the measures. Ultimately, in WinBUGS, it is possible to employ the *spatial.exp, spatial.pred* and *spatial.unipred* functions which provide Bayesian kriging estimators and predictions. These functions allow the specification of a parameterized spatial covariance function for the measures (see Section 6.2). We do not pursue this option here for a variety of reasons. First, the functions can be very slow to iterate due to the need to invert a covariance matrix at each iteration. In addition we have found that the unusual grids found in small area centroid meshes can easily lead to failure of the algorithms due to the need for inversion of a singular covariance matrix. It has also been found that large prediction meshes cause failures. Hence, these functions may not be practical in many applications.

Instead we consider localized smoothing models based on the neighbourhood structure of the data. In our examples we choose to employ the following model specification. Denote the measure at location x as $z(x)$, and the set of observed measures as $\{z(x_k)\}, k = 1, \ldots, n$, and $z_k \equiv z(x_k)$ for short. Also denote the counts of disease within m small areas as $\{y_i\}$ and associated expected counts as $\{e_i\}$. Here we assign nearest neighbour measurement sites to provide an approximate local estimate of the measures at a centroid: $\widehat{z}_i$. In general, we define

$$\widehat{z}_i = \frac{1}{card(\delta_i)} \sum_{l \in \delta_i} z_l,$$

which is a local estimate of the field based on a defined neighbourhood δ_i. The neighbourhoods can be defined conveniently by the nearest neighbour principle. The neighbourhood of the first nearest neighbour leads to the adoption of the value at the nearest neighbour. While a second nearest neighbour field leads to the averaging of the values at the first and second nearest neighbours and so on. Clearly, the larger the neighbourhood the smoother the resulting estimate.

Here we assume that our estimate of the field also contains an error: $\widehat{s}_i = \widehat{z}_i + \epsilon_i$. In this way it is possible to incorporate either or both uncorrelated heterogeneity and correlated heterogeneity within the model for the measurement field. Our model is then:

$$\log(\theta_i) = \log(e_i) + \alpha_0 + \beta\widehat{s_i} + u_i + v_i,$$

where the final terms are the correlated and uncorrelated random effects. At the next level of the hierarchy the terms ϵ_i, u_i, and v_i can be defined by the usual normal prior distributions. In fact ϵ_i could also consist of correlated and uncorrelated components. A correlated component in ϵ_i allows for the correlation between the local means which is implicit in the formulation.

In this approach the question arises as to whether the covariate should be included with error. If the error is purely measurement error then there is a case to exclude the error in the regression formulation because it could be argued that the influence of the covariate itself and not the error-contaminated form is experienced by the small area population. However, the error may also be considered to include some frailty effects and make allowance for the induced correlation in the neighbourhood averaging process. We have included the error in the models fitted here.

Figure 8.20 displays the centroids of the South Carolina county system and the network of groundwater uranium sampling sites. This network is quite irregular, as is commonly the case when sampling has not been developed to estimate spatial features but to assess particular local characteristics. Figure 8.21 displays a needle plot of the uranium measurements (μg/l) found at the sites. The analysis of this data consisted of the following models:

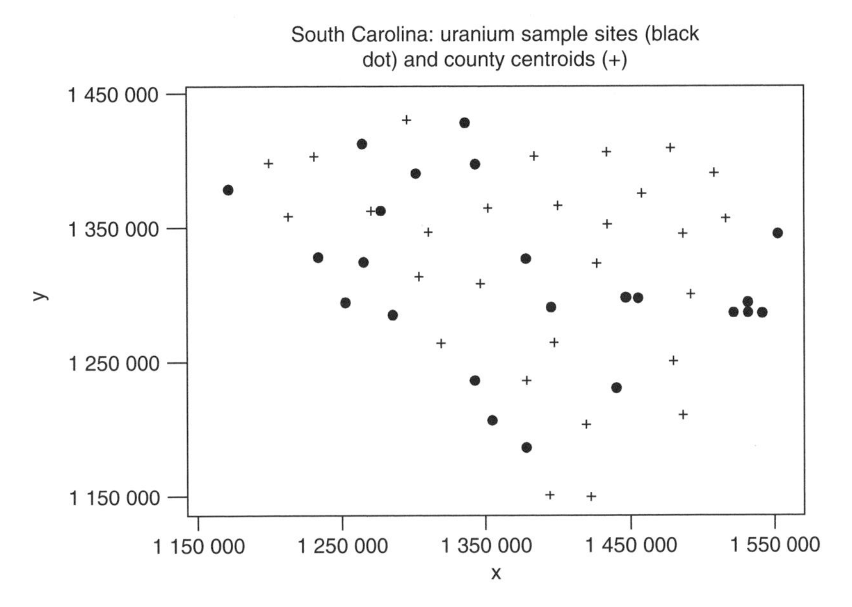

Figure 8.20 Sample sites and county centroids for the uranium example.

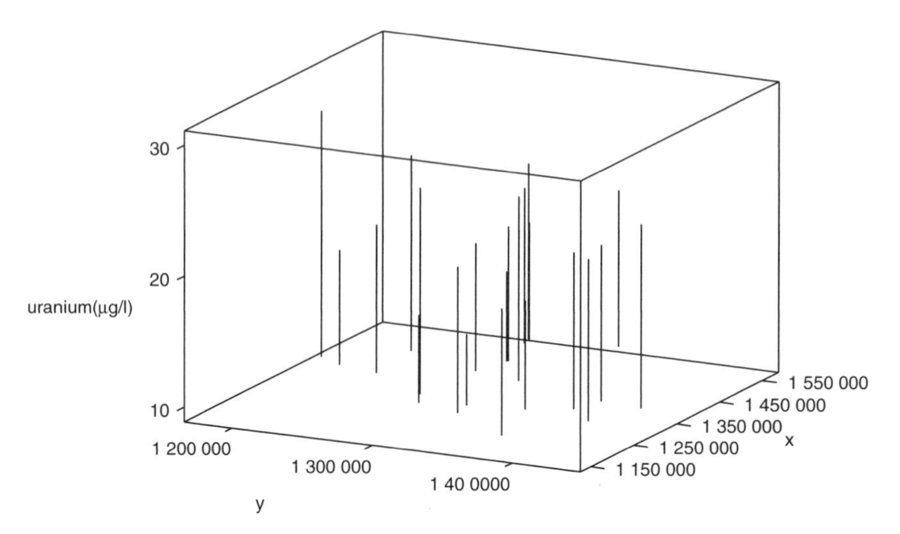

Figure 8.21 Needle plot of groundwater uranium levels measured at sample sites.

(1) An overall mean model: $\widehat{z_i} = \bar{z}$ with $\widehat{s_i} \sim N(\widehat{z_i}, \tau_s)$.

(2) Models based on first, third and fifth nearest neighbour (NN) sets but with:

$$\widehat{z_i} = \frac{1}{card(\delta_i)} \sum_{l \in \delta_i} z_l,$$

where δ_i includes the first or third or fifth NN sets, and $\widehat{s_i} = \widehat{z_i}$.

(3) Models based on first, third and fifth nearest neighbour (NN) sets but with uncorrelated error:

$$\widehat{z_i} = \frac{1}{card(\delta_i)} \sum_{l \in \delta_i} z_l,$$

where δ_i includes the first or third or fifth NN sets, and $\widehat{s_i} = \widehat{z_i} + v_{si}$, where v_{si} has a zero mean normal prior distribution. The code for the latter model with 5NN is given in Figure 8.22.

The results of fitting these different models are given in table 8.7. From this table it is clear that there is little difference in the models based on DIC, although the overall mean and the final first NN models are lowest. The displays below are for the model 2, first NN model, although the overall mean model shows similar results and may be preferred out of parsimony. This effect may be due to the relatively low spatial variability in the simulated data.

```
#this model takes nearest neighbors for the uranium prediction( 1,3, 5 NN can be specified)


model
{

for( i in 1: 46)
{
mhat1[i]<-uran[NNlist5[i,1]]
mhat3[i]<-(uran[NNlist5[i,1]]+uran[NNlist5[i,2]]+uran[NNlist5[i,3]])/3
mhat5[i]<-(uran[NNlist5[i,1]]+uran[NNlist5[i,2]]+uran[NNlist5[i,3]]+uran[NNlist5[i,4]]+uran[NNlist5[i,5]])/5
}


#
#  nowthe Poisson bit

  for(i in 1 : 46) {
O[i]  ~ dpois(mu[i])
hu[i] ~ dnorm(0, tau.h)        # Unstructured random effects
urpred[i]<-mhat5[i]
     log(mu[i]) <- log(E[i]) + alpha0+  alpha1*urpred[i] + hu[i]
     RR[i]  <- exp(alpha0 +  alpha1*urpred[i] + hu[i])   # Area-specific relative risk (for maps)

  }



  # Other priors:
#tau~dgamma(0.001,0.001)
  alpha0  ~ dflat()
alpha1~dnorm(0.0,1.0)
 tau.h  ~ dgamma(0.001,  0.001)

  }
```

Figure 8.22 WinBUGS code for the nearest neighbour model.

Table 8.7 Uranium model fitting results.

Model	$\bar{D}$	$\widehat{D}$	DIC	$\widehat{\beta}^*$
(1) overall mean	166.7	165.7	167.6	−0.0470
(2) 5NN	164.8	158.9	170.7	−0.0530
(2) 3NN	164.6	158.4	170.7	−0.0440
(2) 1NN	163.8	158.7	168.9	−0.0379
(3) 5NN	164.2	157.3	171.0	−0.0481
(3) 3NN	164.2	155.9	172.6	−0.0391
(3) 1NN	163.7	156.2	170.3	−0.0337

*Posterior expected value.

Figures 8.23, 8.24, 8.25 and 8.26, display the SMR map and, for the first NN model, the relative risk surface, the uncorrelated heterogeneity surface and the uranium prediction surface.

In conclusion, in this section, we have tried to demonstrate the wide range of possible types of models which we can fit within the limitations of the currently available WinBUGS functions. The difficulty of using kriging functions is balanced by the flexibility of specifying distance-based smoothing models with

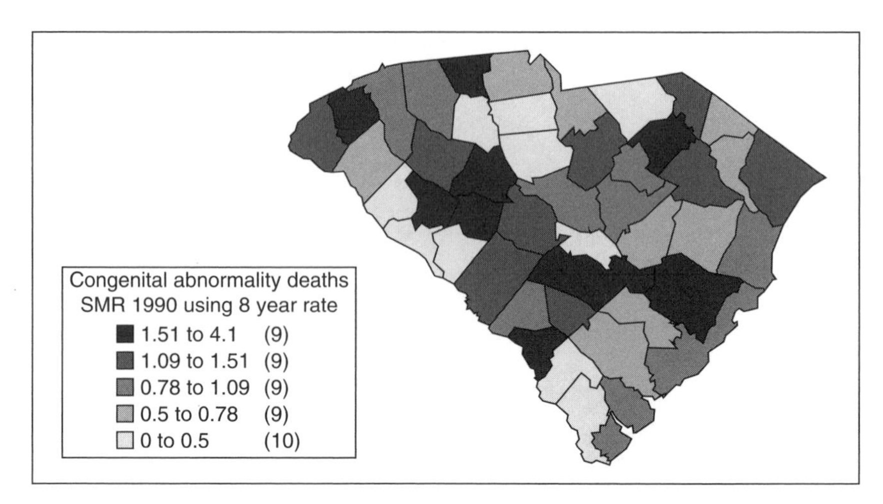

Figure 8.23 SMR map of South Carolina congenital abnormality deaths, 1990.

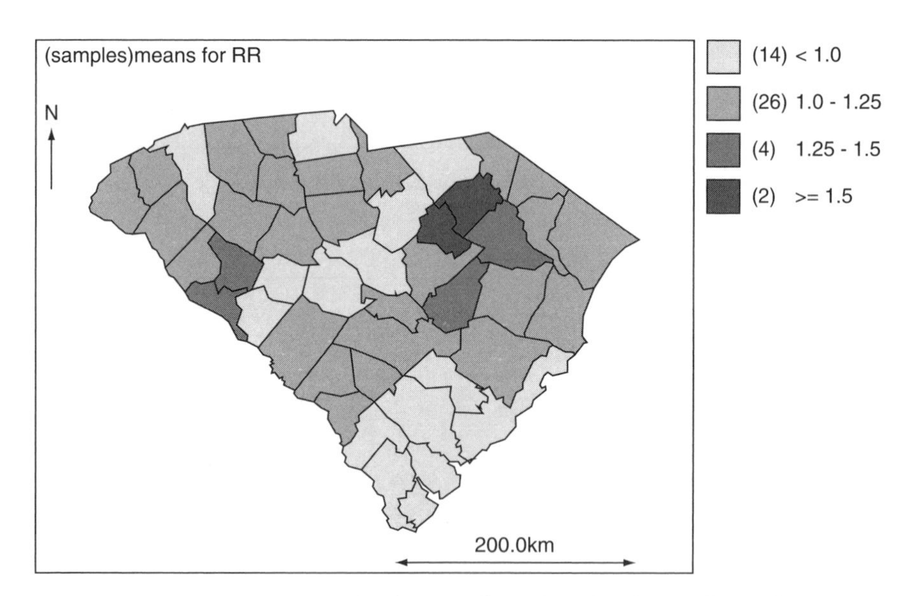

Figure 8.24 Posterior expected relative risk surface for the model 2, first nearest neighbour.

associated error terms. While these models are approximate, they do provide a relatively simple means of incorporating misaligned data. Other possibilities which might be feasible within WinBUGS could lie with the use of site topology to provide interpolants; another variant might be to consider a nonparametric smoothing function as an interpolator. We have not explored these options here.

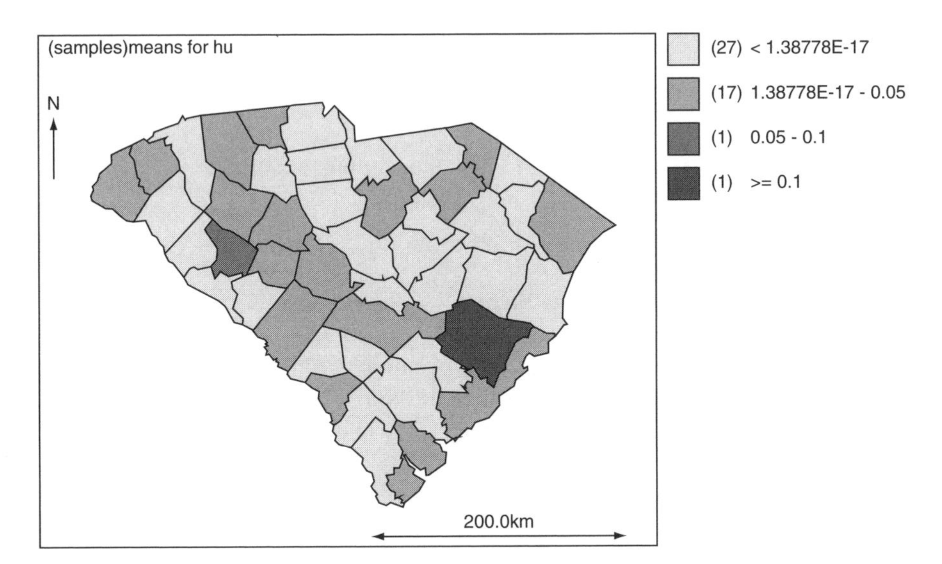

Figure 8.25 Posterior expected uncorrelated heterogeneity for the model 2, first nearest neighbour.

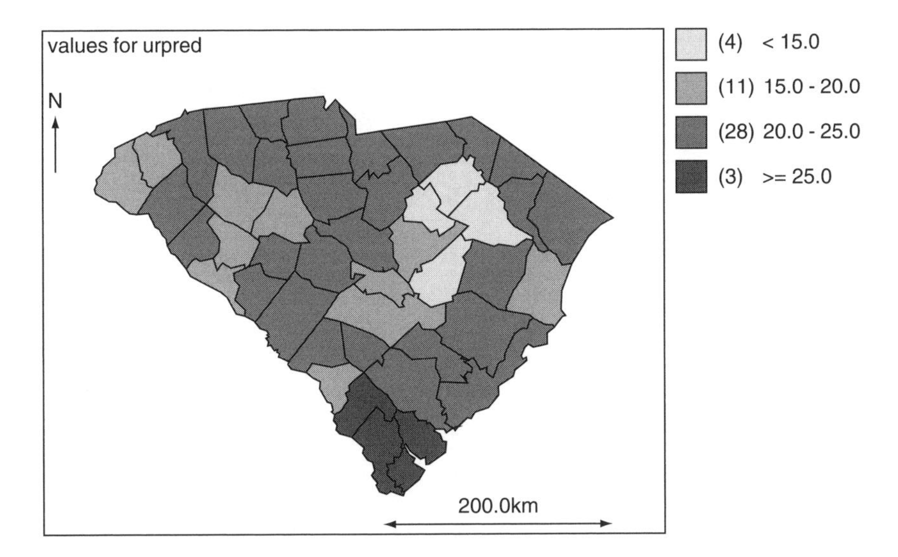

Figure 8.26 Uranium prediction surface for model 2, first nearest neighbour.

8.4 MLwiN ANALYSES OF ECOLOGICAL DATASETS

As we have seen in earlier sections of this chapter, ecological analysis is concerned with describing the relationship between geographical variation in disease incidence in the presence of other explanatory covariates. Fitting models with many predictor variables is common practice in multilevel modelling in general and so is commonly done in MLwiN. In this section we will briefly consider the same datasets that have been considered in earlier sections of this chapter and reanalyse the datasets using MLwiN. We will include slightly less detail of how to set up the models in MLwiN as we will assume the user will now be familiar with the **Equations** window interface.

8.4.1 Cancer mortality in South Carolina

The first example is concerned with a set of data containing 8080 deaths due to malignant neoplasms in the 46 counties in South Carolina in 1999. Here the ecological covariate is population density as it is commonly believed that the higher the population density, the greater risk of cancer. The dataset is stored in the worksheet '*sc_cancer.ws*' which looks as follows:

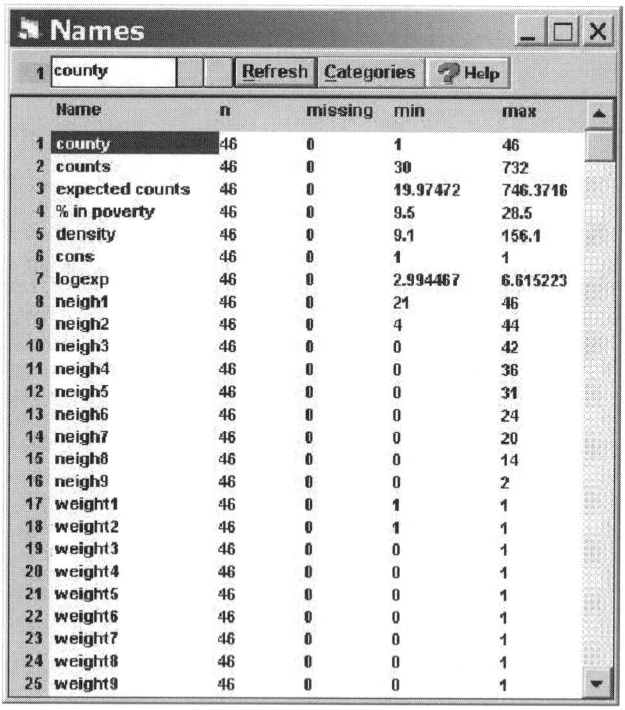

Names

1 county | Refresh | Categories | Help

	Name	n	missing	min	max
1	county	46	0	1	46
2	counts	46	0	30	732
3	expected counts	46	0	19.97472	746.3716
4	% in poverty	46	0	9.5	28.5
5	density	46	0	9.1	156.1
6	cons	46	0	1	1
7	logexp	46	0	2.994467	6.615223
8	neigh1	46	0	21	46
9	neigh2	46	0	4	44
10	neigh3	46	0	0	42
11	neigh4	46	0	0	36
12	neigh5	46	0	0	31
13	neigh6	46	0	0	24
14	neigh7	46	0	0	20
15	neigh8	46	0	0	14
16	neigh9	46	0	0	2
17	weight1	46	0	1	1
18	weight2	46	0	1	1
19	weight3	46	0	0	1
20	weight4	46	0	0	1
21	weight5	46	0	0	1
22	weight6	46	0	0	1
23	weight7	46	0	0	1
24	weight8	46	0	0	1
25	weight9	46	0	0	1

A preliminary analysis would consist of comparing the predictor '*density*' with the SMR which we can create in c26 by typing the commands `CALC c26 = c2/c3` and `NAME c26 'SMR'` in the **Command Interface** window. Then if we use the **Customised Graph** window we get the following:

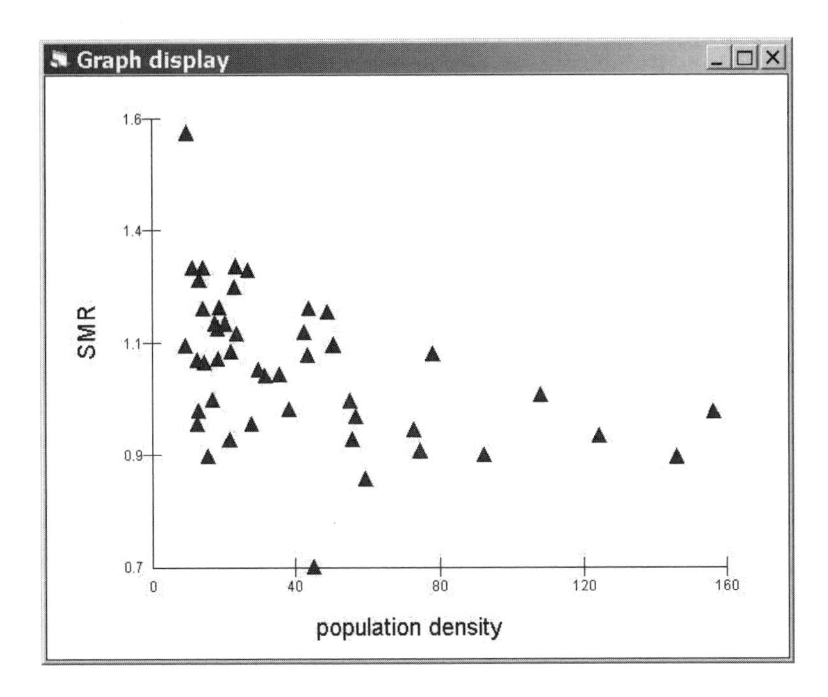

Here we see that generally in South Carolina we observe a negative relationship between SMR and population density (the correlation between the two variables is −0.46.) We would expect greater variability of SMR towards the left of the graph (as we observe) because the number of expected cases will be related to population density (in fact the correlation between the two is +0.96) and so the outlier to the top left is not that surprising. This county is McCormick which had 31 deaths but was expected to have only 20 deaths so had 11 more deaths than expected. Perhaps the more unusual observation is the county with the lowest SMR. This county is Berkeley and was expected to have 296 deaths but only 196 deaths occurred resulting in 100 fewer deaths than expected. This might be worth investigating further but for now we will simply fit models to the whole dataset. The first model we will fit is a standard Poisson regression model and consists of fitting an intercept plus density as a fixed effect. This can be set up in the **Equations** window by declaring '*counts*' as the response variable, *Poisson* as the distribution type, both '*cons*' and '*density*' as fixed effects, '*logexp*' as an offset column and '*county*' as the level 1 identifier. The model when set up and run using first-order MQL will produce the following window:

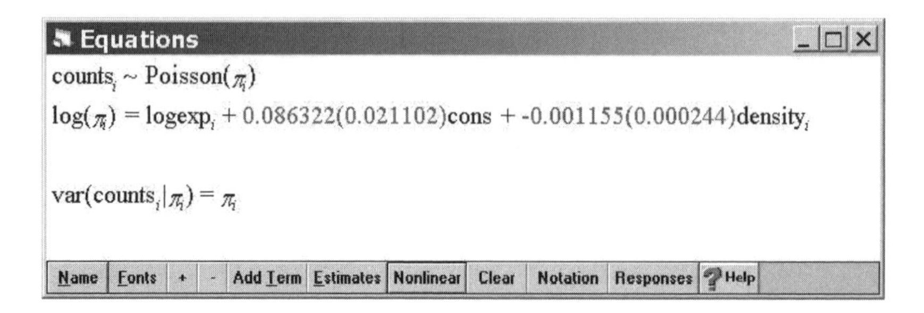

Here we see the similar estimate for the density effect that was observed in the BUGS analysis of this dataset. (Note that in MLwiN we have increased the displayed decimal places from 3 to 6 in the **Options** window.) If we allow extra Poisson variation we observe the following:

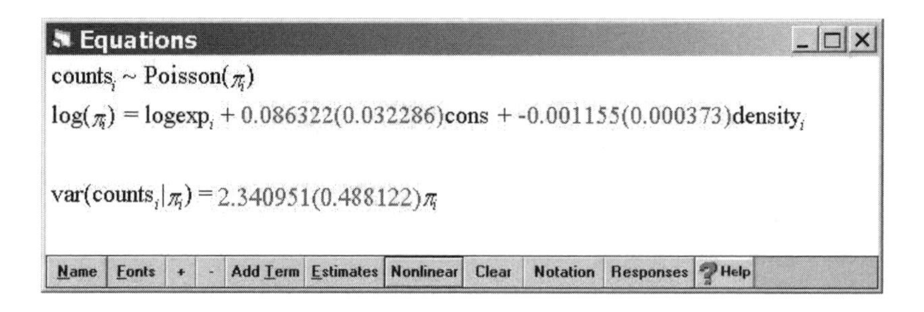

Here we see that there is a large amount of extra variation, suggesting that we need to include some random effects to account for this variation. We will here use MCMC (in MLwiN) to fit simple random effects, multiple-membership (MM) models and CAR models. These can be set up as we did with the lip cancer dataset in the earlier chapters. Note that the columns '*weight 1*'–'*weight9*' are set up for a CAR model and so for a MM model we need to create a new variable in a free column, for example *c27*, that would contain the number of neighbours. This can be achieved via the command

```
CALC c27 = c17 + c18 + c19 + c20 + c21 + c22 + c23 + c24 + c25.
```

We then need to divide each of the weight columns by column *c27*. The following table contains the density effect estimates (plus standard errors) along with the DIC diagnostic and deviance for each model.

Model	β	SE(β)	Deviance	DIC
Density	−0.001167	2.43E−4	418.70	420.65
Density + unstructured	−0.001479	5.13E−4	346.90	372.05
Density + CAR	−0.001594	4.88E−4	349.65	372.69
Density +MM	−0.001699	4.01E−4	351.49	372.83
Density + unst. +MM	−0.001588	5.12E−4	346.73	371.99

Here we see that all the models suggest a negative relationship between population density and deaths from cancer. Of the models fitted here the lowest DIC is given by the model with both unstructured residuals and multiple-membership residuals although there is little to choose between this model and three of the other models. Typically not fitting the population density measure as a fixed effect resulted in a DIC that was larger by around three points over the equivalent model with the density measure giving evidence that population density is an important predictor. If we removed the Berkeley county observation that we highlighted earlier we will find that the negative effect of population density increases slightly so this observation is not very influential in the conclusions we can make from the model.

8.4.2 Low birth weight babies in South Carolina

Our second example concerns a second dataset on the same 46 counties of South Carolina this time in the year 2000. This dataset is, however, different as the at-risk population is not all people in the state but only the newborn children in the year 2000, a population of size 55 962. The response that plays the role of 'disease' in this scenario is whether the babies had a low birth weight, which was true in roughly 10% of births. Each baby was classified by race as to whether they were white, black or 'other' and here we will consider only the white and black babies (the 'other' category only contained around 1100 births in total across the 46 counties.) As low birth weight is a much more common affliction than the cancers we have looked at so far and we know the exact counts for each region, we do not have to use a Poisson assumption but can instead model the proportions of low birth weight babies in each county as a binomial distribution.

We can firstly load up the worksheet '*sc_lbw.ws*' and we will see the names in the **Names** window as follows:

Names

1 county | Refresh | Categories | Help

	Name	n	missing	min	max
1	county	92	0	1	46
2	% in poverty	92	0	9.5	28.5
3	cons	92	0	1	1
4	lbw	92	0	1	335
5	ab	92	0	28	4141
6	race	92	0	1	2
7	%lbw	92	0	3.521127E-02	0.2131148
8	neigh1	92	0	21	46
9	neigh2	92	0	4	44
10	neigh3	92	0	0	42
11	neigh4	92	0	0	36
12	neigh5	92	0	0	31
13	neigh6	92	0	0	24
14	neigh7	92	0	0	20
15	neigh8	92	0	0	14
16	neigh9	92	0	0	2

In the dataset we have for each county two records, one for the white babies and one for the black babies and so we have a two-level structure with '*race*' nested within '*county*'. The response variable we will use is '*%lbw*' which has been formed by dividing the numbers of low birth weight babies, '*lbw*' by the total births '*ab*'. Our predictor variables of interest are firstly the effect of race and secondly for each county we have the variable '*% in poverty*'. We can firstly plot the response against the poverty predictor for each of the white and black groups and look for any patterns that may be observed:

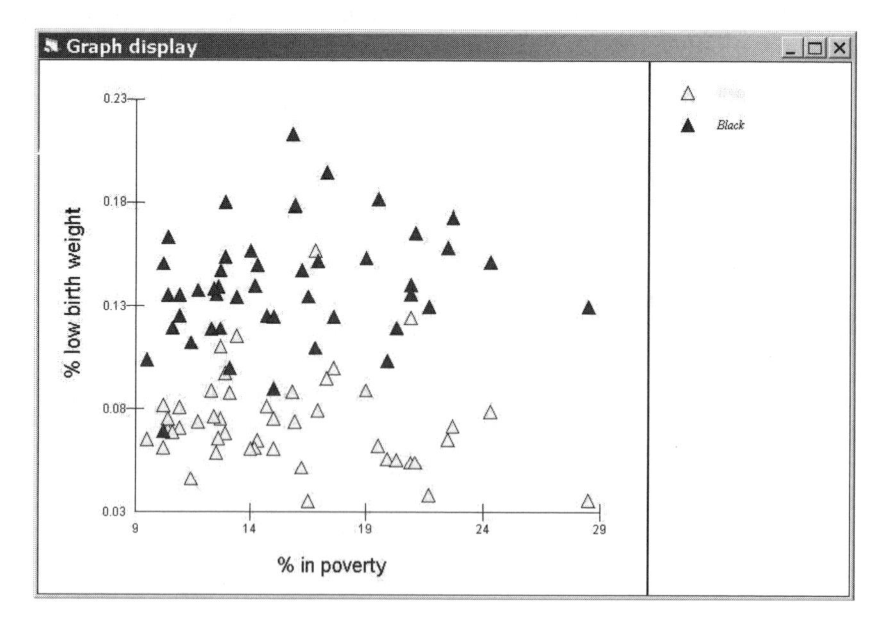

Here we see that there is strong evidence for a link between low birth weight and race, as in nearly every county there are proportionally more low birth weight black babies. There is, however, no evident link between poverty and low birth weight (correlation 0.035) but plotting the data as we have may be misleading as it does not take account of the relative proportions of black and white babies in each county. We can, however, construct variables (via the **Multilevel Data Manipulations** window and the sum operation) that contain the total births and low birth weight babies in each county, and from this information create a new variable that gives the percentage of total births that are low birth weight for each county. We can also do these operations via the following commands in the **Command Interface** window:

```
MLSUm "county" "lbw" C31
MLSUm "county" "ab" C32
CALC C33 = (c31 / c32) * 100
NAME C33 "%lbwall"
```

Note that the new column will contain the percentage low birth weight babies in each county repeated twice so that the column is the same length as the other columns. Next we can plot our newly created variable "*%lbwall*" against the poverty variable as shown below:

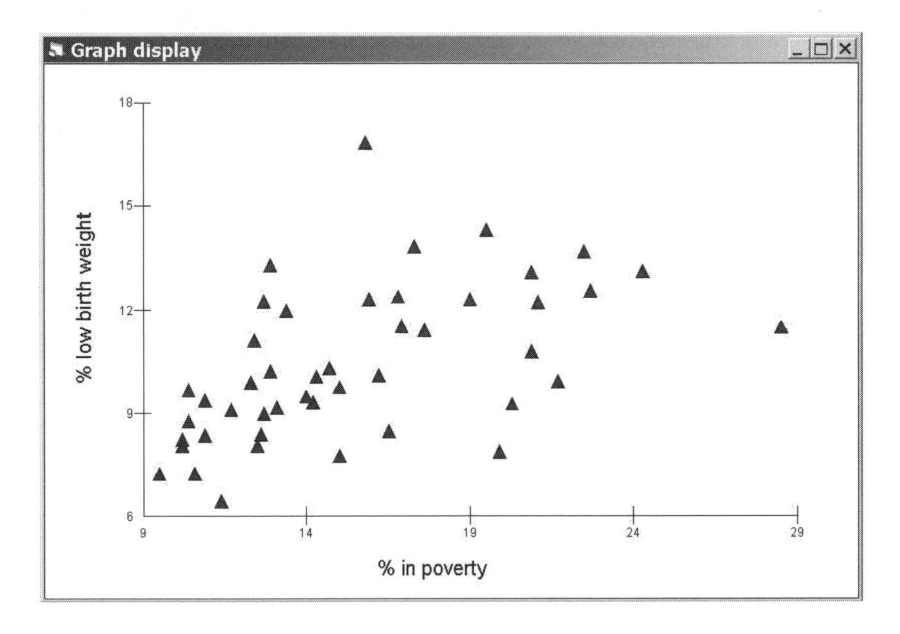

Here we see that there appears to be a positive relationship between the percentage of low birth weight babies and the percentage of poverty in the area (correlation 0.52) which was not obvious from the earlier graph where the races were treated separately. We will now consider fitting some models to the dataset.

Firstly we need to set up a basic binomial model in the **Equations** window. To do this we require to do the following. Firstly set the response variable as '*%lbw*' and the data structure as two levels with '*race*' at level 1 and '*county*' at level 2. To specify the binomial response we need to click on the *N* and choose *binomial* from the window that appears. By default the link function will be set to logit which will fit a logistic regression. We next need to click on the red n_{ij} and set the denominator variable to '*ab*' (all births). Finally we will click on the red x_0 and select '*cons*' to fit a model with just a constant term. If we now run the model using the first-order MQL estimation method we will get the following:

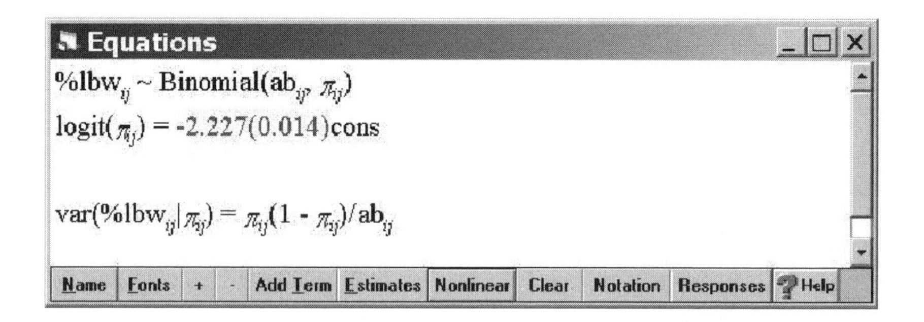

Here we see an estimate of -2.227 for β_0 which can be translated into a probability via the antilogit function (in the **Command interface** window typing CALC ALOG(-2.227)) and corresponds to an overall probability of a low birth weight of 0.097. If we want to allow effects for both race and poverty we can add fixed effects via the **Add Term** window. A model with these additional terms will give the following estimates:

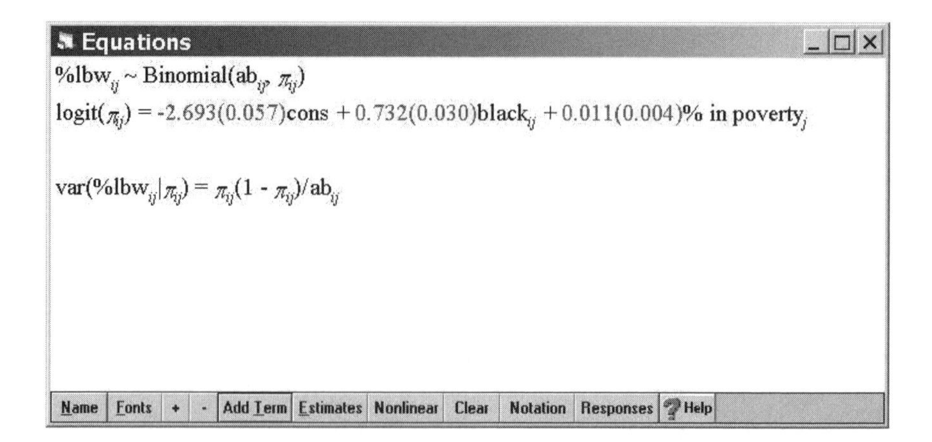

Here we see that there are a greater percentage of low birth weight babies for blacks than whites and the greater the poverty in the district the greater percentage of low birth weight babies. We can use the anti-logit function to work out specific probabilities; for example, the following table gives expected probabilities for four scenarios:

White/black	% Poverty	$X\beta$	Predicted %
White	0	-2.693	6.3%
Black	0	$-2.693 + 0.732$	12.3%
White	9.5	$-2.693 + 9.5^{*}0.011$	7.0%
Black	28.5	$-2.693 + 0.732 + 28.5^{*}0.011$	16.1%

Here we see that, depending on race and poverty level, the probability of a low birth weight baby can range from 7% to 16%. Interestingly we can extend the model to include an interaction term between race and percentage in poverty. If we do this we find that the poverty main effect becomes almost zero whilst the interaction has a positive effect. This suggests that the effect of poverty in the neighbourhood only effects the birth weight of black babies. Using the DIC diagnostic and MCMC estimation we find that the model with the interaction is no better than the model with a single poverty effect, but a model with just a race effect and an effect of poverty and black babies gives a DIC which is two points lower and hence a better model. If we now consider this final fixed effect model we can use quasi-likelihood estimation and, as with the Poisson models, we can relax the binomial assumption and include extra binomial variation. This will result in the following model:

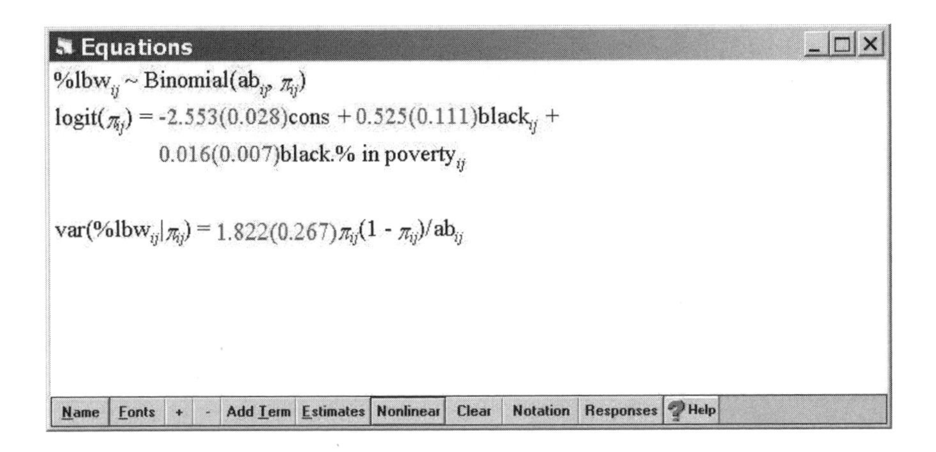

Here we see that again we have more variation than would be expected under a binomial assumption and this is probably due to the fact that we have not accounted for variation between counties apart from that due to the differences in their levels of poverty. We can therefore transform our fixed effect model into a multilevel model by declaring '*cons*' as random at level 2 (county). This will result in the following model:

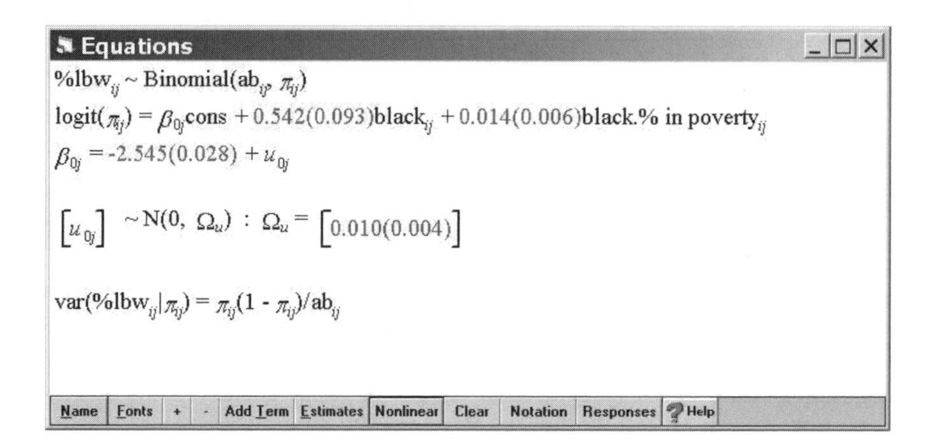

Here we see evidence of some variability captured between the counties in terms of percentage low birth weight babies. As we have more than one measure per county we can extend this model further by allowing the race effect also to vary between counties. This is done by clicking on the '*black*' variable in the model and ticking the '*(j)county*' tickbox. The resulting model, as shown below, suggests that the effect of race also varies between counties.

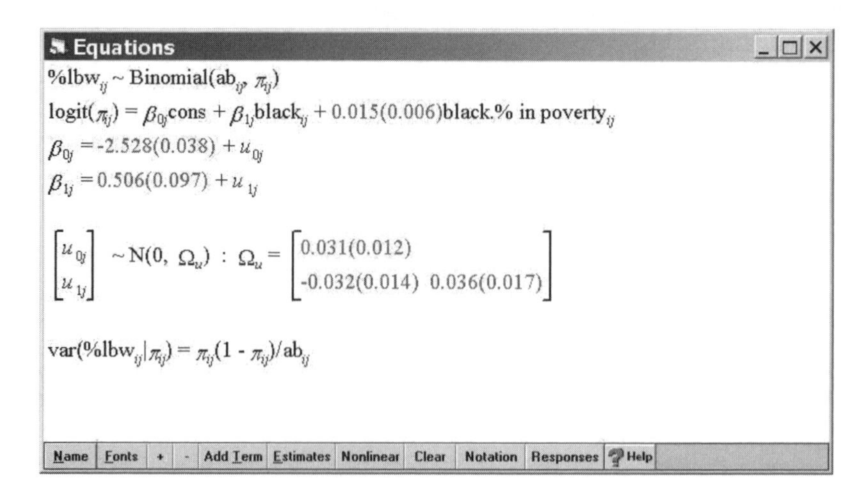

To compare the last three models it is easiest to change estimation method to MCMC (with suitable default priors) and compare the DIC diagnostic as shown in the following table:

Model	Deviance	DIC
Fixed effects	823.51	826.50
Random county effects	785.63	808.23
Random county + race effects	756.99	790.30

Here we see that fitting both random effects for the counties and for the effect of race on each county gives a significantly better model. Up to this point we have not used any spatial information in our models, but this was already done in the WinBUGS analysis of this dataset.

8.4.3 Lip cancer in the former East Germany

In this section we consider a slightly larger dataset of lip cancer mortality in the years 1980–1989 in the former East Germany. The East Germany dataset is geographically interesting as some of the regions, e.g. Berlin, are excluded or isolated and do not have any nearest neighbours. There are in total seven such regions out of the 219 regions (municipalities) some of which are islands which, unlike the islands in the Scottish lip cancer dataset that we looked at in Chapter 5, are also defined as having no neighbours. The covariate that we are interested in, as in the Scottish dataset, is the percentage of the population that are working in agriculture, fisheries or forestry. This is often thought to be a surrogate for exposure to the sun of the general population as such employment will involve long hours outdoors. If we open the worksheet '*lipseg.ws*' we will see the following in the **Names** window:

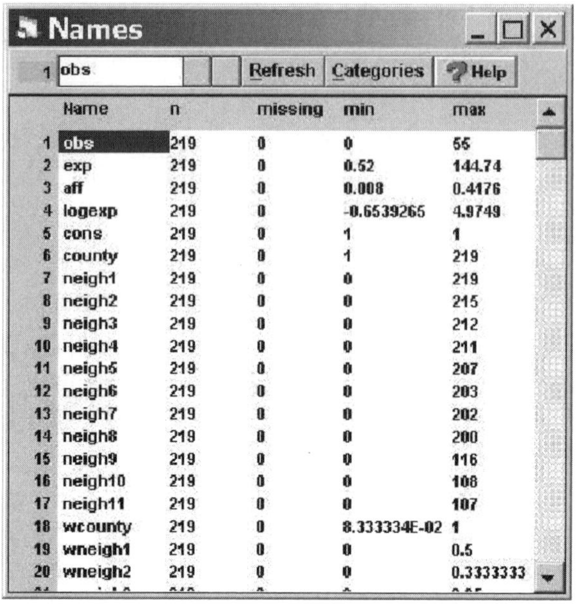

Names

1 obs | Refresh | Categories | Help

	Name	n	missing	min	max
1	obs	219	0	0	56
2	exp	219	0	0.52	144.74
3	aff	219	0	0.008	0.4176
4	logexp	219	0	-0.6539265	4.9749
5	cons	219	0	1	1
6	county	219	0	1	219
7	neigh1	219	0	0	219
8	neigh2	219	0	0	215
9	neigh3	219	0	0	212
10	neigh4	219	0	0	211
11	neigh5	219	0	0	207
12	neigh6	219	0	0	203
13	neigh7	219	0	0	202
14	neigh8	219	0	0	200
15	neigh9	219	0	0	116
16	neigh10	219	0	0	108
17	neigh11	219	0	0	107
18	wcounty	219	0	8.333334E-02	1
19	wneigh1	219	0	0	0.5
20	wneigh2	219	0	0	0.3333333

Here we have for each region an observed count of lip cancer deaths '*obs*', an expected count, '*exp*' and the covariate, the percentage of people employed in agriculture, fishing and farming, '*aff*' which ranges from 0% to 42%. Each region has between 0 and 11 neighbouring regions which are stored in columns, '*neigh1*' to '*neigh11*'. As always we should firstly do some preliminary analysis and so we will firstly plot the SMRs (stored in column *c30*) against the covariate '*aff*'. Such a plot can be seen below.

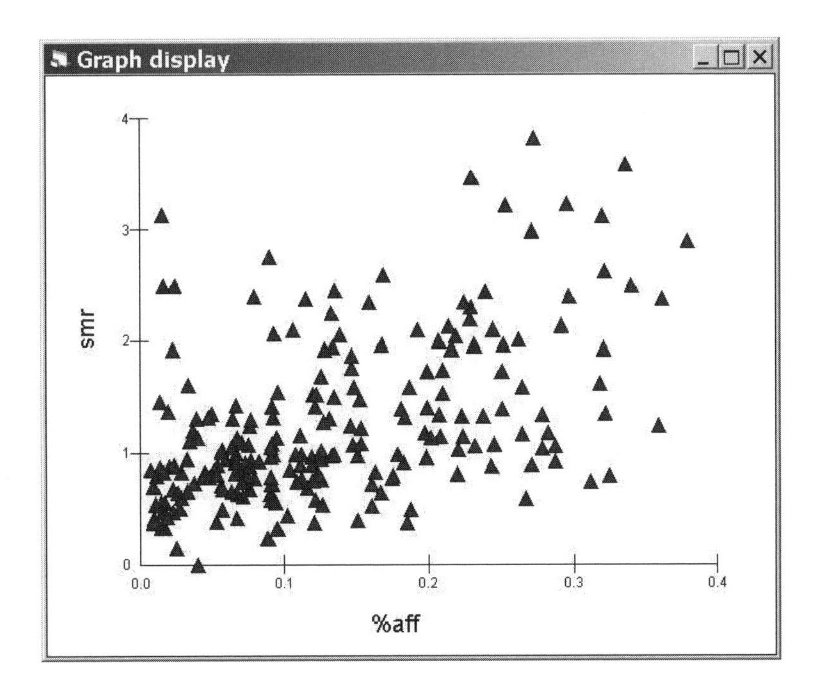

Here we see that there appears to be a positive association between SMR and the covariate (correlation of 0.5) and there also does not appear to be any obvious outlying observations. We can therefore firstly fit a simple Poisson regression in MLwiN and test for overdispersion by fitting the model using quasi-likelihood methods and allowing extra-Poisson variation. The results can be seen below:

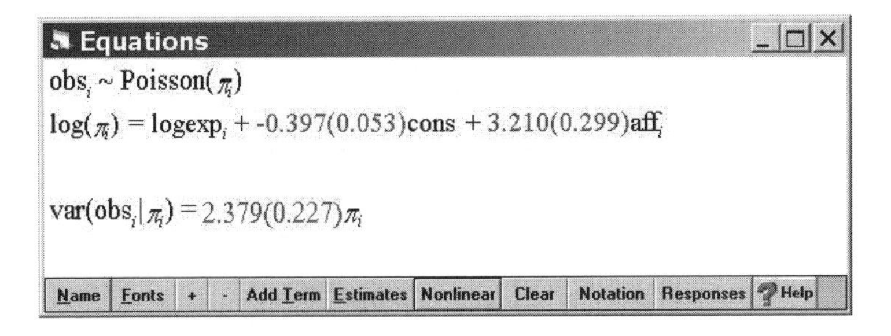

Here we see the strong positive effect of the covariate and also that there is a large amount of unexplained extra-Poisson variation, suggesting we need to fit some additional random effects in the model. One of the motivations of this dataset is to consider fitting models where the effect of the covariate also varies between the regions in the model. We will do this in two ways firstly by considering a simple random slopes type model where we fit a different (unstructured) effect for each region but with these effects assumed to follow a normal distribution. Following the notation of v for unstructured effects we could write such a model as

$$\log(\theta_i) = \beta_0 + \beta_1 x_i + v_{0i} + v_{1i} x_i, \qquad v_i \sim MVN(0, \Sigma_v) \tag{8.1}$$

Here we assume that the rate and the effect of the covariate are different in each region and that the two variables are correlated. This may be a bit ambitious as we are estimating two random effects for each region for which we have only one data point! The alternative is to consider a geographically weighted regression (GWR) approach where for each region we assume the effect of the covariate is dependent on the data in the region itself and its nearest neighbouring regions. We will fit a model that is similar to the GWR models described in the earlier sections by considering a special case of the multiple membership model:

$$\log(\theta_i) = \beta_0 + \beta_1 x_i + \sum_{j \in Neigh(i)} w_{i,j}(u_{0j} + u_{1j} x_j), \qquad u_j \sim MVN(0, \Sigma_u) \tag{8.2}$$

As written above, this is the standard multiple-membership model with the SMR being influenced by all of the neighbouring regions to a particular region, plus in this case the covariate in each of the neighbouring regions. Here, however, rather than also including random effects for the regions we will redefine the set $Neigh(i)$ to be the area i plus its nearest neighbours. The weights will be defined as is typical for multiple-membership models as 1/number of elements in $Neigh(i)$. To see more clearly how the weights are defined we can look at the neighbour and weight columns in the **View/Edit Data** window:

Data

goto line 1 | view | Help | Font

	county(219)	neigh1(219)	neigh2(219)	wcounty(219)	wneigh1(219)	wneigh2(219)
1	1	15	11	.2	.2	.2
2	2	37	7	.2	.2	.2
3	3	28	27	.2	.2	.2
4	4	23	16	.25	.25	.25
5	5	37	29	.125	.125	.125
6	6	37	17	.1428571	.1428571	.1428571
7	7	12	5	.25	.25	.25
8	8	24	23	.1428571	.1428571	.1428571
9	9	0	0	1	0	0
10	10	0	0	1	0	0
11	11	6	1	.3333333	.3333333	.3333333
12	12	7	5	.3333333	.3333333	.3333333
13	13	8	0	.5	.5	0
14	14	3	0	.5	.5	0
15	15	24	17	.1666667	.1666667	.1666667
16	16	23	18	.25	.25	.25
17	17	39	37	.1428571	.1428571	.1428571
18	18	23	19	.25	.25	.25
19	19	95	23	.1666667	.1666667	.1666667
20	20	54	51	.125	.125	.125

To fit this model in MLwiN we have had to ensure that the neighbour columns follow directly the county column as this is now part of the function. There are 12 weight columns that have been named '*wcounty*' and '*wneigh1*' through to '*wneigh11*' which also have to be contiguous. Above we see the first two neighbour columns and the first three weight columns. Here we see that counties 9 and 10 have no neighbours and so the sets $Neigh(9) = \{9\}$ and $Neigh(10) = \{10\}$ and $w_{9,9} = w_{10,10} = 1$ whilst region 13 has one neighbour (region 8) and so $Neigh(13) = \{8, 13\}$ and $w_{13,8} = w_{13,13} = \frac{1}{2}$. We will now use MCMC estimation in MLwiN to fit the two models (8.1) and (8.2) along with several submodels. The models and their DIC values can be seen in the following table:

Model	Corr. RE	Deviance	DIC
$\beta_0 + \beta_1 x_i$	—	1356.3	1358.3
$\beta_0 + \beta_1 x_i + v_{0i}$	—	1104.2	1213.0
$\beta_0 + \beta_1 x_i + v_{0i} + v_{1i} x_i$	No	1103.9	1212.8
$\beta_0 + \beta_1 x_i + v_{0i} + v_{1i} x_i$	Yes	1107.8	1211.4
$\beta_0 + \beta_1 x_i + \sum\limits_{j \in Neigh(i)} w_{i,j}(u_{0j} + u_{1j} x_j)$	No	1088.3	1160.7
$\beta_0 + \beta_1 x_i + \sum\limits_{j \in Neigh(i)} w_{i,j}(u_{0j} + u_{1j} x_j)$	Yes	1086.2	1161.6

Here when the sets of random effects are assumed uncorrelated then inverse gamma priors are used for the random effect variances, and when they are correlated inverse Wishart priors are used. We can see from the table that fitting random effects in the model improves the DIC by over 150, but also that moving from the unstructured random slopes regression model to the model with neighbourhood effects also improves the fit by 50 and that this model is a better fit than the best model fitted in WinBUGS earlier by 20. The choice of correlated or uncorrelated random effects does not make a great difference to the DIC or the model fit.

9

Spatially-correlated Survival Analysis

When individuals are monitored for their time to an end-point, survival methods are often used. Instead of modelling counts of incidence in small areas, these methods model the times to the end-point for individuals. In this situation we can include individual covariates within models as well as higher-level effects. In fact it is possible to include within the model specification random effects relating to individual heterogeneity (frailty) and spatial correlation terms.

9.1 SURVIVAL ANALYSIS IN WinBUGS

The following code (supplied by Sudipto Banerjee) allows the specification of uncensored and censored time failure models coupled with such random effect terms. For details of this approach see Banerjee and Carlin (2003) and Banerjee *et al.* (2003). Further examples of WinBUGS code for nonspatial longitudinal and survival examples are given in Congdon (2003), Chapter 9. Spatially-correlated survival WinBUGS model code is available at *http:/www.biostat.umn.edu/brad* for a range of models.

Cancer registry data consists of records of diagnosis for cancer at the individual level. Usually these data contain a range of diagnosis information, including dates of end-points (diagnosis or progression markers) and covariates, including geo-reference information. The following describes a model for the time to death from breast cancer for a subset of a cohort of females within the state of Iowa for the period 1973 to 1998. This data comes from the SEER registry program. The data set modelled here consists of time to death or censoring. The model includes a failure time distribution for the times to end-point. In this case a Weibull distribution has been assumed for the time distribution. In addition,

Disease Mapping with WinBUGS and MLwiN A. Lawson, W. Browne and C. Vidal Rodeiro
 ISBN: 0-470-85604-1 (HB)

Table 9.1 DIC statistics for the three survival models.

Model	$\bar{D}$	$\widehat{D}$	DIC
Full	8284.0	8251.5	8316.5
W/O other	8283.7	8251.9	8315.5
W/O unknown	8285.5	8254.2	8316.8

Table 9.2 Parameter estimates for the three models.

	Full model		W/O other		W/O unknown	
Parameter	Value	sd	Value	sd	Value	sd
ρ	1.120	0.039	1.115	0.043	1.119	0.041
β_0	−8.734	0.478	−8.742	0.532	−8.758	0.491
β_1	0.296	0.069	0.279	0.239	0.277	0.235
β_2	0.037	0.003	0.038	0.004	0.037	0.003
β_3	1.131	0.309	1.124	0.306	1.130	0.302
β_4	−1.183	1.157	—	—	—	—
β_5	−32.470	8.865	−20.710	15.020	—	—
σ	0.491	0.087	0.486	0.086	0.485	0.084
τ	4.557	1.616	4.631	1.634	4.636	1.612

a number of covariates at the individual level are available (age, race, number of primary cancers). At a higher geographical level, the county within which the individual was diagnosed, is also recorded. Within this framework the Weibull time-to-endpoint model is defined as:

$$f(t_i) = \rho\mu_i t_i^{\rho-1} \exp(-\mu_i t_i^{\rho}),$$

where μ_i is modelled as

$$\log(\mu_i) = \beta_0 + \beta_1 x_{1i} + \beta_2 x_{2i} + \beta_3 x_{3i} + \beta_4 x_{4i} + \beta_5 x_{5i} + W_i.$$

Hence the spatial frailty formulation is

$$f(t_i) \exp(\beta^T x_i + W_i),$$

where $x_{1i} =$ number of primary cancers, $x_{2i} =$ age, $x_{3i} =$ black, $x_{4i} =$ other, $x_{5i} =$ unknown and W has a CAR normal distribution supporting the possibility of correlated random effects at the county level.

The results of this model fit and the results of removing the *other* (x_{4i}) and *unknown* (x_{5i}) covariates are given in Table 9.1 and 9.2.

```
model
{
   for(i in 1:Nsubj){
      obs.t[i] ~ dweib(rho, mu[i])I(t.cen[i],)
      log(mu[i])<- beta0 + beta[1]*Primaries[i]+ beta[2]*Age[i] +beta[3]*Black[i]+ beta[4]*Other[i]+beta[5]*Unknwn[i] +
W[CoRes[i]]
   }

   for(i in 1:nsum){weights[i] <- 1}

   W[1:regions] ~ car.normal(adj[],weights[], num[],tau)
   W.mean <- mean(W[])

   beta0 ~ dnorm(0.0,0.001)
   for(i in 1:5) {beta[i]~ dnorm(0.0,0.001)}

   rho ~ dgamma(1,1)
   tau ~ dgamma(1,1)
   sigma <- 1 / sqrt(tau)

}
```

Figure 9.1 WinBUGS code for the full Weibull frailty model.

The full model fit suggests that for most parameters the model is a reasonably good fit. The exception to this is the covariate *other* which appears to not be well estimated. We examined the model without this covariate and the results are also displayed above. The removal of this covariate makes little difference to the model fit, although its removal leaves a number of other variables poorly estimated (*primaries* and *unknown*). Removal of *unknown* has little effect on the model fit also. The WinBUGS code in Figure 9.1 is for the full regression model.

9.2 SURVIVAL ANALYSIS IN MLwiN

In this chapter we will consider disease data in which we have data for the individual cases of disease rather than counts for small areas. The data consists of survival times after diagnosis of breast cancer for a cohort of 2122 women who resided in the US state of Iowa. This dataset is a subset of a larger dataset analysed by Banerjee and Carlin (2003) of the period 1973–1998. The dataset has two levels as the individual cases are nested within counties within the state of Iowa. There is a large amount of censoring in the dataset and for each woman we have a time after diagnosis (in months) when they died or when they were censored. To be classified as a failure (death) the woman has to have died from metastasis of cancerous nodes in the breast. Any other cause of death is included in the censored observations. In the subset of the data 68% of women were considered censored observations. We have several predictor variables for each woman including the age of the woman when breast cancer was detected, the number of primaries, i.e. cancers including breast cancer, and the race of the woman (white, black, other, unknown). At the level of the county we have identifiers for the neighbouring counties. The data is stored in the worksheet '*iowa.ws*' and, as we can see below, we have a combination of a county level datafile and an individual level datafile.

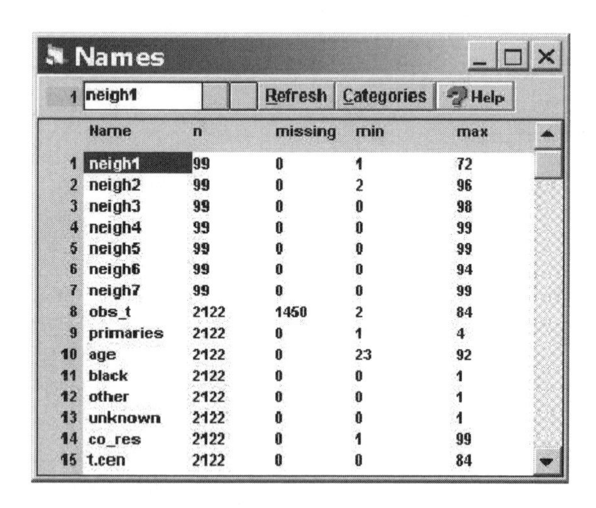

	Name	n	missing	min	max
1	neigh1	99	0	1	72
2	neigh2	99	0	2	96
3	neigh3	99	0	0	98
4	neigh4	99	0	0	99
5	neigh5	99	0	0	99
6	neigh6	99	0	0	94
7	neigh7	99	0	0	99
8	obs_t	2122	1450	2	84
9	primaries	2122	0	1	4
10	age	2122	0	23	92
11	black	2122	0	0	1
12	other	2122	0	0	1
13	unknown	2122	0	0	1
14	co_res	2122	0	1	99
15	t.cen	2122	0	0	84

Here we see that columns *c1–c7* contain the neighbouring county identifiers and are of length 99 (one record per county) whilst the columns *c8–c15* are of length 2122 (one record per individual).

9.2.1 Preliminary data manipulations

The data has not been sorted on county identifiers so before we start we should firstly sort the data and expand the neighbour identifiers to be one record for each individual. Firstly we will sort the data in columns *c8–c15* so that all records for each county are contiguous. The county of residence identifiers are stored in column '*co_res*' and so we can select the **Sort** window from the **Data Manipulation** window and set it up as follows:

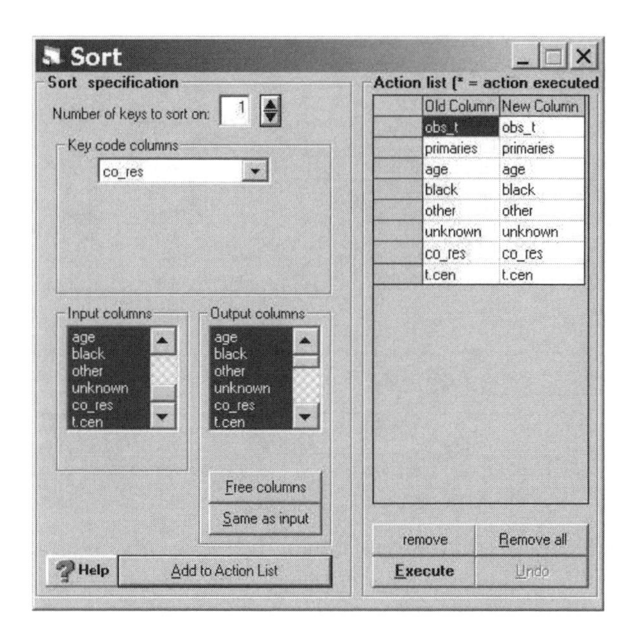

If we press the **Execute** button we will sort the data on the county identifiers. Next we need to match the counties with their neighbours and expand the neighbour columns. To do this we will firstly need a column of length 99 (the number of counties) that identifies the counties. This we can produce using the **Generate Vector** window as follows:

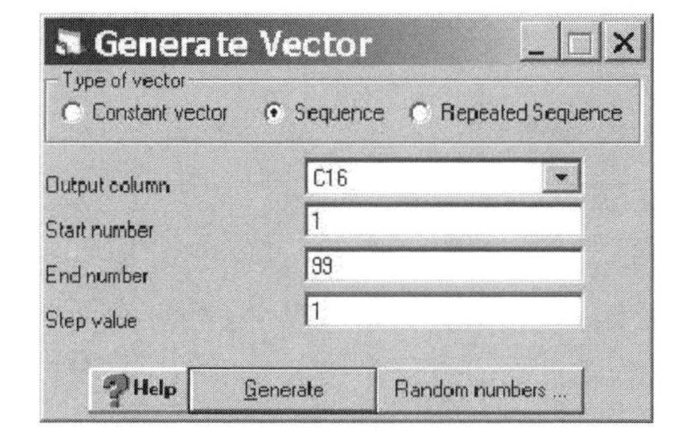

We can name column *c16* as '*county*' via the **Names** window and next we can use the **Merge** window in the **Data Manipulation** menu to match the neighbourhood identifiers to the individual level data. Here we see that we need to define '*county*' as the identifiers to merge from and '*co_res*' as the identifiers to merge to. We will merge the columns '*neigh1*'–'*neigh7*' into the first seven empty columns as shown below:

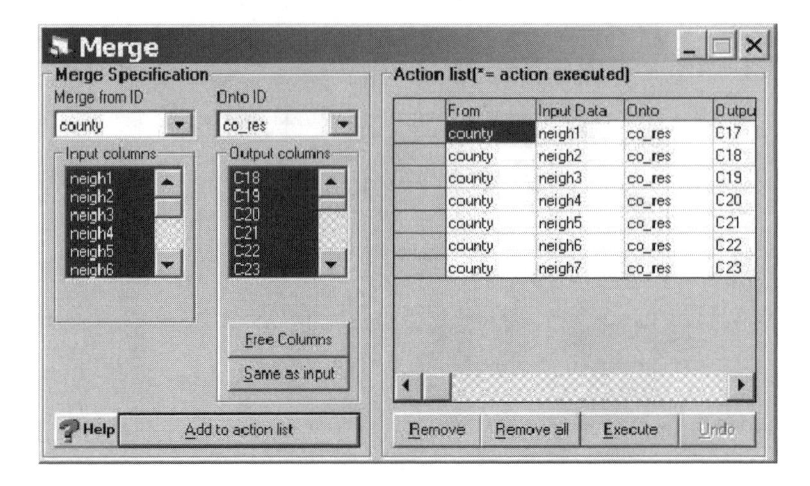

Clicking on the **Execute** button will create the seven new columns in *c17–c23*. We finally need to remove the shorter neighbourhood and county columns and name the new longer neighbourhood columns as '*neigh1*' – '*neigh7*'. To remove the columns and start the columns at *c1* we can use the following commands in the **Command Interface** window: `ERASE c1-c7` and `ERASE c16` will delete the short columns and `MOVE` will move all the columns to fill in the gaps. If we rename the columns then the **Names** window should look as follows:

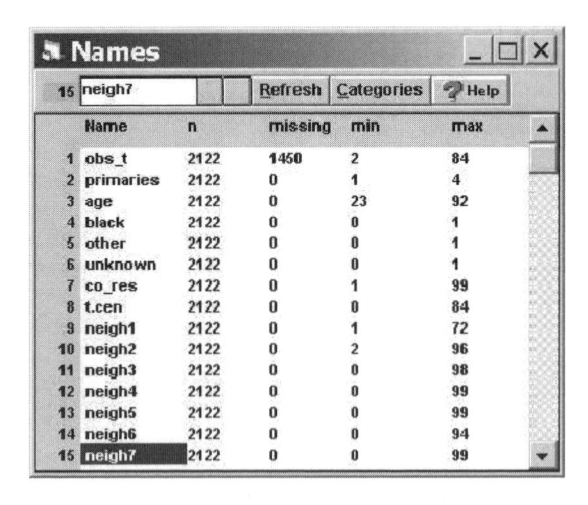

Now we are ready to perform some initial data examinations.

9.2.2 Preliminary data examination

The column '*obs_t*' contains the time in months after diagnosis that a patient dies of breast cancer (a missing value is given if they are censored). The column '*t.cen*' contains the censoring time in months for the censored observations with zeros representing non-censored data. For consistency we will convert these zeros to missing data via the **Recode** window. This can be set up as follows:

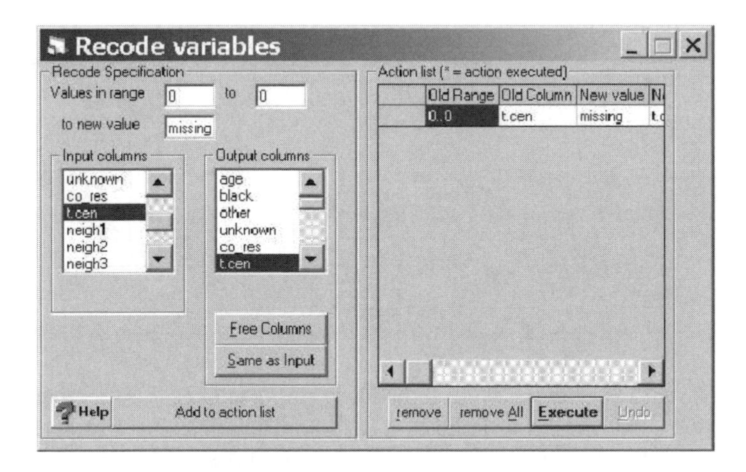

Now we can look at the data by plotting histograms of both the observed deaths and the censoring times as follows:

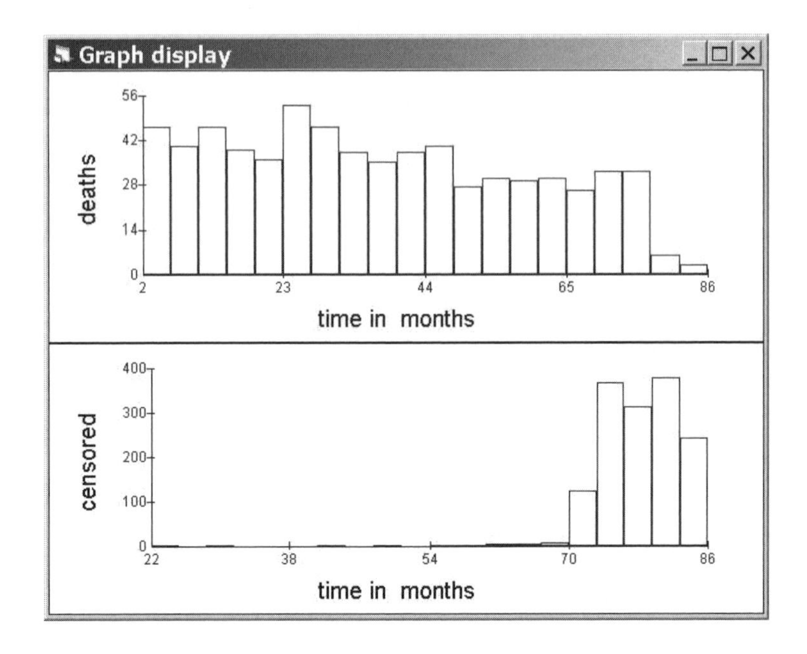

Here we see that the deaths seem to be fairly uniform over the time period with no particular peaks and a little bit of a reduction towards the end, although at this point there will be a smaller population at risk. The censored times, however, seem to generally be at times greater than 70 months, which suggests that most of the censored observations can be considered as individuals who have survived breast cancer for the observation period. We can construct a variable that indicates whether an observation is censored or not and we will do this via the command `CALC c16 = 'obs_t' == missing`. We can now name column *c16* as '*censored*' and explore correlations between this variable and our predictors. If we look at correlations via the **Average and Correlations** window available from the **Basic Statistics** menu, we see negative correlations between censored observations and the number of primaries (−0.12), the age of the patient (−0.26) and the black group dummy variable (−0.05). There is therefore less chance of being censored if you are older or have more cancers. It should be noted that of the 2122 patients, 2085 were white and so any effects for the other race groups will be based on small sample sizes (22 black, 10 other and 5 unknown).

If we instead consider the observed times of death for the 672 deaths in the dataset we find fairly small correlations, positive with the number of primaries (0.05) and negative with age (−0.01) so there does not seem to be as strong a relationship between the actual lengths of survival times of these cases and the predictor variables. We will firstly consider treating whether a patient dies or is censored as a response variable and fit binomial response models to this response then we will consider how we can fit a survival model to the data.

9.2.3 Binary response models for mortality

Before fitting models to the dataset we need to add two additional columns, which we can produce via the **Generate Vector** window as shown previously. Firstly we need a column that contains the patient identifiers (which we will name '*id*') and for this we will create a sequence from 1 to 2122. Then we need a constant column consisting of a column of 2122 1s which we will name '*cons*' and this will be the intercept column. Before fitting any models we will save this model (using the **Save Worksheet** option available in the **File** menu) as '*iowa2.ws*'. We will return to the worksheet in this state later in the chapter.

We will consider 3 models: a fixed effect model, a model with county random effects and a multiple-membership model with both county and neighbour effects. Firstly we will set up a fixed effect logistic regression model. This can be set up in the **Equations** window by specifying '*censored*' as the response variable, and three as the number of levels (for use with the multiple-membership model later) with '*id*' at level 1, '*co res*' at level 2 and '*neigh1*' at level 3. We then click on the N and choose instead *Binomial* as the response type. The Binomial distribution has a red n_{ijk} which can be clicked on and '*cons*' selected as we are fitting binary data and hence have denominators of 1 for each observation. To finish the model we will add '*cons*', '*primaries*', '*age*' and '*black*' as fixed effects, noting that the other race categories have too little data to use in the model. If we run this model with first-order MQL estimation we will get the following estimates:

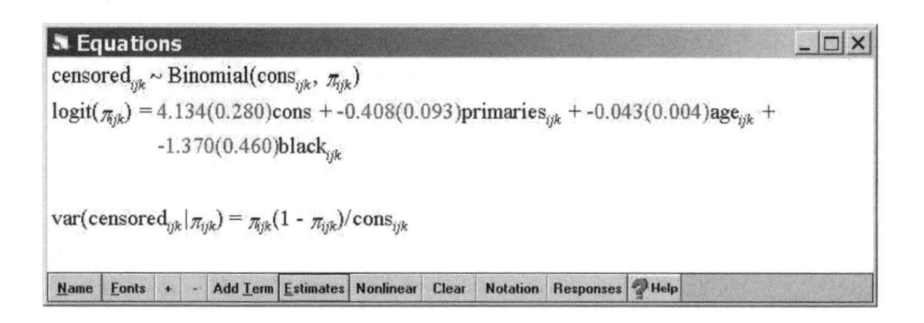

We see here that (as the correlations predicted) the more cancers (primaries) present and the older the patient is the less chance of being censored. The black patient group has a smaller probability of being censored but this group is very small (22 patients) and as we see the standard error for this coefficient is reasonably large. We can now progress on to fitting random effects and a multiple-membership model. For these models we will use MCMC estimation so that we can compare the models via the DIC diagnostic. It should be noted that weight columns for the multiple membership model should be constructed in a similar way to the Ohio dataset example in the relative risk chapter. The results of fitting a multiple-membership model for 50 000 iterations using MCMC are as follows:

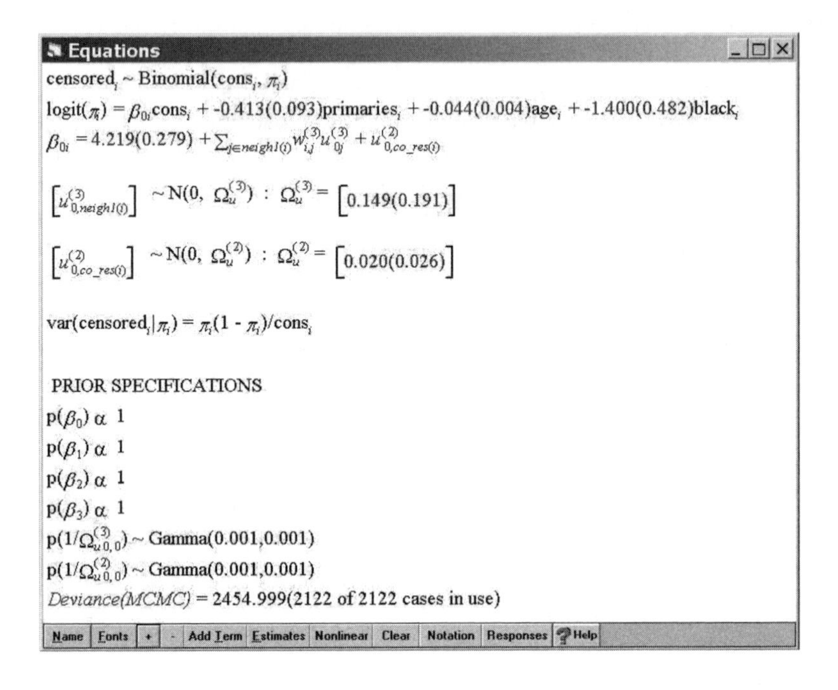

As we can see, this model gives very similar fixed effects as the earlier model. The variances for both the county and neighbour effects have large standard errors; however, if we compare the DIC difference between this model and the fixed effect model, we do see a marginal reduction in DIC.

Model	Deviance	DIC
Fixed effects	2472.2	2476.2
Random county effects	2459.8	2474.8
Random county + MM effects	2455.0	2473.2

In this section we have fitted models to the censoring indicator variable. We are, however, more concerned with general survival models. Both the survival and censoring time are recorded to the nearest month but, as we will see in the WinBUGS analysis of the dataset, we can assume that this is a continuous measure and fit a Weibull model. MLwiN has many options through which we can fit survival models (both continuous and discrete) but currently many of these are still only available via the MLwiN macro language (see Yang *et al.*, 2000 for more details) and it is difficult to combine these models with spatial information. The alternative approach is to fit a discrete time logistic model which involves expanding the dataset into a sequence of zeros followed by a 1 for each observation, and analysing this dataset using a binomial response model. We will investigate this approach in the next section.

9.2.4 Discrete survival models for mortality

Here we can consider each of our individual breast cancer patients as having a probability at each of a sequence of time points of failing (dying). Then we observe at each time point that either the patient survives (0) or dies (1) in which case the observations end. If we are also considering censoring then a censored sequence simply ends without a death and so is simply a set of zeros. Therefore for each individual we can convert their response from a time to a sequence of zeros followed by a 1 or simply a sequence of zeros. This conversion will result in a far larger dataset, particularly if we treat each month as a unit for which we require a 0 or 1. As our observed times (both deaths and censorings) range from 0 to 84 we could instead discretize our dataset to units of years and for each individual we then have (up to) seven measures. The transformation of the dataset from a time to a sequence of failure indicators is complex and so we will use the following set of macro commands to perform the transformation. We will firstly load up the saved worksheet '*iowa2.ws*' so that we start at the correct place. Then we can run the following macro file by opening the macro '*iowa.txt*' via the **File** menu and clicking on the **Execute** button.

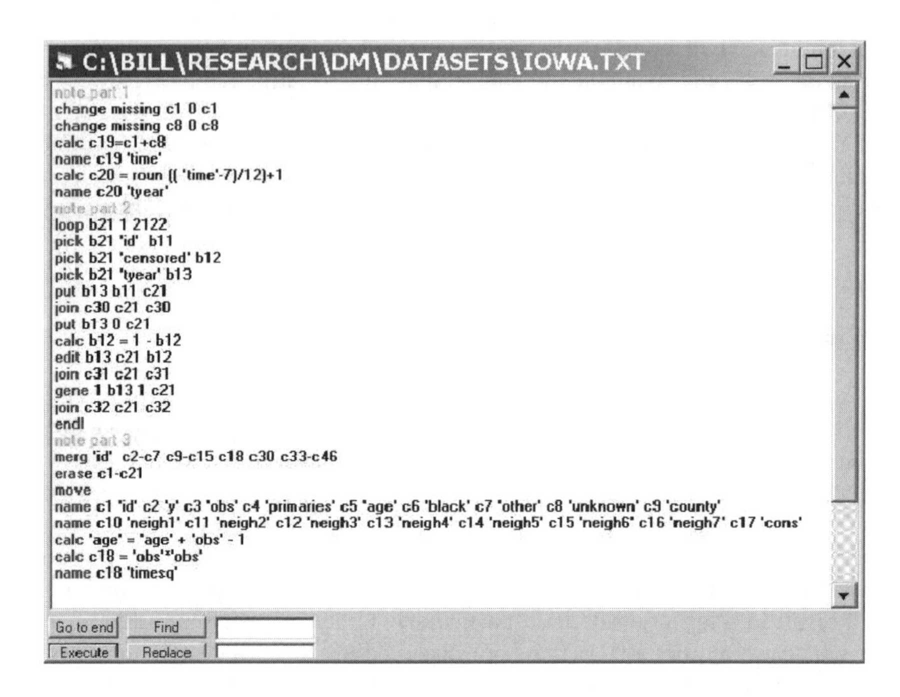

C:\BILL\RESEARCH\DM\DATASETS\IOWA.TXT
note part 1
change missing c1 0 c1
change missing c8 0 c8
calc c19=c1+c8
name c19 'time'
calc c20 = roun (('time'-7)/12)+1
name c20 'tyear'
note part 2
loop b21 1 2122
pick b21 'id' b11
pick b21 'censored' b12
pick b21 'tyear' b13
put b13 b11 c21
join c30 c21 c30
put b13 0 c21
calc b12 = 1 - b12
edit b13 c21 b12
join c31 c21 c31
gene 1 b13 1 c21
join c32 c21 c32
endl
note part 3
merg 'id' c2-c7 c9-c15 c18 c30 c33-c46
erase c1-c21
move
name c1 'id' c2 'y' c3 'obs' c4 'primaries' c5 'age' c6 'black' c7 'other' c8 'unknown' c9 'county'
name c10 'neigh1' c11 'neigh2' c12 'neigh3' c13 'neigh4' c14 'neigh5' c15 'neigh6' c16 'neigh7' c17 'cons'
calc 'age' = 'age' + 'obs' - 1
calc c18 = 'obs'*'obs'
name c18 'timesq'
Go to end
Find
Execute
Replace

We have here split the macro into three parts. In the first part we construct a variable that measures the time in years until either death or censoring and label this variable '*tyear*'. Here any death/censoring between 1 and 12 months is defined as 1 year, 13–24 months as 2 years and so on. This results in a variable that takes values between 1 and 7 for all individuals. The second part of the macro takes this new variable and from it constructs the larger dataset with three columns, one that contains the repeated patient number (*c30*), one that gives a survival indicator variable (*c31*) and a final column that contains the period identifiers (*c32*). The third part of the macro then expands the other variables of interest to fit the larger data size. We then remove all the old columns, move the new columns to the start of the worksheet and rename all the columns. Finally we correct the '*age*' variable to account for ageing by adding the number of years the individual has survived to each observation. We also construct a variable '*timesq*' which consists of the squares of the observation times. This will allow a quadratic effect to influence the survival probability when included in the model. When executed this macro file will produce a worksheet as follows:

Names

	Name	n	missing	min	max
1	id	12501	0	1	2122
2	y	12501	0	0	1
3	obs	12501	0	1	7
4	primaries	12501	0	1	4
5	age	12501	0	23	97
6	black	12501	0	0	1
7	other	12501	0	0	1
8	unknown	12501	0	0	1
9	county	12501	0	1	99
10	neigh1	12501	0	1	72
11	neigh2	12501	0	2	96
12	neigh3	12501	0	0	98
13	neigh4	12501	0	0	99
14	neigh5	12501	0	0	99
15	neigh6	12501	0	0	94
16	neigh7	12501	0	0	99
17	cons	12501	0	1	1
18	timesq	12501	0	1	49

We now have three levels to our dataset with observations ('*obs*') nested within individuals ('*id*') nested within counties ('*county*') and if we wish to specify spatial effects we will need to specify four classifications. We will fit a Binomial model with response '*y*' and the four level identifiers defined as '*obs*', '*id*', '*county*' and '*neigh1*' respectively. We will fit an intercept ('*cons*') and fixed effects for the variables '*primaries*', '*age*', '*black*', '*obs*' and '*timesq*'. The results of fitting a fixed effects only model can be seen below:

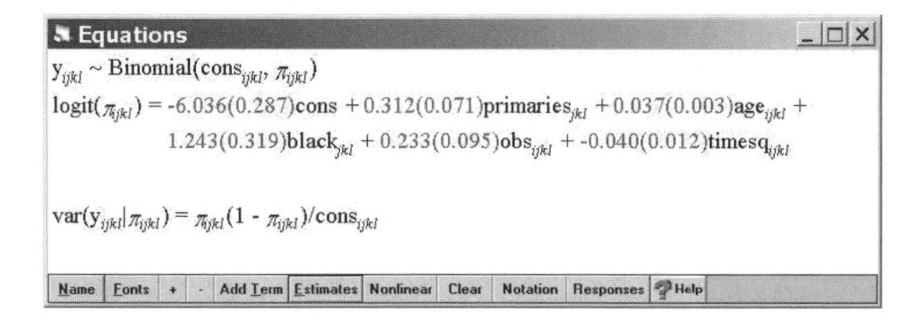

Here we can see that the probability of death increases with the number of primaries, the age of the patient and if the patient is black. Also the probability of death is quadratically related to the time since diagnosis, with the probability of death increasing in the first three years and then reducing. This suggests that if a patient survives the first few years then there is a good chance they will survive breast cancer which is backed up by the large number of censored observations.

We can once again fit more complex models including county effects, individual effects and spatial neighbour effects at the county level. The table below shows the DIC values for four possible models that we fitted using MCMC estimation.

Model	Deviance	DIC
Fixed effects	5033.0	5045.3
Random county + individual effects	5022.0	5042.7
Random county effects only	5023.9	5042.7
Random county + MM effects	5019.4	5041.8

As with the modelling of the censoring response variable we see that fitting random effects at higher levels gives a marginally better model. The final model actually took over an hour to run for 50 000 iterations using MCMC estimation and gave the estimates given below:

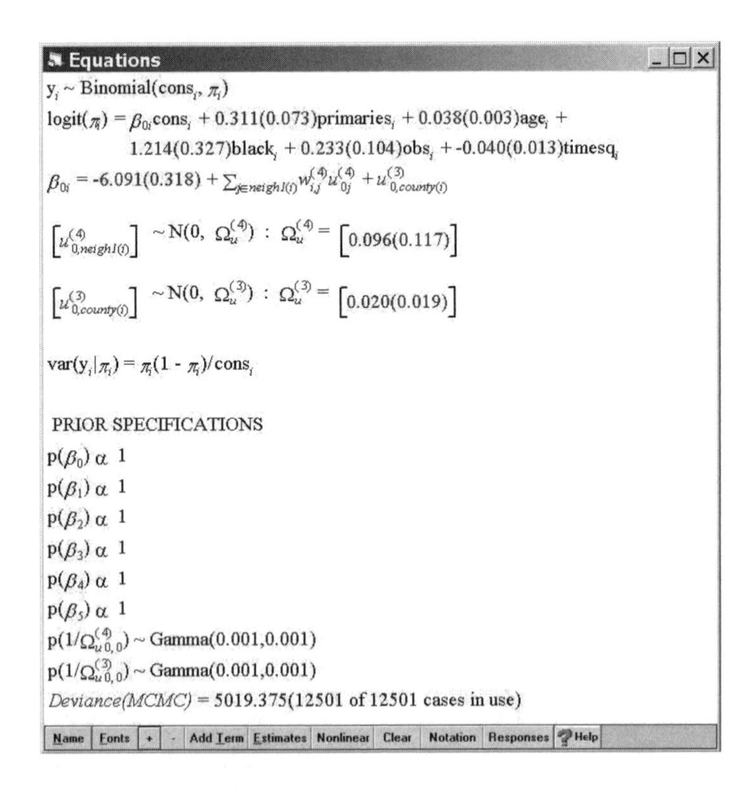

Here we see that the fixed effect estimates are very similar to the fixed effect model and the variance estimates again have large standard errors. The spatial multiple-membership effects have a larger variance than the county random effects suggesting they are more important. In this section we have studied how to fit both multilevel and spatial effects into a discrete time survival model in the MLwiN software package.

10
Epilogue

Besides the introduction of powerful statistical tools to a wider audience, in this volume we have tried to achieve two basic aims. First, we have attempted to highlight the range and depth of the capabilities of both MLwiN and WinBUGS. Second we have attempted to provide a comparison of the methods in application to specific datasets. In some cases, WinBUGS has greater flexibility, for instance in the specification of nonlinear and spatial correlation models. In other cases, MLwiN provides an easier vehicle for analysis, particularly when a clear multilevel structure is apparent in the data. We have not attempted to provide an exhaustive account of possible applications. Indeed there are a number of application areas which we have not examined, although the main areas have been covered. In addition, a number of features of the packages have been excluded from consideration.

The areas of application which have not been addressed are:

- *Edge effects*
  Edge effects have an effect on the estimation of relative risk. In many US states around 40% of counties lie on the state boundary and the added censoring at these locations can have an effect on relative risk estimation and on the variance of the relative risks. Proximity to a boundary can have an impact in a variety of ways. One option is to regard the neighbourhoods of edge regions as missing and to estimate external regions' relative risks as missing parameters within MCMC runs. This could also have been applied in the spatial prediction example instead of using Bayesian kriging (Section 6.2). Another approach could be to employ weighting schemes which depend on boundary proximity. Both these approaches are feasible within WinBUGS and MLwiN.
- *Multivariate disease mapping*
  The analysis of multiple disease maps could be of great importance both within epidemiology and public health. There are a number of examples within public health of the analysis of groups of diseases, when, for example,

Disease Mapping with WinBUGS and MLwiN A. Lawson, W. Browne and C. Vidal Rodeiro
 ISBN: 0-470-85604-1 (HB)

a putative health hazard is to be assessed for its environmental impact. Maps can be modelled individually of course, but currently it is not easily possible to fit multivariate CAR models (Carlin and Banerjee, 2003) in the packages. Macros exist for fitting multivariate multiple-membership models in MLwiN using the IGLS algorithm (see Leyland *et al.*, (2000)). A bivariate approach to mapping has been proposed by Knorr-Held and Best (2001), although this is not easily extended to more than two diseases.

- *Specific cluster modelling*
  While we have presented the analysis of focused clustering in Chapter 7, the analysis of non-focused specific clustering is more complex as the location and number of clusters is usually unknown in these applications (see, for example, Lawson and Denison, 2002). The intensity of *spatially-specific* computation required (e.g. dynamic nearest neighbour or hidden process updating) is likely to make this computation prohibitive in these packages.
- *Real-time map surveillance*
  Retrospective spatiotemporal analysis is certainly possible in both WinBUGS or MLwiN. Examples of such analyses have been presented in Sections 6.1.6 and 7.7.2. However, if analyses are to be carried out in real-time or near real-time then these packages are not ideally suited to the task. In map surveillance, monitoring of changes in the spatial and spatiotemporal distribution of disease requires the detection to be made at any given time point. This is, in essence, prospective analysis and must be capable of handling an expanding data dimension and parameter set enlargement too. This is the concern of sequential Monte Carlo estimation estimation (Berzuini *et al.*, 1997; Doucet *et al.*, 2001). WinBUGS and MLwiN were originally designed to provide analyses of fixed dimension problems, and extension into dynamic dimensions could be difficult within the current versions.
- *Case-event data analysis*
  Finally we have not examined models for case-event data in this volume. Usually point process models are valid for such data. Specifically, it is usually assumed that a modulated non-homogeneous Poisson process likelihood model governs the case-event distribution. Point process models do not appear within the remit of the packages as they are defined, although computational tricks can be used to provide (inefficient) algorithms for sampling some simple cases. Often these models require the evaluation of spatial integrals which can only be inefficiently computed. One case where integrals are not required is the binary conditional logistic regression model, where it is possible to model an additive link useful in focused clustering. The WinBUGS code for a version of this model is given by Congdon (2003, Section 7.11).

Appendix 1

WinBUGS Code for Focused Clustering Models

A.1 FALKIRK EXAMPLE

A.1.1. Model 1.

```
model{
for (i in 1 : 26) {
num[i] ~dpois(mu[i] )
log(mu[i] ) <-log(e[i] ) + alpha0
RR[i] <-mu[i] / e[i]}
alpha0 ~dflat()}
list(alpha0=0.01)
```

A.1.2. Model 2.

```
model{
for (i in 1 : 26) {
num[i] ~dpois(mu[i] )
log(mu[i] ) <-log(e[i] ) + alpha0 + alpha1 * dep[i]
RR[i] <-mu[i] / e[i]
}
alpha1 ~dnorm(0.0, 1.0E-5)
alpha0 ~dflat()
}
list(alpha0=0.01, alpha1=0.01)
```

A.1.3. Model 3.

```
model{
for (i in 1 : 26) {
num[i] ~dpois(mu[i] )
f[i] <-1+exp( -alpha2* dist[i] )
log(mu[i] ) <-log(e[i] )+ alpha0 + alpha1 * dep[i] +log(f[i] )
RR[i] <-mu[i] / e[i]
}
```

Disease Mapping with WinBUGS and MLwiN A. Lawson, W. Browne and C. Vidal Rodeiro
 ISBN: 0-470-85604-1 (HB)

```
alpha2 ~dnorm(0.0,1.0)
alpha1 ~dnorm(0.0, 1.0E-5)
alpha0 ~dflat()
}
list(alpha0=0.01, alpha1=0.01, alpha2=0.01)
```

A.1.4. Model 3b.

```
model{
for (i in 1 : 26) {
num[i] ~dpois(mu[i])
ab[i] <-alpha2*dist[i]
f[i] <-1+exp( -ab[i] * ab[i] )
log(mu[i]) <-log(e[i])+alpha0 + alpha1 * dep[i] +log(f[i])
RR[i] <-mu[i] / e[i]
}
alpha2 ~dnorm(0.0,1.0)
alpha1 ~dnorm(0.0, 1.0E-5)
alpha0 ~dflat()
}
list(alpha0=0.01, alpha1=0.01, alpha2=0.01)
```

A.1.5. Model 4.

```
model{
for (i in 1 : 26) {
num[i] ~dpois(mu[i])
f[i] <-1+exp( alpha3* log(dist[i])-alpha2*dist[i])
log(mu[i]) <-log(e[i])+alpha0 + alpha1 * dep[i] +log(f[i])
RR[i] <-mu[i] / e[i]
}
alpha3 ~dnorm(0.0,0.5)
alpha2 ~dnorm(0.0,1.0)
alpha1 ~dnorm(0.0,1.0E-5)
alpha0 ~dflat()
}
list(alpha0=0.01, alpha1=0.01, alpha2=0.01,alpha3=0.01)
```

A.1.6. Model 5.

```
model{
for (i in 1 : 26) {
num[i] ~dpois(mu[i])
f[i] <-exp(alpha4*cos(ang[i])+alpha5*sin(ang[i]))
log(mu[i]) <-log(e[i])+alpha0 + alpha1 * dep[i] +log(f[i])
RR[i] <-mu[i] / e[i]
}
alpha5 ~dnorm(0.0,1.0)
alpha4 ~dnorm(0.0,1.0)
alpha2 ~dnorm(0.0,1.0)
alpha1 ~dnorm(0.0, 1.0E-5)
alpha0 ~dflat()
}
list(alpha0=0.01,
alpha1=0.01,alpha2=0.01,alpha4=0.01,alpha5=0.01)
```

A.1.7. Model 6.

```
model{
for (i in 1 : 26) {
num[i] ~dpois(mu[i])
f[i] <- (1+exp(-alpha2*dist[i]))*exp(alpha4*
cos(ang[i])+alpha5*sin(ang[i]))
log(mu[i]) <-log(e[i])+alpha0 + alpha1 * dep[i] +log(f[i])
RR[i] <-mu[i] / e[i]
}
alpha5 ~dnorm(0.0,1.0)
alpha4 ~dnorm(0.0,1.0)
alpha2 ~dnorm(0.0,1.0)
alpha1 ~dnorm(0.0, 1.0E-5)
alpha0 ~dflat()
}
list(alpha0=0.01, alpha1=0.01,
alpha2=0.01,alpha4=0.01,alpha5=0.01)
```

A.1.8. Model 7.

```
model{
for (i in 1 : 26) {
v[i] ~dnorm(0.0,dstar)
num[i] ~dpois(mu[i])
f[i] <- (1+exp(-alpha2*dist[i]))
log(mu[i]) <-log(e[i])+alpha0 + alpha1 *
dep[i] +log(f[i])+v[i]
RR[i] <-mu[i] / e[i]
}
dstar ~dgamma(0.001,0.001)
alpha2 ~dnorm(0.0,1.0)
alpha1 ~dnorm(0.0, 1.0E-5)
alpha0 ~dflat()
}
list(alpha0=0.01, alpha1=0.01, alpha2=0.01,dstar=0.001,
v=c(0,0,0,0,0,0,0,0,0,0,0,0,0,0,0,0,0,0,0,0,0,0,0,0,
0,0))
```

A.1.9. Model 8.

```
model{
u[1 : 26] ~car.normal(adj[],wei[],nn[],tau)
for (i in 1 : 26) {
v[i] ~dnorm(0.0,dstar)
num[i] ~dpois(mu[i])
f[i] <- (1+exp(-alpha2*dist[i]))
log(mu[i]) <-log(e[i])+alpha0 + alpha1 *
dep[i] +log(f[i])+v[i] +u[i]
RR[i] <-mu[i] / e[i] }
tau ~dgamma(0.001,0.001)
dstar ~dgamma(0.001,0.001)
alpha2 ~dnorm(0.0,1.0)
alpha1 ~dnorm(0.0, 1.0E-5)
alpha0 ~dflat()
```

```
for (i in 1 : 104) { wei[i] <-1}
}
#inits
list(
alpha0=0.01, alpha1=0.01,
alpha2=0.01,dstar=0.001,tau=0.01,
v=c(0,0,0,0,0,0,0,0,0,0,0,0,0,0,0,0,0,0,0,0,0,0,0,
0,0,0),
u=c(0,0,0,0,0,0,0,0,0,0,0,0,0,0,0,0,0,0,0,0,0,0,0,0,
0,0))
#adjacency data
list( nn = c(1, 8, 3, 4, 5, 4, 3, 5, 1, 2,
1, 3, 4, 3, 3, 2, 6, 7, 6, 6,
4, 4, 5, 6, 5, 3
),
adj = c(
2,
20, 19, 10, 6, 5, 4, 3, 1,
22, 4, 2,
8, 5, 3, 2,
8, 7, 6, 4, 2,
8, 7, 5, 2,
8, 6, 5,
9, 7, 6, 5, 4,
8,
11, 2,
10,
19, 18, 13,
18, 17, 14, 12,
17, 15, 13,
17, 16, 14,
17, 15,
25, 18, 16, 15, 14, 13,
25, 24, 21, 19, 17, 13, 12,
24, 21, 20, 18, 12, 2,
24, 23, 22, 21, 19, 2,
24, 20, 19, 18,
26, 23, 20, 3,
26, 25, 24, 22, 20,
25, 23, 21, 20, 19, 18,
26, 24, 23, 18, 17,
25, 23, 22
)
)
```

A.2 OHIO EXAMPLE

A.2.1. Model 1.

```
model{
for (i in 1:m)
```

```
{
for (k in 1:T)
{
# Poisson likelihood for observed counts
y[i,k] ~dpois(mu[i,k])
log(mu[i,k])<-log(e[i,k])+alpha+v[i]+tt[k]
# Relative Risk in each area and period of time
theta[i,k]<-exp(alpha+v[i]+tt[k])
}
theta_area[i]<-exp(v[i])
}
for(k in 1:T){ time[k]<-exp(tt[k])}
# Prior distributions for the Uncorrelated Heterogeneity
for(i in 1:m)
{
v[i]~dnorm(0,tau.v)
}
tt[1]~dnorm(0,tau.delta)
for(k in 2:T){
tt[k] ~dnorm(tt[k-1],tau.delta)
}
alpha~dflat()
# Hyperprior distributions on inverse variance parameter of
random effects
tau.v~dgamma(0.5,0.0005)
tau.delta~dgamma(0.5,0.0005)
}
INITS
list(alpha=0,tau.v=1,tau.delta=1,
v=c(0,0,0,0,0,0,0,0,0,0,0,0,0,0,0,0,0,0,0,0,0,0,0,0,
0,0,0,0,0,0,0,0,0,0,0,0,0,0,0,0,0,0,0,0,0,0,0,0,0,
0,0,0,0,0,0,0,0,0,0,0,0,0,0,0,0,0,0,0,0,0,0,0,0,0,
0,0,0,0,0,0,0,0,0,0,0,0,0,0,0),
tt=c(0,0,0,0,0,0,0,0,0,0))
```

A.2.2. Model 2.

```
model{
for (i in 1:m)
{
f[i]<-1+exp(-alpha1*dist[i])
for (k in 1:T)
{
# Poisson likelihood for observed counts
y[i,k]~dpois(mu[i,k])
log(mu[i,k])<-log(e[i,k])+alpha+v[i]+tt[k]+log(f[i])
# Relative Risk in each area and period of time
theta[i,k]<-exp(alpha+v[i]+f[i]+tt[k])
}
theta_area[i]<-exp(v[i]+f[i])
}
for(k in 1:T){ time[k]<-exp(tt[k])}
# Prior distributions for the Uncorrelated Heterogeneity
```

```
for(i in 1:m)
{
v[i] ~dnorm(0,tau.v)
}
tt[1] ~dnorm(0,tau.delta)
for(k in 2:T) {
tt[k] ~dnorm(tt[k-1] ,tau.delta)
}
alpha~dflat()
# Hyperprior distributions on inverse variance parameter of
random effects
tau.v~dgamma(0.5,0.0005)
tau.delta~dgamma(0.5,0.0005)
alpha1~dnorm(0,1.0E-5)
}
INITS
list(alpha=0,alpha1=0,tau.v=1,
tau.delta=1,v=c(0,0,0,0,0,0,0,0,0,0,0,0,
0,0,0,0,0,0,0,0,0,0,0,0,0,0,0,0,0,0,0,0,0,0,
0,0,0,0,0,0,0,0,0,0,0,0,0,0,0,0,0,0,0,0,0,0,
0,0,0,0,0,0,0,0,0,0,0,0,0,0,0,0,0,0,0,0,0,0,
0,0,0,0,0,0,0,0,0,0),tt=c(0,0,0,0,0,0,0,0, 0,0))
```

A.2.3. Model 3.

```
model{
for (i in 1:m)
{
f[i] <-1+exp(-alpha1*dist[i] )
for (k in 1:T)
{
# Poisson likelihood for observed counts
y[i,k] ~dpois(mu[i,k] )
log(mu[i,k] )<-log(e[i,k] )+
alpha+v[i] +u[i] +tt[k] +log(f[i] )
# Relative Risk in each area and period of time
theta[i,k] <-exp(alpha+v[i] + u[i] +f[i] +tt[k] )
}
theta_area[i] <-exp(v[i] +u[i] + log(f[i] ) )
}
for(k in 1:T){ time[k] <-exp(tt[k] )}
# CAR prior distribution for spatial correlated
heterogeneity
u[1:m] ~car.normal(adj[],weights[],num[], tau.u)
# Prior distributions for the Uncorrelated Heterogeneity
for(i in 1:m)
{
v[i] ~dnorm(0,tau.v)
}
tt[1] ~dnorm(0,tau.delta)
for(k in 2:T) {
tt[k] ~dnorm(tt[k-1] ,tau.delta)
}
```

```
# Weights
for(k in 1:sumNumNeig)
{
weights[k] <-1
}
alpha~dflat()
# Hyperprior distributions on inverse variance parameter of
random effects
tau.v~dgamma(0.5,0.0005)
tau.u~dgamma(0.5,0.0005)
tau.delta~dgamma(0.5,0.0005)
alpha1~dnorm(0,1.0E-5)
}
INITS
list(alpha=0,
tau.v=1,
tau.u=1,
tau.delta=1,alpha1=0,
u=c(0,0,0,0,0,0,0,0,0,0,0,0,0,0,0,0,0,
0,0,0,0,0,0,0,0,0,0,0,0,0,0,0,0,0,0,0,0,0,0,
0,0,0,0,0,0,0,0,0,0,0,0,0,0,0,0,0,0,0,0,0,0,
0,0,0,0,0,0,0,0,0,0,0,0,0,0,0,0,0,0,0,0,0,0,
0,0,0,0,0,0),
v=c(0,0,0,0,0,0,0,0,0,0,0,0,0,0,0,0,0,
0,0,0,0,0,0,0,0,0,0,0,0,0,0,0,0,0,0,0,0,0,0,
0,0,0,0,0,0,0,0,0,0,0,0,0,0,0,0,0,0,0,0,0,0,
0,0,0,0,0,0,0,0,0,0,0,0,0,0,0,0,0,0,0,0,0,0,
0,0,0,0,0),tt=c(0,0,0,0,0,0,0,0,0, 0))
```

2.4. Model 4.

```
model{
for(i in 1:m)
{
f[i] <-1+exp(-alpha1*dist[i])
for(k in 1:T)
{
# Poisson likelihood for observed counts
y[i,k] ~dpois(mu[i,k])
log(mu[i,k])<-log(e[i,k])+ alpha+v[i] +u[i] +tt[k] +
log(f[i])+nu[i,k]
# Relative Risk in each area and period of time
theta[i,k] <-exp(alpha+v[i] + u[i] +log(f[i])+tt[k])
}
theta_area[i] <-exp(v[i] +u[i] + log(f[i]))
}
for(k in 1:T){ time[k] <-exp(tt[k])}
# CAR prior distribution for spatial correlated heterogeneity
u[1:m] ~car.normal(adj[],weights[],num[], tau.u)
# Prior distributions for the Uncorrelated Heterogeneity
for(i in 1:m)
{
v[i] ~dnorm(0,tau.v)
```

```
for(k in 1:T){
nu[i,k]~dnorm(0,tau.nu)}
}
tt[1]~dnorm(0,tau.delta)
for(k in 2:T){
tt[k]~dnorm(tt[k-1],tau.delta)
}
# Weights
for(k in 1:sumNumNeig)
{
weights[k]<-1
}
alpha~dflat()
# Hyperprior distributions on inverse variance parameter of
random effects
tau.nu~dnorm(0,1.0E-5)
tau.v~dgamma(0.5,0.0005)
tau.u~dgamma(0.5,0.0005)
tau.delta~dgamma(0.5,0.0005)
alpha1~dnorm(0,1.0E-5)
}
INITS
list(alpha=0,
tau.v=1,
tau.u=1,tau.nu=1,
tau.delta=1,alpha1=0,
u=c(0,0,0,0,0,0,0,0,0,0,0,0,0,0,0,0,0,
0,0,0,0,0,0,0,0,0,0,0,0,0,0,0,0,0,0,0,0,0,0,
0,0,0,0,0,0,0,0,0,0,0,0,0,0,0,0,0,0,0,0,0,0,
0,0,0,0,0,0,0,0,0,0,0,0,0,0,0,0,0,0,0,0,0,0, 0,0,0,0,0),
v=c(0,0,0,0,0,0,0,0,0,0,0,0,0,0,0,0,0,0,
0,0,0,0,0,0,0,0,0,0,0,0,0,0,0,0,0,0,0,0,0,0,
0,0,0,0,0,0,0,0,0,0,0,0,0,0,0,0,0,0,0,0,0,0,
0,0,0,0,0,0,0,0,0,0,0,0,0,0,0,0,0,0,0,0,0,0,
0,0,0,0,0),tt=c(0,0,0,0,0,0,0,0,0, 0))
```

A.2.5. Model 4b.

```
Model{
for (i in 1:m)
{
f[i]<-1+exp(-alpha1*dist[i])
for (k in 1:T)
{
# Poisson likelihood for observed counts
y[i,k]~dpois(mu[i,k])
log(mu[i,k])<-log(e[i,k])+
alpha+v[i]+tt[k]+log(f[i])+nu[i, k]
# Relative Risk in each area and period of time
theta[i,k]<-exp(alpha+v[i] + log(f[i])+tt[k])
}
theta_area[i]<-exp(v[i] + log(f[i]))
}
```

```
for(k in 1:T){ time[k] <-exp(tt[k] )}
# Prior distributions for the Uncorrelated Heterogeneity
for(i in 1:m)
{
v[i] ~dnorm(0,tau.v)
for(k in 1:T){
nu[i,k] ~dnorm(0,tau.nu)}
}
tt[1] ~dnorm(0,tau.delta)
for(k in 2:T) {
tt[k] ~dnorm(tt[k-1] ,tau.delta)
}
alpha~dflat()
# Hyperprior distributions on inverse variance parameter of
random effects
tau.nu~dnorm(0,1.0E-5)
tau.v~dgamma(0.5,0.0005)
tau.delta~dgamma(0.5,0.0005)
alpha1~dnorm(0,1.0E-5)
}
INITS
list(alpha=0,
tau.v=1,
tau.nu=1,
tau.delta=1,
alpha1=0,
v=c(0,0,0,0,0,0,0,0,0,0,0,0,0,0,0,0,0,0,
0,0,0,0,0,0,0,0,0,0,0,0,0,0,0,0,0,0,0,0,0,0,0,
0,0,0,0,0,0,0,0,0,0,0,0,0,0,0,0,0,0,0,0,0,0,0,
0,0,0,0,0,0,0,0,0,0,0,0,0,0,0,0,0,0,0,0,0,0,0,
0,0,0,0,0),tt=c(0,0,0,0,0,0,0,0,0, 0))
```

Appendix 2

S-Plus Function for Conversion to GeoBUGS Format

An S-Plus function has been developed for the conversion of map polygons produced by the S-Plus function *map*( ) into the correct format for use with GeoBUGS. The resulting file is in S-Plus input format for GeoBUGS. The function was written by Lance Waller with additions by Brad Carlin. It is called *poly.S* and can be downloaded from *http://www.biostat.umn.edu/~brad/software.html.* The function is also given below:

```
##
## poly.S
##
####################
# S-plus polygon extractor v. 2.1 Lance Waller 8/22/01
# (including some of Brad Carlin' s output formatting)
# This converts S-plus map() format boundary files into
# a list suitable for input to GeoBUGS v. 1.4
# Uses mapgetl, mapgetg, and makepoly functions.
# first: source("poly.S")
# then (e.g.): mkpoly("idaho")
# This will generate idaho.txt, suitable for reading into GeoBUGS.
####################
mkpoly < - function(state)
{
toradians < - atan(1)/45
radiusearth < -0.5*(6378.2 + 6356.7)
sine51 -sin (51.5* toradians)
## in lieu of a call to X11():
{
outfile2 < - paste(state,".ps",sep="")
```

Disease Mapping with WinBUGS and MLWiN A. Lawson, W. Browne and C. Vidal Rodeiro
 ISBN: 0-470-85604-1 (HB)

```
pscript(outfile2)
}
##
# First, list state for which county boundaries are desired:
# state < - "idaho"
#state < - "minnesota"
#state < - "connecticut"
library(maps)
outfile < - paste(state,".txt",sep="")
# mnpolys < - map ("county",state,fill=T)
namesvec < - map("county",state,namesonly=T)
write(paste("map:",length(namesvec),"\n"),outfile)
for (i in 1:length(namesvec)) {
write(paste( i, paste("grid",i,sep="")),outfile,append=T)
}
for (i in 1:length(namesvec)) {
exact < - T
gon < - mapname("county",namesvec[i] ,exact)
line < - mapgetg ("county", gon, fill=T)
coord < - mapgetl("county", linenumber)
gonsize < - linesize
# Brad tweaks, next 3 lines:
color < - 1
color < - rep(color, length = length (gonsize))
keep < - !is.na(color)
coord[ c("x", "y")] < - makepoly(coord, gonsize, keep)
# Add repeat of first point to close polygon
# Brad tweak: Comment these out; not needed by GeoBUGS!
# coordx < - c(coordx,coordx[1] )
# coordy < - c(coordy,coordy[1] )
# Lance rough conversion of US lat/long to km (used by GeoBUGS):
# (see also forum.swarthmore.edu/dr.math/problems/
longandlat.html)
# radius of earth:
# r = 3963.34 (equatorial) or 3949.99 (polar) mi
# = 6378.2 or 6356.7 km
# which implies: km per mile = 1.609299 or 1.609295
# a change of 1 degree of latitude corresponds to the same number
# of km, regardless of longitude. arclength=r* theta, so the
# multiplier for coordy should probably be just the radius of
# earth.
# On the other hand, a change of 1 degree in longitude corresponds
# to a different distance, depending on latitude. (at N pole,
# the change is essentially 0. at the equator, use equatorial
# radius.
# Perhaps for U.S., might use an "average" latitude, 30 deg is
# roughly Houston, 49deg is most of N bdry of continental 48 states.
# 0.5(30+49)=39.5 deg. so use r approx 6378.2* sin(51.5)
## Let's try these new multipliers:
# coordx < - (coordx* atan(1)/45)* 6416.6
# coordy < - (coordy* atan(1)/45)* 6416.6
coordx < - (coordx* toradians)* radiusearth* sine51
```

```
coordy < - (coordy* toradians)* radiusearth
##
coordmat < - cbind(
rep(paste("grid",i,sep=" "),length(coordx)),
round(coordx,5),
round(coordy,5))
if (i == 1) {
write(" ",outfile,append =T)}
else{
write(c(NA,NA,NA),outfile,append=T)
}
write(t(coordmat),outfile,append= T,ncol=3)
# Uncomment if you want the "movie" of county maps . . .
plot(coord,type="l")
polygon(coord)
# A counter for the loop, comment out if unnecessary.
print( paste("i = ",i))
} ## for i
write("END",outfile,append=T)
}
```

Bibliography

Aitkin, M. (1996a). Empirical Bayes shrinkage using posterior random effect means from nonparametric maximum likelihood estimation in general random effect models. In A. Forcina, G. M. Marchetti, R. Hatzinger, and G. Galmacci (Eds), *Proceedings of the 11th International Workshop on Statistical Modelling*, pp. 87–92. Graphos, Citta di Castello.

Aitkin, M. (1996b). A general maximum likelihood analysis of overdispersion in generalised linear models. *Statistics and Computing* **6**, 251–262.

Aitkin, M. and N. Longford (1986). Statistical modelling in school effectiveness studies (with discussion). *Journal of the Royal Statistical Society* **149**, 1–43.

Aitkin, M., D. Anderson, and J. Hinde (1981). Statistical modelling of data on teaching styles (with discussion). *Journal of the Royal Statistical Society* **144**, 148–161.

Assuncao, R. (2003). Space varying coeffcient models for small area data. *Environmetrics*. To appear.

Assuncao, R., J. Potter, and S. Cavenghi (2002). A Bayesian space varying parameter model applied to estimating fertility schedules. *Statistics in Medicine* **21**, 2057–2075.

Banerjee, S. and B. P. Carlin (2003). Semiparametric spatio-temporal frailty modeling. *Environmetrics*. To appear.

Banerjee, S., M. M. Wall, and B. P. Carlin (2003). Frailty modeling for spatially-correlated survival data, with application to infant mortality in Minnesota. *Biostatistics* **4**, 123–142.

Bernardinelli, L., D. G. Clayton, C. Pascutto, C. Montomoli, M. Ghislandi, and M. Songini (1995). Bayesian analysis of space-time variation in disease risk. *Statistics in Medicine* **14**, 2433–2443.

Bernardinelli, L., D. G. Clayton, C. Montomoli (1995b) Bayesian estimates of disease maps: how important are priors? *Statistics in Medicine* **14**, 2411–2431.

Bernardinelli, L., C. Pascutto, C. Montomoli, J. Komakec, and W. Gilks (1999). Ecological regression with errors in covariates. In A. B. Lawson, D. Boehning, E. Lasaffree, A. Biggeri, J. F. Viel, and R. Bertolline (Eds), *Disease Mapping and Risk Assessment for Public Health*. Chichester: John Wiley & Sons, Ltd.

Bernardinelli, L., C. Pascutto, C. Montomoli, and W. Gilks (2000). Investigating the genetic association between diabetes and malaria: an application of Bayesian ecological regression models with errors in covariates. In P. Elliott, J. Wakefield, N. G. Best, and D. Briggs (Eds), *Spatial Epidemiology. Methods and Applications*. Oxford: Oxford University Press.

Bernardo, J. M. and A. F. M. Smith (1994). *Bayesian Theory*. New York: John Wiley & Sons, Inc.

Disease Mapping with WinBUGS and MLwiN A. Lawson, W. Browne and C. Vidal Rodeiro
 ISBN: 0-470-85604-1 (HB)

Berzuini, C., N. G. Best, W. R. Gilks, and C. Larissa (1997). Dynamic conditional independence models and Markov Chain Monte Carlo methods. *Journal of the American Statistical association* **92**, 1403–1412.

Besag, J. and P. J. Green (1993). Spatial statistics and Bayesian computation. *Journal of the Royal Statistical Society, Series B* **55**, 25–37.

Besag, J., J. York, and A. Mollié (1991). Bayesian image restoration with two applications in spatial statistics. *Annals of the Institute of Statistical Mathematics* **43**, 1–59.

Best, N. G. and J. C.Wakefield (1999). Accounting for inaccuracies in population counts and case registration in cancer mapping studies. *Journal of the Royal Statistical Society* **162**, 363–382.

Best, N. G., M. K. Cowles, and K. Vines (1995). *CODA: Convergence Diagnosis and Output Analysis Software for Gibbs Sampling Output*. Cambridge, England: MRC Biostatistics Unit.

Biggeri, A., F. Divino, A. Frigessi, A. B. Lawson, D. Boehning, E. Lesaffree, and J. F. Viel (1999). Introduction to spatial models in ecological analyses. In A. B. Lawson, D. Boehning, E. Lasaffree, A. Biggeri, J. F. Viel, and R. Bertolline (Eds), *Disease Mapping and Risk Assessment for Public Health*. Chichester: John Wiley & Sons, Ltd.

Bock, R. D. and M. Aitkin (1981). Marginal maximum likelihood estimation of item parameters: an application of an EM algorithm. *Psychometrika* **46**, 443–459.

Boehning, D., E. Dietz, and P. Schlattmann (2000). Space-time mixture modelling of public health data. *Statistics in Medicine* **19**, 2333–2344.

Bowman, A. and A. Azzalini (1997). *Applied Smoothing Techniques for Data Analysis: The Kernel Approach with S-Plus Illustrations*. London: Oxford University Press.

Breslow, N. and D. G. Clayton (1993). Approximate inference in generalized linear mixed models. *Journal of the American Statistical Association* **88**, 9–25.

Breslow, N. and N. Day (1987). *Statistical Methods in Cancer Research*, Vol. 2: *The Design and Analysis of Cohort Studies*. Lyon: International Agency for Research on Cancer.

Brooks, S. and A. E. Gelman (1998). General methods for monitoring convergence of iterative simulations. *Journal of Computational and Graphical Statistics* **7**, 434–455.

Browne, W. (1998). Applying MCMC methods to multilevel models. PhD thesis, University of Bath.

Browne, W. (2003). *MCMC Estimation in MLwiN*. London: Institute of Education.

Browne, W. and D. Draper (2000). Implementation and performance issues in the Bayesian and likelihood fitting of multilevel models. *Computational Statistics* **15**, 391–420.

Browne, W., H. Goldstein, and J. Rasbash (2001). Multiple membership multiple classification (MMMC) models. *Statistical Modelling* **1**, 103–124.

Brunsdon, C., A. E. Fotheringham, and M. Charlton (1999). Some notes on parametric significance tests for geographically weighted regression. *Journal of Regional Science* **39**, 497–524.

Bryk, A. and S. Raudenbush (1992). *Hierarchical Linear Models*. Newbury Park: Sage.

Bryk, A., S. Raudenbush, M. Seltzer, and R. Congdon (1988). *An Introduction to HLM: Computer Program and User's Guide* (2nd edn). Chicago: University of Chicago, Department of Education.

Carlin, B. P. and S. Banerjee (2003). Hierarchical multivariate CAR models for spatio-temporally correlated survival data. In J. M. Bernardo *et al.* (Eds), *Bayesian Statistics 7*. Oxford: Oxford University Press.

Carlin, B. P. and T. A. Louis (1996). *Bayes and Empirical Bayes Methods for Data Analysis*. London: Chapman & Hall.

Carstairs, V. (1981). Small area analysis and health service research. *Community Medicine* **3**, 131–139.

Casella, G. and E. I. George (1992). Explaining the Gibbs Sampler. *The American Statistician* **46**, 167–174.

Chen, M., Q. Shao, and J. Ibrahim (2000). *Monte Carlo Methods in Bayesian Computation.* New York: Springer Verlag.

Clayton, D. G. (1991). A Monte Carlo method for Bayesian inference in frailty models. *Biometrics* **47**, 467–485.

Clayton, D. G. and L. Bernardinelli (1992). Bayesian methods for mapping disease risk. In P. Elliott, J. Cuzick, D. English, and R. Stern (Eds), *Geographical and Environmental Epidemiology: Methods for Small-Area Studies*. Oxford: Oxford University Press.

Clayton, D. G. and J. Kaldor (1987). Empirical Bayes estimates of age-standardised relative risks for use in disease mapping. *Biometrics* **43**, 671–691.

Clayton, D. G., L. Bernardinelli, and C. Montomoli (1993). Spatial correlation in ecological analysis. *International Journal of Epidemiology* **22**, 1193–1202.

Cliff, A. D. and J. K. Ord (1981). *Spatial Processes: Models and Applications*. London: Pion.

Congdon, P. (2003). *Applied Bayesian Modelling*. Clinchester: John Wiley & Sons, Ltd.

Cressie, N. A. C. (1993). *Statistics for Spatial Data* (revised edn). New York: John Wiley & Sons, Inc.

Cuzick, J. and M. Hills (1991). Clustering and clusters-summary. In G. Draper (Ed.), *Geographical Epidemiology of Childhood Leukaemia and Non-Hodgkin Lymphomas in Great Britain 1966–1983*, pp. 123–125. London: HMSO.

Daley, D. and D. Vere-Jones (1988). *An Introduction to the Theory of Point Processes*. New York: Springer Verlag.

Devine, O. and T. Louis (1994). A constrained empirical Bayes estimator for incidence rates in areas with small populations. *Statistics in Medicine* **13**, 1119–1133.

Diggle, P. J., J. Tawn, and R. Moyeed (1998). Model-based geostatistics. *Journal of the Royal Statistical Society C* **47**, 299–350.

Diggle, P. J., S. Morris, and J. C. Wakefield (2000). Point-source modelling using matched case-control data. *Biostatistics* **1**, 1–17.

Doucet, A., N. de Freitas, and N. Gordon (Eds) (2001). *Sequential Monte Carlo Methods in Practice*. New York: Springer Verlag.

Elliott, P., J. C. Wakefield, N. G. Best, and D. J. Briggs (Eds) (2000). *Spatial Epidemiology: Methods and Applications*. London: Oxford University Press.

Esman, N. A. and G. M. Marsh (1996). Applications and limitations of air dispersion modeling in environmental epidemiology. *Journal of Exposure Analysis and Environmental Epidemiology* **6**, 339–353.

Ferreira, J., D. Denison, and C. Holmes (2002). Partition modelling. In A. B. Lawson and D. Denison (Eds), *Spatial Cluster Modelling*, Chapter 7, pp. 125–145. New York: CRC Press.

Fotheringham, A. E., M. Charlton, and C. Brunsdon (1998). Geographically weighted regression: a natural evolution of the expansion method for spatial data analysis. *Environment and Planning A* **30**, 1905–1927.

Gelman, A. E. and D. Rubin (1992). Inference from iterative simulation using multiple sequences (with discussion). *Statistical Science* **7**, 457–511.

Gelman, A., J. Carlin, H. Stern, and D. Rubin (1995). *Bayesian Data Analysis*. London: Chapman & Hall.

Gilks, W. and P. Wild (1992). Adaptive rejection sampling for Gibbs Sampling. *Applied Statistics* **41**, 337–348.

Gilks, W. R., D. G. Clayton, D. J. Spiegelhalter, N. G. Best, A. J. McNeil, L. D. Sharples, and A. J. Kirby (1993). Modelling complexity: applications of Gibbs Sampling in medicine. *Journal of the Royal Statistical Society B* **55**, 39–52.

Gilks, W. R., S. Richardson, and D. J. Spiegelhalter (Eds) (1996). *Markov Chain Monte Carlo in Practice*. London: Chapman & Hall.

Glick, B. J. (1979). The spatial autocorrelation of cancer mortality. *Social Science and Medicine*. **13D**, 123–130.

Goldstein, H. (1986). Multilevel mixed linear model analysis using iterative generalized least squares. *Biometrika* **73**, 43–56.

Goldstein, H. (1989). Restricted unbiased iterative generalised least squares estimation. *Biometrika* **76**, 622–623.

Goldstein, H. (1991). Nonlinear multilevel models with an application to binary response data. *Biometrika* **78**, 45–51.

Goldstein, H. (1995). *Multilevel Statistical Models* (2nd edn). London: Edward Arnold.

Goldstein, H. and J. Rasbash (1996). Improved approximations for multilevel models with binary response. *Journal of the Royal Statistical Society* **159**, 505–513.

Hedeker, D. and R. Gibbons (1994). A random-effects ordinal regression model for multilevel analysis. *Biometrics* **50**, 933–944.

Hill, P. and H. Goldstein (1998). Multilevel modelling of educational data with cross-classifications and missing identification of units. *Journal of Educational and Behavioral Statistics* **23**, 117–128.

Kelsall, J. and P. Diggle (1998). Spatial variation in risk of disease: a nonparametric binary regression approach. *Applied Statistics* **47**, 559–573.

Knorr-Held, L. (2000). Bayesian modelling of inseparable space-time variation in disease risk. *Statistics in Medicine* **19**, 2555–2567.

Knorr-Held, L. and J. Besag (1998). Modelling risk from a disease in time and space. *Statistics in Medicine* **17**, 2045–2060.

Knorr-Held, L. and N. G. Best (2001). A shared component model for detecting joint and selective clustering of two diseases. *Journal of the Royal Statistical Society* **164**, 73–85.

Knorr-Held, L. and G. Rasser (2000). Bayesian detection of clusters and discontinuities in disease maps. *Biometrics* **56**, 13–21.

Langford, I., G. Bentham, and A. McDonald (1998). Multilevel modelling of geographically aggregated health data: a case study on malignant melanoma mortality and uv exposure in the European Community. *Statistics in Medicine* **17**, 41–58.

Langford, I., A. Leyland, J. Rasbash, and H. Goldstein (1999). Multilevel modelling of the geographical distribution of rare diseases. *Journal of the Royal Statistical Society* **48**, 253–268.

Lawson, A. B. (1993). On the analysis of mortality events around a prespecified fixed point. *Journal of the Royal Statistical Society A* **156**, 363–377.

Lawson, A. B. (1994). On using spatial Gaussian priors to model heterogeneity in environmental epidemiology. *The Statistician* **43**, 69–76. Proceedings of the Practical Bayesian Statistics Conference.

Lawson, A. B. (1997). Some spatial statistical tools for pattern recognition. In A. Stein, F. W. T. P. de Vries, and J. Schut (Eds), *Quantitative Approaches in Systems Analysis*, Vol. 7, pp. 43–58. C. T. de Wit Graduate School for Production Ecology, Wageningen, Netherlands.

Lawson, A. B. (2001). *Statistical Methods in Spatial Epidemiology*. Chichester: John Wiley & Sons, Ltd.

Lawson, A. B. and A. Clark (2002). Spatial mixture relative risk models applied to disease mapping. *Statistics in Medicine* **21**, 359–370.

Lawson, A. B. and D. Denison (2002). Spatial cluster modelling: an overview. In A. B. Lawson and D. Denison (Eds), *Spatial Cluster Modelling*, Chapter 1, pp. 1–19. New York: CRC Press.

Lawson, A. B., A. Biggeri, and C. Lagazio (1996). Modelling heterogeneity in discrete spatial data models via MAP and MCMC methods. In A. Forcina, G. Marchetti, R.

Hatzinger, and G. Galmacci (Eds), *Proceedings of the 11th International Workshop on Statistical Modelling*, pp. 240–250. Graphos, Citta di Castello.

Lawson, A. B., D. Böhning, E. Lessafre, A. Biggeri, J. F. Viel, and R. Bertollini (Eds) (1999). *Disease Mapping and Risk Assessment for Public Health*. Chichester: John Wiley & Sons, Ltd.

Lawson, A. B., A. Biggeri, D. Boehning, E. Lesaffre, J.-F. Viel, A. Clark, P. Schlattmann, and F. Divino (2000). Disease mapping models: an empirical evaluation. *Statistics in Medicine* **19**, 2217–2242. Special issue: Disease Mapping with Emphasis on Evaluation of Methods.

Lesaffre, E. and B. Spiessens (2001). On the effect of the number of quadrature points in a logistic random-effects model: an example. *Journal of the Royal Statistical Society C*, **50**, 325–336.

Leyland, A. and H. Goldstein (Eds) (2001). *Multilevel Modelling of Health Statistics*. Chichester: John Wiley and Sons, Ltd.

Leyland, A. and A. McLeod (2000). Mortality in England and Wales, 1979–1992: An introduction to multilevel modelling using MLwiN. Occasional Paper no. 1, MRC Social and Public Health Sciences Unit.

Leyland, A., I. Langford, J. Rashbash, and H. Goldstein (2000). Multivariate spatial models for event data. *Statistics in Medicine* **19**, 2469–2478.

Longford, N. (1987). A fast scoring algorithm for maximum likelihood estimation in unbalanced mixed models with nested random effects. *Biometrika* **74**, 817–827.

Longford, N. (1988). *VARCL – software for variance components analysis of data with hierarchically nested random effects (maximum likelihood)*. Princeton, NJ: Educational Testing Service.

MacEachren, A. M. (1995). *How Maps Work: Representation, Visualization and Design*. New York: Guilford Press.

MacNab, Y. C. and C. B. Dean (2002). Spatio-temporal modelling of rates for the construction of disease maps. *Statistics in Medicine* **21**, 347–358.

Manton, K., M. Woodbury, and E. Stallard (1981). A variance components approach to categorical data models with heterogeneous mortality rates in North Carolina counties. *Biometrics* **37**, 259–269.

Marshall, R. (1991). Mapping disease and mortality rates using empirical Bayes estimators. *Applied Statistics* **40**, 283–294.

Mollié, A. (2000). Bayesian mapping of Hodgkin's disease in France. In P. Elliott, J. Wakefield, N. G. Best, and D. Briggs (Eds), *Spatial Epidemiology. Methods and Applications*. Oxford: Oxford University Press.

Monmonier, M. (1996). *How to Lie with Maps* (2nd edn). London: University of Chicago Press.

Neal, R. (1998). Suppressing random walks in Markov chain Monte Carlo using ordered over-relaxation. In M. I. Jordan (Ed.), *Learning in Graphical Models*. Dordrecht: Kluwer Academic Publishers.

Neal, R. M. (2003). Slice sampling. *Annals of Statistics* **31**, 1–34.

Panopsky, M. A. and J. A. Dutton (1984). *Atmospheric Turbulence*. New York: John Wiley & Sons, Inc.

Pickle, L. W. and D. J. Hermann (1995). Cognitive aspects of statistical mapping. Technical Report 18, NCHS O.ce of Research and Methodology, Washington, DC.

Pickle, L., M. Mungiole, G. Jones, and A. White (1999). Exploring spatial patterns of mortality: the new atlas of United States mortality. *Statistics in Medicine* **18**, 3211–3220.

Rabe-Hesketh, S., T. Toulopoulou, and R. Murray (2001). Multilevel modeling of cognitive function in schizophrenic patients and their first degree relatives. *Multivariate Behavioural Research* **36**, 279–298.

Rasbash, J. and H. Goldstein (1994). Efficient analysis of mixed hierarchical and crossed random structures using a multilevel model. *Journal of Behavioral Statistics* **19**, 337–350.

Rasbash, J., W. Browne, H. Goldstein, M. Yang, I. Plewis, M. Healy, G. Woodhouse, D. Draper, I. Langford, and T. Lewis (2000). *A User's Guide to MLwiN*. London: Institute of Education.

Raudenbush, S., M. Yang, and M. Yosef (2000). Maximum likelihood for generalized linear models with nested random effects via high-order, multivariate Laplace approximation. *Journal of Computational and Graphical Statistics* **9**, 141–157.

Richardson, S. (2003). Spatial models in epidemiological applications. In P. Green, N. Hjort, and S. Richardson (Eds), *Highly Structured Stochastic Systems*. London: Oxford University Press.

Richardson, S., C. Monfort, M. Green, G. Draper, and C. Muirhead (1995). Spatial variation of natural radiation and childhood leukaemia incidence in Great Britain. *Statistics in Medicine* **14**, 2487–2501.

Ripley, B. D. (1987). *Stochastic Simulation*. New York: John Wiley & Sons, Inc.

Robert, C. P. and G. Casella (1999). *Monte Carlo Statistical Methods*. New York: Springer Verlag.

Schlattman, P. and D. Boehning (1993). Mixture models and disease mapping. *Statistics in Medicine* **12**, 1943–1950.

Smith, A. F. M. and A. E. Gelfand (1992). Bayesian statistics without tears: a sampling-resampling perspective. *American Statistician* **46**, 84–88.

Smith, A. F. M. and G. Roberts (1993). Bayesian computation via the Gibbs Sampler and related Markov chain Monte Carlo methods. *Journal of the Royal Statistical Society B*, **55**, 3–23.

Smith, B. (1999). *BOA, BUGS Output Analysis Program*. The University of Iowa College of Public Health.

Snow, J. (1854). *On the Mode of Communication of Cholera* (2nd edn). London: Churchill Livingstone.

Spiegelhalter, D. J., N. G. Best, W. R. Gilks, and H. Inskip (1996). Hepatitis B: a case study in MCMC methods. In W. R. Gilks, S. Richardson, and D. J. Spiegelhalter (Eds), *Markov Chain Monte Carlo in Practice*. London: Chapman & Hall.

Spiegelhalter, D. J., N. G. Best, B. P. Carlin, and A. van der Linde (2002a). Bayesian deviance, the effective number of parameters and the comparison of arbitrarily complex models. *Journal of the Royal Statistical Society B*, **64**, 583–640.

Spiegelhalter, D. J., A. Thomas, N. G. Best, and D. Lunn (2002b). *WinBugs User Manual. Version 1.4*. Cambridge, England: MRC Biostatistics Unit.

Stern, H. S. and N. A. Cressie (1999). Inference for extremes in disease mapping. In A. B. Lawson, D. Böhning, E. Lasaffree, A. Biggeri, J. F. Viel, and R. Bertolline (Eds), *Disease Mapping and Risk Assessment for Public Health*. Chichester: John Wiley & Sons, Ltd.

Stern, H. S. and N. A. C. Cressie (2000). Posterior predictive model checks for disease mapping models. *Statistics in Medicine* **19**, 2377–2397.

Tanner, M. A. (1996). *Tools for Statistical Inference* (3rd edn). New York: Springer Verlag.

Thomas, A., N. G. Best, R. Arnold, and D. J. Spiegelhalter (2002). *GeoBUGS User Manual. Version 1.1*. London: Imperial College School of Medicine.

Tsutakawa, R. (1988). Mixed model for analysing geographic variability in mortality rates. *Journal of the American Statistical Association* **83**, 37–42.

Wakefield, J. C. and S. Morris (2001). The Bayesian modeling of disease risk in relation to a point source. *Journal of the American Statistical Association* **96**, 77–91.

Waller, L. A., B. P. Carlin, H. Xia, and A. E. Gelfand (1997). Hierarchical spatiotemporal mapping of disease rates. *Journal of the American Statistical Association* **92**, 607–617.

Walter, S. D. (1993). Visual and statistical assessment of spatial clustering in mapped data. *Statistics in Medicine* **12**, 1275–1291.

Yang, M., J. Rasbash, H. Goldstein, and M. Barbosa (2000). *MLwiN Macros for Advanced Multilevel Modelling*. London: Institute of Education.

Yu, B. and P. Mykland (1998). Looking at Markov samplers through cusum path plots: A simple diagnostic idea. *Statistics and Computing* **8**, 275–286.

Zia, H., B. P. Carlin, and L. A. Waller (1997). Hierarchical models for mapping Ohio lung cancer rates. *Environmetrics* **8**, 107–120.

Index

Disease Mapping with WinBUGS and MLWiN A. Lawson, W. Browne and C. Vidal Rodeiro
 ISBN: 0-470-85604-1 (HB)

Statistics in Practice

Human and Biological Sciences

Brown and Prescott – Applied Mixed Models in Medicine
Ellenberg, Fleming and DeMets – Data Monitoring in Clinical Trials: A Practical Perspective
Lawson, Browne and Vidal Rodeiro – Disease Mapping with WinBUGS and MLwiN
Marubini and Valsecchi – Analysing Survival Data from Clinical Trials and Observation Studies
Parmigiani – Modeling in Medical Decision Making: A Bayesian Approach
Senn – Cross-over Trials in Clinical Research, Second Edition
Senn – Statistical Issues in Drug Development
Whitehead – Design and Analysis of Sequential Clinical Trials, Revised Second Edition
Whitehead – Meta-analysis of Controlled Clinical Trials

Earth and Environmental Sciences

Buck, Cavanagh and Litton – Bayesian Approach to Intrepreting Archaeological Data
Glasbey and Horam – Image Analysis for the Biological Sciences
Webster and Oliver – Geostatistics for Environmental Scientists

Industry, Commerce and Finance

Aitken – Statistics and the Evaluation of Evidence for Forensic Scientists
Lehtonen and Pahkinen – Practical Methods for Design and Analysis of Complex Surveys, Second Edition
Ohser and Mücklich – Statistical Analysis of Microstructures in Materials Science